SYMBOLIC
AND KNOWLEDGE-BASED
SIGNAL PROCESSING

PRENTICE HALL SIGNAL PROCESSING SERIES

Alan V. Oppenheim, Series Editor

ANDREWS AND HUNT   *Digital Image Restoration*
BRIGHAM   *The Fast Fourier Transform*
BRIGHAM   *The Fast Fourier Transform and Its Applications*
BURDIC   *Underwater Acoustic System Analysis, 2/E*
CASTLEMAN   *Digital Image Processing*
COWAN AND GRANT   *Adaptive Filters*
CROCHIERE AND RABINER   *Multirate Digital Signal Processing*
DUDGEON AND MERSEREAU   *Multidimensional Digital Signal Processing*
HAMMING   *Digital Filters, 3/E*
HAYKIN, ED.   *Advances in Spectrum Analysis and Array Processing, Vols. I & II*
HAYKIN, ED.   *Array Signal Processing*
JAYANT AND NOLL   *Digital Coding of Waveforms*
KAY   *Modern Spectral Estimation*
KINO   *Acoustic Waves: Devices, Imaging, and Analog Signal Processing*
LEA, ED.   *Trends in Speech Recognition*
LIM   *Two-Dimensional Signal and Image Processing*
LIM, ED.   *Speech Enhancement*
LIM AND OPPENHEIM, EDS.   *Advanced Topics in Signal Processing*
MARPLE   *Digital Spectral Analysis with Applications*
MCCLELLAN AND RADER   *Number Theory in Digital Signal Processing*
MENDEL   *Lessons in Digital Estimation Theory*
OPPENHEIM, ED.   *Applications of Digital Signal Processing*
OPPENHEIM AND NAWAB, EDS.   *Symbolic and Knowledge-Based Signal Processing*
OPPENHEIM, WILLSKY, WITH YOUNG   *Signals and Systems*
OPPENHEIM AND SCHAFER   *Digital Signal Processing*
OPPENHEIM AND SCHAFER   *Discrete-Time Signal Processing*
QUACKENBUSH ET AL.   *Objective Measures of Speech Quality*
RABINER AND GOLD   *Theory and Applications of Digital Signal Processing*
RABINER AND SCHAFER   *Digital Processing of Speech Signals*
ROBINSON AND TREITEL   *Geophysical Signal Analysis*
STEARNS AND DAVID   *Signal Processing Algorithms*
STEARNS AND HUSH   *Digital Signal Analysis, 2/E*
TRIBOLET   *Seismic Applications of Homomorphic Signal Processing*
VAIDYANATHAN   *Multirate Systems and Filter Banks*
WIDROW AND STEARNS   *Adaptive Signal Processing*

SYMBOLIC AND KNOWLEDGE-BASED SIGNAL PROCESSING

EDITORS

Alan V. Oppenheim
Massachusetts Institute of Technology

S. Hamid Nawab
Boston University

Prentice Hall
Englewood Cliffs, New Jersey 07632

Library of Congress Cataloging-in-Publication Data

SYMBOLIC AND KNOWLEDGE-BASED SIGNAL PROCESSING / Alan V. Oppenheim and
  Hamid Nawab [editors].
     p. cm. — (Prentice-Hall signal processing series)
   Includes bibliographical references and index.
   ISBN 0-13-880444-3
     1. Signal processing—Equipment and supplies—Design. 2. Signal
processing—Digital techniques—Data processing. 3. Expert systems
(Computer science) 4. Logic, Symbolic and mathematical.
  I. Oppenheim, Alan V. II. Nawab, Hamid. III. Series.
  TK5102.5.S959 1992   91-47100
  621.382′2′0285—dc20   CIP

Editorial/production supervisor
   and interior design: **Karen Bernhaut and Barbara Marttine**
Cover design: **Wanda Lubelska**
Prepress buyer: **Mary McCartney**
Manufacturing buyer: **Susan Brunke**

 © 1992 by Prentice-Hall, Inc.
A Simon & Schuster Company
Englewood Cliffs, New Jersey 07632

ILS and nthPOWER are trademarks of Signal Technology, Inc. IMSL is a trademark of
IonSL, Inc. Macintosh is a trademark of Apple Computer, Inc. MACSYMA and
Symbolics Common Lisp are trademarks of Symbolics, Inc. *Maple* is a trademark of
Waterloo Maple Software. *Mathematica* is a trademark of Wolfram Research, Inc.
MATLAB is a trademark of MathWorks, Inc. *MILO* is a trademark of Paracomp. *NeXT* is
a trademark of Next Computer. *nodal* is a trademark of Alfred Riddle. *Photoshop* is a
trademark of Adobe Systems, Inc. POSTSCRIPT is a trademark of Adobe Systems, Inc.
QM is a trademark of the MIT AI Lab. SPICE is a trademark of the University of
California at Berkeley. TEX is a trademark of American Mathematical Society.

All rights reserved. No part of this book may be
reproduced, in any form or by any means,
without permission in writing from the publisher.

Printed in the United States of America
10 9 8 7 6 5 4 3 2 1

ISBN   0-13-880444-3

Prentice-Hall International (UK) Limited, *London*
Prentice-Hall of Australia Pty. Limited, *Sydney*
Prentice-Hall Canada Inc., *Toronto*
Prentice-Hall Hispanoamericana, S.A., *Mexico*
Prentice-Hall of India Private Limited, *New Delhi*
Prentice-Hall of Japan, Inc., *Tokyo*
Simon & Schuster Asia Pte. Ltd., *Singapore*
Editora Prentice-Hall do Brasil, Ltda., *Rio de Janeiro*

CONTENTS

PREFACE — vii
Alan V. Oppenheim and S. Hamid Nawab

FOREWORD — xv
Randy Davis

1 SIGNAL REPRESENTATIONS FOR NUMERICAL PROCESSING — 1
Gary E. Kopec

2 COMPUTER-AIDED ALGORITHM DESIGN AND REARRANGEMENT — 30
Michele M. Covell, Cory S. Myers, and Alan V. Oppenheim

3 SYMBOLIC ANALYSIS OF SIGNALS AND SYSTEMS — 88
Brian L. Evans and James H. McClellan

4	**THE SYMBOLIC MANIPULATION AND ANALYSIS OF MORPHOLOGICAL ALGORITHMS**	**142**
	Craig H. Richardson and Ronald W. Schafer	
5	**INTERACTIVE SIGNAL PROCESSING DOCUMENTS**	**173**
	Malcolm Slaney	
6	**BLACKBOARD SYSTEMS FOR KNOWLEDGE-BASED SIGNAL UNDERSTANDING**	**205**
	Norman Carver and Victor Lesser	
7	**INTEGRATED PROCESSING AND UNDERSTANDING OF SIGNALS**	**251**
	S. Hamid Nawab and Victor Lesser	
8	**SIGNAL ABSTRACTION CONCEPT FOR SIGNAL INTERPRETATION**	**286**
	Evangelos E. Milios and S. Hamid Nawab	
9	**KNOWLEDGE-BASED SIGNAL PROCESSIING APPLICATIONS**	**303**
	Erkan Dorken, Evangelos E. Milios, and S. Hamid Nawab	
	INDEX	**331**

PREFACE

Signal processing has its traditional roots in mathematics, physics and engineering. A significant component of the field focuses on generating one signal from another through an appropriately designed analog or digital system. The computer-related aspects of this activity, which have evolved and grown through sampled data systems and digital signal processing have generally been concerned with numerical processing issues such as computational efficiency, finite precision effects, modularity, recursive characteristics of algorithms, and so on. This has also been accompanied by the development of computer-based tools to aid the design of components of signal processing systems such as digital filters. However for the most part, the scope of such design tools has been limited to problems for which the design activity itself mostly involves the use of numeric processing algorithms.

There has also been considerable activity in signal understanding systems for such tasks as speech recognition and understanding, target classification, signal sorting and so on. Such systems are most typically configured to use signal processing for front-end feature extraction to map the signals to a symbolic feature set, followed by symbolic analysis of these features utilizing techniques and principles from the field of artificial intelligence and expert systems. Because the traditional paradigms and cultures of signal processing and computer science are historically different, in such systems the front end signal processing and higher level symbolic processing have tended to be treated as separate stages without reprocessing of the data based on higher level inferences.

With the continuing sophistication of signal processing systems and techniques, and the rapid development of techniques and technology in computer science, there are significant opportunities for more effectively exploiting computer science in the design and implementation of signal processing systems and in more closely integrating the signal and symbolic processing stages in signal understanding systems. A key goal of this book is to distill and summarize some of the research that has been carried out in this direction and to set the stage for a next phase of the research. It is our expectation and hope that these exploratory contributions are suggestive of the wealth of ideas, concepts and techniques that are still to emerge from these areas.

The material in this book can be roughly categorized into two basic themes. The first, addressed primarily in Chapters 1–5, considers the use of symbolic and knowledge-based techniques to provide software environments for system design and for processing signals and systems symbolically as well as numerically. The second theme is directed at the design of signal understanding systems in which the signal processing and symbolic processing are more highly integrated than has been traditional in the past. An inherent difficulty in integrating numerical and symbolic processing relates to the fact that numerical representations of signals and symbolic representation of features are fundamentally different as are the mechanisms and properties associated with processing them. Integrating numerical and symbolic processing in an effective way rather than simply using isolated stages is a significant and difficult objective, and provides the main focus for the last four chapters of this book.

As addressed in the first five chapters, the development of software environments to support signal processing system design and for the symbolic manipulation and processing of signals and systems is motivated by a number of factors. One is the clear need for more sophisticated tools in designing signal processing systems. Signal processing requirements and options are becoming increasingly complex and the tradeoffs more subtle. Also significant is the fact that the design space is essentially unbounded, requiring efficient and effective search and evaluation techniques. Consequently it has become difficult or even impossible for a designer to effectively and thoroughly explore design alternatives without the aid of sophisticated computer-based tools. Exploring the design options for a given signal processing objective naturally involves algorithm rearrangement and evaluation on the basis of various cost metrics. Exploiting a variety of computer science paradigms in the symbolic representation and manipulation of signals and systems and the development of innovative methods for searching the design space can lead to significantly enhanced software environments for designing signal processing systems.

As with most complex engineering tasks, signal processing system design in its most general form is highly intuitive, and in many aspects the principles for guiding rearrangements are not highly structured. Fortunately, however, there are also significant components of signal processing technology that are highly structured with

Preface

well established rearrangement rules. These rules and principles, in fact, are the core of most textbooks and courses in signal processing. While it is obviously a large task to effectively incorporate these rules into a software environment for symbolically exploring algorithm rearrangement, with appropriate research and development effort it should be achievable. In some, but not all respects, an analogy can be drawn with software environments such as *MACSYMA* which symbolically manipulate mathematical expressions through well established mathematical rules. The concept of a signal processing design environment as proposed in a broad sense in the first five chapters of this book requires considerably more global rearrangement, search, inference, cost evaluation, etc. than is incorporated in currently available symbolic mathematics packages. Also, as noted in Chapter 3, *MACSYMA* consists of over 300,000 lines of compiled Lisp code and represents more than 100 man-years of development. In contrast, the work represented in this book on symbolic signal processing represents only a few man years on a topic with considerably more complexity. However these chapters describe a direction and strongly suggest the feasibility of symbolic analysis and rearrangement of signal processing expressions.

Although only alluded to in several of the chapters, symbolic representation and manipulation of signals and systems can potentially become an important aspect of on-line signal processing as well as off-line design. Specifically, by representing and manipulating signals as data objects and exploiting delayed evaluation as the signals pass through a system or sequence of transformation stages, a processor with symbolic capabilities could potentially generate an output signal object, the numerical values for which, when eventually required, could be more efficiently and accurately obtained than by the successive numerical processing specified in the algorithm. As a perhaps trivial example to illustrate the point, if a processor requires the DFT of the DFT of the input signal, delayed evaluation and symbolic representation can lead to efficiently and accurately recognizing and defining this signal object as the delayed and time reversed input signal. While on-line reconfigurable symbolic signal processing with delayed evaluation is even more ambitious than the use of symbolic processing in the context of a system design environment, many of the issues and principles are similar and are identified and discussed throughout this book.

Some of the specific issues that are addressed in the first five chapters of the book are:

- The criteria to be used to evaluate the appropriateness of signal representations utilized in the environment;
- representation of signals, systems and algorithms in the environment in order that their various properties (e.g. stability of systems, region of support of signals, computational efficiency of algorithms) and the ways in which they interact with each other may be conveniently analyzed;
- ways in which a software environment can support transform analysis of signals and systems;

- mechanisms needed in the environment to enable the automatic derivation of detailed algorithms for signal processing transformations specified at a high level of abstraction;
- mechanisms needed to incorporate strategies for searching for the most appropriate algorithm for a signal processing transformation in accordance with various cost measures (such as computational efficiency or compatibility with a particular implementation technology);
- the types of algebraic manipulations of signal-processing expressions required and the mechanisms needed to support them.

In Chapter 1, Kopec develops and discusses the issues of signal representation and establishes requirements, such as immutability in which the observable properties of a signal remain invariant after the signal is created, and deferred evaluation whereby the representation of a signal can be generated without necessarily computing its values. Signal representations such as the use of arrays, streams and abstract data objects, and associated signal processing languages are discussed in the context of these requirements. Chapter 2 by Covell, Myers and Oppenheim builds from many of the concepts introduced in Chapter 1 and describes a particular software environment implemented in Lisp. This environment, referred to as ADE (A Design Environment) was developed as a feasibility demonstration of the basic concepts associated with carrying out symbolic algorithm rearrangement and evaluation. In addition to outlining some basic requirements for the system, the use of the system in two design contexts is illustrated. In both cases the software environment arrived at efficient and novel implementations. Through these examples, Chapter 2 suggests the clear potential for such an environment while also underscoring the difficulties associated with carrying it to a practical level.

Chapter 3 is also concerned with the symbolic manipulation of signal processing expressions. In this chapter, Evans and McClellan begin with a currently available symbolic algebra environment, specifically *Mathematica,* and extend it to manipulate algebraic forms of signals and systems. A primary feature of the current packages is to solve linear systems problems in both the digital and analog domains using convolution and transforms, complete with justification of answers. A future goal of these packages is to provide the system design capabilities present in the Lisp-based environments described in Chapter 2. Although some features have been implemented, it appears to be extremely difficult for a signal processing environment implemented in a currently available symbolic mathematics program to evaluate and compare relative costs of different expression rearrangements.

An immediate use of the signal processing packages for *Mathematica* is to teach students concepts in linear systems. This is enhanced by *Mathematica*'s support of an interactive learning environment called the *notebook* facility which provides a sophisticated interface to its symbolic computational engine. In Chapter 5, Slaney explores the use of notebooks as interactive scientific documents in the context of his own signal processing extensions. Notebooks, like technical reports, include text,

Preface

tables, figures, and graphs. Readers, however, can gain a richer understanding of the material by interacting with computer models, animations, and sound in a notebook.

As with Chapter 2, Chapter 4 describes an environment for algorithm manipulation and design built on Lisp. Whereas the systems in chapters 2 and 3 focus primarily on linear operations such as transforms, convolution, and so on, and incorporate only simple nonlinearities, Chapter 4 is entirely directed at a nonlinear class of algorithms referred to as morphological operations. The system described by Richardson and Schafer in this chapter incorporates cost measures and analysis very similar to those in the system described in Chapter 2. Mathematical morphology basically views signals as sets and involves nonlinear superpositions of signals using set operations such as union and intersection. As with the earlier chapters, this chapter presents a convincing demonstration of the feasibility of a design environment to aid in the symbolic rearrangement of a class of algorithms including simplification, cost analysis and comparison.

The second theme in this book, covered in chapters 6 through 9, involves the sophisticated integration of signal processing into signal understanding systems. Broadly speaking, a signal understanding system is designed to generate assertions about the real-world environment from which the system receives its input signal data. Furthermore, an important characteristic of a signal understanding system is that it should be able to adjust its future actions on the basis of the assertions it has already formed at any given time. Most signal understanding systems to date can perform such situation-dependent adjustment of their problem-solving at all levels except the signal processing level. A major reason for this is that researchers in the signal understanding area have generally limited their investigations to the exploration of engineering principles involved in the strictly symbolic and heuristic processing aspects of signal understanding. Consequently, not much attention was given to designing signal understanding systems that can adjust their largely numeric and mathematically based signal processing in response to changing contexts. Recent work, a large part of which is represented in the last four chapters of this book, has begun to systematically address the issues that arise in designing signal understanding systems whose signal processing activities need to be controlled in a sophisticated manner.

The capability for situation-dependent adjustment of signal processing is desirable in a variety of signal understanding applications. Consider, for example, the design of a robot that can monitor the sounds produced in a household environment to determine the nature of the events taking place in that environment. Furthermore, suppose the robot has to be able to take appropriate actions in response to important events such as a ringing fire alarm, a ringing telephone, a crying baby, or the sounding of an oven buzzer. Since the individual characteristics of particular sounds produced in a household environment vary enormously and since it is not uncommon for several sound-producing events to occur simultaneously, signal processing requirements for transient analysis, steady-state analysis, noise suppression, and signal detection also tend to vary enormously over time. It therefore appears desirable that the adjustment of signal processing be performed in a manner that can take advantage of any abstract and largely symbolic descriptions of the sound generating environment that are available to the robot through its higher level interpretation activities. Another important factor that leads to the need for adjustable signal processing in this robot is due to the goal-directed aspects of the robot's activities. For example, when the robot hears a burglar alarm, it may decide to focus its signal processing activity toward the goal of detecting the sounds of an intruder's footsteps.

The last four chapters of this book address issues that are pertinent to applications such as robotic hearing as well as a variety of other applications alluded to in those chapters. The context of these applications serves to illustrate how the design of signal understanding systems can involve a careful blend of ideas from signal processing and artificial intelligence. More specifically, some of the important issues addressed in these chapters include:

- The types of characteristics that a signal understanding application must have in order to require a complex integration of signal processing and signal understanding;
- modifications or additions needed in existing signal understanding architectures to enable such systems to use signal processing in a more sophisticated manner than the traditional fixed front-end approach;
- ways in which the mathematical theory underlying signal processing algorithms can be utilized to help design signal understanding systems that must select signal processing strategies in a situation-dependent manner;
- mechanisms required for a signal understanding system to be able to assess the reliability of evidence based on signal processing output data;
- the degree to which the higher-level constraints in the interpretation process alleviate the need to utilize computationally expensive signal processing for extracting detailed characteristics of the input signal. Conversely, the degree to which judicious use of signal processing alleviates the need for computationally expensive problem-solving in the higher levels of the interpretation process;
- possible new design-criteria for signal-processing techniques which may be elicited from considerations of complex integration of signal processing and signal understanding.

In Chapter 6, Carver and Lesser present essential elements of the blackboard model in artificial intelligence and discuss how the blackboard model has been utilized in the design of knowledge-based signal understanding systems. In particular, they critically analyze those aspects of the blackboard model that are deemed to be especially relevant to signal understanding applications. In this vein, they identify the nature of the control mechanisms built into a blackboard system as being crucial for giving a signal understanding system the capability to change its problem-solving strategies in a context-dependent manner. In particular, they discuss a specific control framework which they have developed for signal understanding systems of this type. This framework endows a signal understanding system with two basic problem-solving modes: *evidence aggregation* and *differential diagnosis*. Evidence aggregation problem-solving seeks data for increasing or decreasing the uncertainty of one particular interpretation, whereas differential diagnosis problem-solving seeks data for resolving ambiguities that produced competing interpretations. Utilizing these methods, a signal understanding system can decide when to carry out adjusted *reprocessing* of signal data for resolving uncertainties in the corresponding interpretations. Throughout the chapter, an aircraft monitoring application is used to illustrate the nature of the issues discussed.

In Chapter 7, Nawab and Lesser discuss a particular architecture for integrating signal processing and signal understanding within the context of a blackboard model developed in Chapter 6. They introduce the concept of "discrepancy detection" as the means for checking consistency between different signal processing outputs for the same signal data or between a signal processing output and the results obtained from higher level interpretation processes. They also formulate certain knowledge-based reasoning processes that can use the detected discrepancies as the basis for adjusting future signal processing actions of the system. A key feature of the resulting design is that it takes advantage of the underlying theory of signal processing algorithms in the process of deciding how to adjust the signal processing in response to the detected discrepancies. This decision process involves mechanisms that are akin to the knowledge-based manipulation of symbolic expressions that was alluded to in some of the earlier chapters on symbolic signal processing.

An alternative architecture for integrating signal processing and signal understanding is presented in Chapter 8. In contrast to the previous two chapters, the work of Milios and Nawab reported in this chapter does not make use of the blackboard problem-solving model. They point out that in some applications it is more appropriate to use an architecture in which there is much more structured control of the problem-solving activity than that which is typically associated with blackboard problem-solving models. To illustrate their points, they describe a particular application system that carries out adaptive spectral estimation. This application differs from the aircraft monitoring application of Chapter 6 and the sound understanding application of Chapter 7 in that its symbolic processing objectives are relatively simpler.

In Chapter 9, Dorken, Milios and Nawab describe and discuss some of the salient features of systems that have been developed over the last decade for addressing specific signal understanding applications. Each of these systems utilizes sophisticated mechanisms for the integration of signal processing into the overall signal understanding architectures. One of the systems, the sound understanding testbed, uses the architecture described in Chapter 7. Another system, the helicopter signal tracker, utilizes the architecture described in Chapter 8. The pitch detector's assistant is the third system described in this chapter. This last system is unique in that its architecture makes use of the fact that in addition to the speech signal input it is also provided with a symbolic input in the form of a transcript corresponding to the speech signal.

The importance of the subject matter in this book can be seen from both a technology perspective and an applications perspective. The material in this book addresses the emergence of a technology which requires expertise in both signal processing and computer science principles. The ultimate payoff in the development of such an integrated technology lies in its applications. The area of software environments for signal processing design is aimed at enhancing the productivity of scientists and engineers working on application problems that involve complex design and experimentation activities. The area of integrated signal processing and signal understanding opens up the possibilities of new solutions to traditional signal processing problems as well as broadening the scope of the problems on which signal processing expertise can be brought to bear. Examples of traditional application areas that stand to benefit from this perspective include signal enhancement, spectral analysis (particularly for non-stationary signals), pitch detection in speech processing, and feature extraction in image processing. The broadening of application areas from the signal processing expert's point of view includes problems such as speaker separation in speech, sound classification in a multi-sound environment, knowledge-based target tracking, and multi-sensor information fusion.

Although there are certain unifying themes that give this book an overall coherence, each of the chapters represents a largely self-contained discussion of issues relevant to the topics discussed in that chapter. The ultimate strength of the book therefore lies in the contributions made by the various chapter authors. We are grateful to them for contributing their respective chapters as well as for the many useful suggestions they made in reviewing the chapters written by other authors. We would also like to thank all the authors for their enthusiastic cooperation and timely responses that made the task of editing this book considerably easier. We also thank Professor Randy Davis for contributing a foreword for the book, and for his involvement in the early aspects of some of the research presented.

Alan V. Oppenheim
S. Hamid Nawab

FOREWORD

- Randy Davis
  Massachusetts Institute of Technology

The collection of chapters in this book examines the intriguing and relatively unexplored territory that lies in the intersection of signals and symbols. With few exceptions, signal processing and AI have historically grown up as distinct undertakings, evolving their own tools, techniques and languages, with signal processing focusing largely on numeric calculation and AI emphasizing symbolic inference.

This collection nicely illustrates several important themes that arise in work at the intersection of these two fields. It shows us, first of all, that there are problems that lie squarely in the intersection, probems that pose substantial challenges for both signal processing and AI. Second, it illustrates that those problems are best attacked with approaches informed by insights from both disciplines. Third, it exemplifies in a number of ways an interesting evolution in our conception of what a computer is and what a program is, a change that arises out of the attempt to work on problems requiring both signals and symbols. The progression of that change is marked by several significant steps:

1. In the beginning, computers were seen as calculators and a program was conceived of as a set of instructions for doing arithmetic. Computers were viewed initially as fast number crunchers whose task was to make short work of arithmetic. Given this conception of the hardware, it followed that software was to be simply a set of instructions that would cause the machine to carry out its calculations.

Since speed was invariably an issue, programs had to be designed to be fast. Since memory space was invariably an issue, programs had to be designed to be small. This led to a kind of minimalist view of programs: they should be the shortest and fastest set of instructions that accomplish the task.

2. Computers soon came to be recognized as symbol processors, and a program correspondingly was viewed as a set of instructions for manipulating symbols. This wider view arose in large measure from the attempt by AI programs to deal with non-numeric data such as algebraic expressions. The success of this effort made it clear that numbers were only one of the kinds of symbols that the machine can handle. It can also deal with symbols like $e^{j\omega n}$ or $cosh[\omega n + b]u[n]$, or `periodic`, `continuous`, `discrete`, or `homogeneous`.

3. The next step was recognizing that a computer can be an intelligence amplifier and that programs can reason as well as calculate. AI has made realistic the notion that some varieties of intelligent thinking can be captured by a general purpose symbol processor, i.e., this machine, with the appropriate programs, can for some tasks produce intelligent problem solving behavior. With this our view of the computer changed yet again: As with many of our tools, we use it to amplify our powers, but where it was once simply a high speed number cruncher used to amplify our ability to calculate, it is now also an intelligence amplifier, augmenting our ability to think. A machine of this sort is usable on a much wider range of tasks, notably those that involve reasoning, not just calculation.

It was in this context that the work on knowledge-based systems was born and gave rise to what is now recognized as the fundamental principle of that subfield: The hypothesis that knowledge is power. That is, programs can be intelligent to the degree that they capture and use knowledge about the world. In programs, as in people, expertise arises largely from what they know.

4. The recognition that programs can reason and that reasoning is based on knowledge led to a change in the view of programming: It became a process of specifying what to know, rather than what to do. Knowledge-based systems made this particularly clear, by demonstrating that human experts could be debriefed, explaining what they knew that made them good at their task. Programs became expert to the extent that they had the same knowledge. As a result, programming became a task of knowledge accumulation rather than instruction specification, telling the machine what to know, rather than what to do.

5. This soon led to the recognition that a program can be considered a repository for knowledge. In this view a program is no longer simply a set of instructions to carry out a calculation. It is now appropriately viewed as a means of cataloging, expressing, and using knowledge about a field. It is in effect an interesting sort of reference text, capable of storing knowledge (as any text can), but capable as well of applying it to a problem, as no ordinary text can.

Efforts to accumulate the knowledge led to additional insights. It became clear, for example, that much of the knowledge of human experts is non-numeric: Much of what an expert knows is not well expressed with numbers, much of the interesting reasoning is not easily modeled as arithmetic. This is clearly true in fields like medical diagnosis, where some of the early knowledge-based systems work was done, but it is equally true in fields that are fundamentally quantitative, like signal processing. Relatively little of the skill of an expert is in the collection of equations he may know; any diligent student can soon memorize the same list. The more elusive expertise is in knowing how and when to use which equation, or what properties of a problem make it susceptible to one or another approach. Put slightly differently, most of the expert-level reasoning is done before the equations ever hit the page.

It also became clear that language has a strong influence on this process. Languages influence how we think; notably, what a language makes difficult to express, we rarely think of. When programs were instructions for number crunching, programming languages gave us a vocabulary for calculation and we typically thought in those terms. Now programs are viewed as repositories of knowledge and appropriate languages for this view have been developed, languages that allow us to express the sort of non-numeric knowledge and reasoning that make up much of human expertise.

6. Viewing programs as knowledge repositories led naturally to the possibility that a program can be considered a medium of communication. That is, in addition to being executed by a machine, programs may also be literary works of a sort, to be read by people. In this view knowledge bases can become valuable documents representing a significant portion of what is known about a particular subfield, as has become true in areas like symbolic mathematics (*MACSYMA, Mathematica*), structure elucidation in chemistry (DENDRAL), and others.

Having programs read by people as well as by computers leads as well to the notion that programs should be written to communicate, not simply to compute. Programs can themselves be textual works, with their own notion of lucidity and elegant style.

7. Recent advances in hardware capability and decreasing cost have illustrated that the digital medium is ubiquitous and polymorphic. The ubiquity of digital information is a notion at the heart of computer science in general. It has been a long time in making its influence widely felt, partly due to economics, but partly due to one of the earliest mistakes to be made in computing, using typewriters as I/O devices. That unfortunate act set the dominant metaphor for man/machine interaction for nearly forty years: We are to communicate with the machine by typing at it, and it to us. Yet as the authors of the chapters in this book perhaps know better than most of us, digital information can in fact be used to capture a remarkable array of media in addition to text. Graphics, animation, sound, shape, and more are now routinely used in multimedia interaction with computers.

Each of these themes in the evolution of our conception of computers and programs is illustrated in the collection of chapters in this book. Gary Kopec describes data abstractions for signals, developing a language for describing signals that allows us to express their important properties. The result is one example of a programming language that describes more than calculation. Even though he is concerned with numerical processing, many of the most useful properties expressed in the language are distinctly non-mathematical.

In their work on computer-aided algorithm design Covell, Meyers, and Oppenheim illustrate the power of an intelligence amplifier. Their system aids the programmer by systematically exploring possible algorithm designs. By searching methodically it was able to produce a design for a non-integer sampling rate conversion algorithm of sufficient interest that it was worthy of publication when independently (re-)discovered by another researcher.

Evans and McClellan's work on symbolic analysis of signals illustrates several of the themes, including accumulating the appropriate language (terms like `associative`, `continuous`, `discrete`, `homogeneous`, and `memoryless`), doing symbolic rather than numeric calculation (e.g., Laplace transforms that whose input and output are both formulas rather than numbers), and programs as accumulated knowledge (e.g., the list of strategies for inverse z-transforms).

Richardson and Schafer show how programs can capture knowledge about the design of algorithms and transforms (the computer as an intelligence amplifier) and how they can employ that knowledge in the symbolic manipulation of code (the computer engaged in non-numeric calculation).

Slaney's work on interactive documents illustrates the power of an environment that is interactive, animated, and multi-media, and explores the change in mindset that comes with considering programs as a medium of communication: The notion of literate programming gives a concrete illustration of the expressive power of code carefully written.

The three articles by Carver and Lesser, Nawab and Lesser, and Milios and Nawab each illustrate the expanding set of problems that accompanies conceiving of computers and programs as more than calculators. When we begin to conceive of them as capable of some forms of reasoning, the scope of tasks they can handle broadens from signal processing to embrace signal interpretation and signal understanding.

Dorken, Milios and Nawab then survey the challenging problems that arise at the intersection of paradigms, the place where numeric and symbolic processing meet, via a historical review of some of the pioneering work in the area.

I am pleased and honored by the opportunity to comment this thought-provoking collection.

1

SIGNAL REPRESENTATIONS FOR NUMERICAL PROCESSING

• Gary E. Kopec
Xerox PARC

1.1 INTRODUCTION

Historically, the need for execution-time efficiency has been a dominant factor in the design of digital signal processing software. More recently, however, the increased availability of low-cost, high-performance hardware has made run-time efficiency a less significant determinant of software quality than factors such as producibility, reliability, and maintainability. As the progress in hardware technology continues, methodologies and tools for enhancing these attributes will become as critically important in signal processing as they already are in less performance-oriented fields.

Research in programming methodology over the past 20 years has led to the view that the fundamental activity in the development of well-engineered software is the recognition of *abstractions* [49, 51]. In particular, three general kinds of abstractions have been identified: abstract *operations*, abstract *data types*, and *control* abstractions. An abstract operation corresponds to the notion of a procedure or subroutine that is supported in most contemporary programming languages. An abstract data type, or *data abstraction*, consists of a set of objects plus a set of primitive operations for creating and manipulating those objects. Examples of common data abstractions include fixed and floating point numbers, characters, strings, arrays, and structures. Control abstractions are concerned with program control flow and are illustrated by looping constructs such as `while` and `for`.

Program development in any particular application area may be characterized by the specific kinds of procedural, data, and control abstractions that are used. Furthermore, a programming language or environment aids software development in some domain to the extent that it provides explicit support for the natural abstractions of the domain.

Digital signal processing is based on a well-established body of mathematical theory that is explicitly used during program development. Perhaps the most basic feature of this theory is the central role of the concepts *signal* and *system*. This suggests that a well-structured signal processing program should be organized as a collection of "signal" and "system" abstractions. Similarly, a signal processing programming language is one that provides explicit support for such abstractions.

Most previous attempts to define signal processing languages have been based on two main models of signal processing computation: *next-state simulation* [7] and *array processing*. Next-state simulation is the underlying computation model in stream-oriented block diagram programming languages [5–20]. Array processing refers to the large and ill-defined collection of implementation techniques that exploit the separability of address arithmetic and "kernel computation" which characterizes many signal processing algorithms [3, 4]. Signal processing algorithms frequently consist of a small number of *computation kernels* which are repeatedly applied to many sets of arguments. The sequencing of the kernels and the data paths among them are independent of the values of the data. This algorithmic structure has been exploited in the design of highly parallel array processing hardware [1, 2] and time-efficient software [3, 4].

Next-state simulation and array processing have each been highly successful within some range of applications. Neither model, however, has supported the design of an effective general-purpose signal processing programming system. The fundamental problem is that block diagram languages and array processing are attempts to model discrete-time systems as procedural abstractions that share a common algorithmic structure [26]. Unfortunately, the class of "signal processing algorithms" is becoming increasingly heterogeneous. This observation, together with the growing recognition of the utility of the data abstraction concept, has motivated exploration of an alternative possibility—signal processing languages in which signals and systems take the form of specialized data abstractions.

This chapter reviews three approaches to the representation of discrete-time signals as data abstractions. Two of these representations—the *streams* of block diagram languages and the *arrays* of array processing—are widely used in contemporary signal processing programming. The third representation—*abstract signal objects*—has been the basis for much of the more recent work in signal processing language design [26–34]. A set of signal representation criteria is developed, based on elementary observations about the mathematics of discrete-time signals and the mathematical notations used to describe signals. These criteria are then used to assess the advantages and limitations of the three types of representation.

Sec. 1.2 Signal Representation Requirements

The discussions of block diagram and array processing languages will be carried out in the context of simple Lisp-based exemplars, called BDL and AOL, respectively.[1] The facilities of these hypothetical languages are intended to be representative of those found in the many "real" block diagram and array processing systems that have been proposed. The discussion of abstract signal objects will focus on the Signal Representation Language (SRL) [27].

The remainder of this chapter is organized as follows. Section 1.2 develops a set of general signal representation criteria. Sections 1.3–1.5 review the use of streams, arrays, and abstract signal objects in signal processing programming, and describe the languages BDL, AOL, and SRL. The three types of signal representations are assessed in terms of the criteria developed in section 1.2. Finally, section 1.6 contains a brief summary.

1.2 SIGNAL REPRESENTATION REQUIREMENTS

In order to assess the relative advantages and limitations of alternate approaches to signal representation, it is necessary to define the dimensions along which the alternatives are to be compared. Possible dimensions of comparison include time or space efficiency, theoretical soundness, and naturalness for expressing some class of algorithms. The criteria developed in this section are based on the mathematical properties of discrete-time signals and are aimed at ensuring that the behavior of a signal representation object in a program reflects the mathematical properties of the represented signal. The objective is to derive a set of desirable attributes for a signal representation from first principles, without specific consideration of practical trade-offs which might be imposed by a particular implementation technology. This point of view is complementary to that manifested in investigations of signal processing software time/space efficiency trade-offs [3, 4]. However, practical experience with signal processing languages based on the criteria developed below suggests that such languages are not necessarily disadvantaged in terms of time or space efficiency.

This section begins with a brief review of elementary signal processing theory in order to establish notation and to provide a point of reference. Five signal representation criteria are then developed from observations about the mathematics of signals and the mathematical notations used to describe signals. Three of these criteria—*uniform reference*, *immutability*, and *dimensional extensibility*—are primarily concerned with the external observable behavior of signal data objects. The remaining two requirements—*deferred evaluation* and *manifest typing*—focus on the way in which signals and systems are specified and are motivated by the use of expressions in mathematical notation. A more complete discussion of signal and system representation criteria may be found in [26, 29].

---

[1] The program examples in this chapter are in Common Lisp [50].

1.2.1 Basic Signal Processing Concepts and Notation

A *discrete-time signal* (*sequence*) is a function $x : D \to R$, where $D \subset I^m$ (m-tuples of integers) and R is some set.[2] If x is a signal from D into R then, fundamentally, x is a subset of $D \times R$ (the Cartesian product of D and R) such that for each $d \in D$ there is a unique $r \in R$ such that $(d, r) \in x$. The notation "$r = x(d)$" means $(d, r) \in x$; in that case (d, r) is called a *sample* of x. If $x : D \to R$, where $D \subset I^m$, then the *domain* of x is D, the *range* of x is R, and the *dimension* of x is m.

This chapter will focus on signals whose domains are Cartesian products of finite intervals, i.e.

$$x : [0, N_1) \times \cdots \times [0, N_m) \to R \quad (1.1)$$

where $[0, N_i)$ is the set of integers $0 \ldots N_i - 1$. The class of all signals of the form (1.1) will be denoted S. The restriction to discrete multidimensional box domains is not essential for the central points of this chapter. It is an expository simplification that avoids a number of technical issues associated with more general domains. The representation of infinitely long signals and functions of a continuous variable are developed in the signal representation languages SPLICE [29] and QuickSig [33, 34]. These languages provide facilities for explicitly representing and manipulating signal domain concepts such as *interval* and *support*.

A particular signal is usually defined by giving an *expression* for the values of its samples. For example, a complex exponential with frequency ω is defined by

$$z_\omega(n) = e^{j\omega n}, \forall n \in I \quad (1.2)$$

The expression "$e^{j\omega n}$" defines the class of all complex exponential signals; a particular complex exponential is characterized by its value for the *parameter* ω.

A *discrete-time system* with M *inputs* and P *outputs* is a function from S^M into S^P. A particular system is usually defined in one of two ways—by its input/output mapping or by a *realization*. An *input/output mapping* is the identity of a system as a mathematical function between signal spaces. In general, it is defined by specifying a predicate which characterizes the input/output pairs of the system. However, since the inputs and outputs of a system are signals, the input/output mapping is usually expressed in terms of the relationships that hold among the samples of the input and output. For example, the first-order recursive filter shown in Figure 1.1 is that system whose input/output pairs are of the form (x, y), where the samples of x and y are related by

$$y(n) = \begin{cases} 0 & \text{if } n < 0 \\ a \cdot y(n-1) + x(n) & \text{if } n \geq 0 \end{cases} \quad (1.3)$$

for all $n \in I$, where it is assumed that $x(n) = 0$ for $n < 0$.

---

[2] In general, R may contain any type of object; most often R is a set of real or complex scalars or vectors.

Sec. 1.2 Signal Representation Requirements

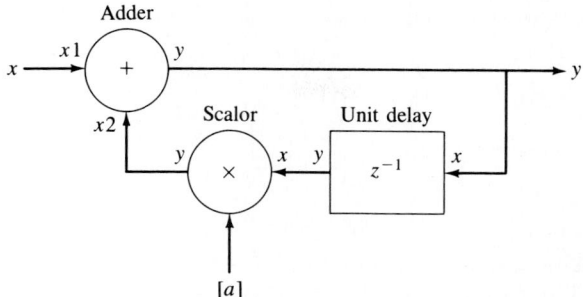

Figure 1.1 Block diagram of a first-order filter.

A *realization* of a system is a representation of its input/output mapping in some prescribed form. For example, a *state-space realization* [45] of the above filter is

$$s(n + 1) = a \cdot s(n) + x(n)$$
$$y(n) = a \cdot s(n) + x(n) \quad (1.4)$$
$$s(0) = 0$$

An important attribute of a state-space realization is that it can be interpreted as a dynamic process for incrementally generating the response of a system to a specified input signal. In general, the focus of realization theory is the dynamic behavior of such generation processes, rather than the abstract relation between the input and output signals [46].

The mathematical definitions just reviewed form the basis for the following signal representation criteria.

1.2.2 Uniform Reference

The most fundamental signal representation requirement concerns the kinds of observable properties that distinguish one signal from another. Mathematically, two discrete-time signals

$$x_1 : [0, N_1^{(1)}) \times \cdots \times [0, N_m^{(1)}) \to R$$
$$x_2 : [0, N_1^{(2)}) \times \cdots \times [0, N_m^{(2)}) \to R \quad (1.5)$$

are *equal* as functions if and only if

$$N_i^{(1)} = N_i^{(2)}, \; i = 1 \ldots m, \quad \text{and}$$
$$x_1(n) = x_2(n), \; \forall n \in D \quad (1.6)$$

where $D = [0, N_1) \times \cdots \times [0, N_m)$ is the common domain of x_1 and x_2. In effect, the identity of a signal is determined solely by the identities of its domain and range and the values of its samples.

A fundamental property of signal equality is that it is preserved by discrete-time systems. That is, if

$$x_1 = x_2 \tag{1.7}$$

then

$$F(x_1) = F(x_2) \tag{1.8}$$

for every discrete-time system F, where equality in (1.7) and (1.8) means (1.6).

The preservation of signal equality reflects the fact that a discrete-time system is typically defined by the relationships that hold among the samples of the input and output signals. For example, the input and output of a one-dimensional signal *scalor* [45] (constant-gain amplifier) are related by

$$y(n) = a \cdot x(n), \; \forall n \in D \tag{1.9}$$

where a is the applied gain. If $x_1 = x_2$ then, by (1.6)

$$x_1(n) = x_2(n), \; \forall n \in D \tag{1.10}$$

Thus, by (1.9)

$$y_1(n) = y_2(n), \; \forall n \in D \tag{1.11}$$

and $y_1 = y_2$ by (1.6).

The data abstraction formalization of the notion of observable property is the concept of an *inquiry operation* [48]. The inquiry operations of a data type are the primitive operations used to extract information from objects of the type. They define the observable attributes by which members of the type are distinguished. The mathematical definition of signal equality suggests that the inquiry operations for a signal abstraction should include functions that identify the domain of a signal and functions that return the values of its samples. A data abstraction for signals of the form (1.1) will satisfy this requirement if it includes two inquiry operations, `dimensions` and `fetch`, which satisfy the following specification: if s is the object that represents signal $x : [0, N_1) \times \cdots \times [0, N_m) \rightarrow R$, then

$$\begin{aligned}&(\text{dimensions } s) = (N_1 \ldots N_m), \text{ and} \\ &(\text{fetch } s \; i_1 \ldots i_m) = x(i_1 \ldots i_m), \; \forall (i_1 \ldots i_m) \in D\end{aligned} \tag{1.12}$$

The requirement that a signal abstraction include generic `dimensions` and `fetch` operations which may be applied to any signal object is called the requirement of *uniform reference*.

The implication of (1.7) and (1.8) is that two signals are mathematically indistinguishable if they are equal as functions. This suggests that it should not be possible to distinguish between two signal objects s_1 and s_2 in a program using *any* operation if

$$\begin{aligned}&(\text{dimensions } s_1) = (\text{dimensions } s_2) \text{ and} \\ &(\text{fetch } s_1 \, i_1 \ldots i_m) = (\text{fetch } s_2 \, i_1 \ldots i_m), \; \forall (i_1 \ldots i_m) \in D\end{aligned} \tag{1.13}$$

One reliable general way to guarantee the complete equivalence of signal objects that represent equal signals is simply to provide no inquiry operations other than `dimensions` and `fetch`.

1.2.3 Immutability

Mathematically, the value of a function at any point of its domain is a unique, well-defined entity. As a result, if $x : D \to R$ is a signal and $n \in D$, then the expression "$x(n)$" denotes the same value in every context in which it occurs. The constancy of discrete-time signals as mathematical entities suggests that they should be represented in programs by objects that are *immutable* [49]. The observable properties of a signal object should be defined when the signal is first created and remain fixed thereafter. For example, if s represents $x : D \to R$ and $(i_1 \ldots i_m) \in D$, then the function invocations

$$(\texttt{dimensions } s)$$
$$(\texttt{fetch } s\ i_1 \ldots i_m) \tag{1.14}$$

should return the same value whenever they are evaluated.

An important corollary of signal immutability is that the inputs and outputs of a discrete-time system should usually be represented as distinct objects. Although systems are loosely described as "transforming" their inputs, the function of a system is actually to create a new signal whose samples are related to those of the inputs according to some specification. This implies that distinct objects should be used to represent the inputs and outputs of a system unless the corresponding functions are mathematically identical.

1.2.4 Dimensional Extensibility

There are numerous theorems in one-dimensional signal processing that do not have multidimensional counterparts [44]. Nevertheless, many classes of multidimensional signals and systems may be derived from one-dimensional prototypes by simply increasing the number of independent "time" variables in a straightforward way. For example, two-input adders for one- and two-dimensional signals are defined, respectively, by the input-output relations

$$y(n) = x_1(n) + x_2(n) \text{ and}$$
$$y(n_1, n_2) = x_1(n_1, n_2) + x_2(n_1, n_2) \tag{1.15}$$

The similarity between the mathematical descriptions of one- and multidimensional systems suggests that abstractions for one- and multidimensional signals should be similar as well. This requirement is called *dimensional extensibility*.

1.2.5 Deferred Evaluation

As previously noted, a particular signal is typically defined by giving an expression for the values of its samples. There is a distinction between the processes of defining a function, by constructing an expression for its values, and computing the function, by evaluating the expression at a specific point of its domain. This suggests that creating the program representation for a discrete-time signal should not imply the immediate computation of its samples. The requirement that it be possible to create a signal without computing its samples is called the criterion of *deferred evaluation*.

In addition to being conceptually appealing, deferred evaluation is absolutely necessary for representing certain important classes of signals [29]. For example, if creating a signal involves computing all of its samples, it would not be possible to represent infinitely long signals or signals defined using feedback loops, such as y in Figure 1.1.

1.2.6 Manifest Typing

The traditional purpose of a signal processing system is the analysis, synthesis, or modification of numerical signal data. In general, numerical signal processing algorithms may be characterized as those for which the inquiry operations `dimensions` and `fetch` provide all of the necessary information about a signal. However, there are important nonnumerical aspects of signal and system analysis which cannot be supported if the only inquiry operations are `dimensions` and `fetch`. For example, it is not possible, using just `dimensions` and `fetch`, to implement operations such as

$$(\text{is-sine? } s) = \begin{cases} t & \text{if } s \text{ was generated} \\ & \text{by a sine oscillator} \\ \text{nil} & \text{otherwise} \end{cases}$$

or

$$(\text{omega } s) = \begin{cases} \omega & \text{if } s \text{ was generated by a sine} \\ & \text{oscillator in the form} \\ & s(n) = \sin(\omega n) \\ \text{nil} & \text{otherwise} \end{cases}$$

It is important to emphasize that `is-sine?` and `omega` return information about the *actual* history of s and are not parameter estimation operators that decide if (a signal equal to the signal represented by) s *could* be generated by a sine oscillator.

Operations such as `is-sine?` and `omega` are "nonmathematical" in that they do not satisfy (1.7) and (1.8). For example, $x_1 = x_2$ does not imply that $(\text{is-sine? } s_1) = (\text{is-sine? } s_2)$, where s_1 and s_2 are representations for x_1 and x_2, since s_1 might be the output of a sine oscillator while s_2 was created by delaying the output of a cosine oscillator.

History-based operations, while not strictly necessary for numerical processing, are very useful during interactive program development and exploratory signal analysis [38]. This observation, in part, motivates recent interest in *symbolic* signal processing [29, 36, 37, 35], the subject of Chapter 2. Symbolic signal processing involves the manipulation of expressions for signal samples rather than numerical sample values. When such manipulations are important, information about the type or history of a signal should be explicitly recorded as part of the signal's representation, and the signal data abstraction should provide inquiry operations that allow this information to be retrieved. This requirement is called the criterion of *manifest typing* [29] or *perspicuity* [26].

1.3 STREAMS AND BLOCK DIAGRAM LANGUAGES

Block diagram languages for discrete-time signal processing were first introduced in the early 1960s with the language BLODI [5]. Since that time, numerous variations have appeared, including BLODIB [6], PATSI [7], LOTUS [8], DARE [9], MITSYN [10], CIRCUS [11], BLOSIM [12, 13], GOSPL [14], and GABRIEL [16].

Although these languages vary greatly in their syntax and implementation, they are all derived from a common model of discrete-time system realization, called *next-state simulation* (n.s.s.) [7], and are based on the use of a *stream* representation for discrete-time signals. This section briefly reviews the basic concepts of next-state simulation and streams, and then describes the hypothetical Lisp-based block diagram language BDL.

1.3.1 Next-State Simulation

Next-state simulation (n.s.s.) is a generalization of the state-space representation of linear dynamical systems. A system is *n.s.s.-realizable* if it can be represented by a set of equations of the form

$$s_k(n+1) = F_k(s_1(n)\ldots s_N(n), x_1(n)\ldots x_M(n)), \quad k = 1\ldots N$$
$$y_p(n) = G_p(s_1(n)\ldots s_N(n), x_1(n)\ldots x_M(n)), \quad p = 1\ldots P \quad (1.16)$$
$$s_k(0) = s_k^0, \quad k = 1\ldots N$$

where $x_1\ldots x_M, y_1\ldots y_P$, and $s_1\ldots s_N$ are *input*, *output*, and *state* sequences, respectively. The functions $F_1\ldots F_N$ and $G_1\ldots G_P$ are *state-update* and *observation* functions, and the values $s_1^0\ldots s_N^0$ are *initial conditions*.

A basic property of n.s.s. realizability is that it is preserved when systems are interconnected according to three fundamental composition rules or productions [45]. These productions, illustrated in Figure 1.2, correspond to the formation of an elementary *direct product*, *cascade*, and *feedback loop*. The importance of these productions is that any block diagram network may be constructed using a finite sequence of them.

A cascade or direct product of two n.s.s.-realizable systems is always n.s.s.-

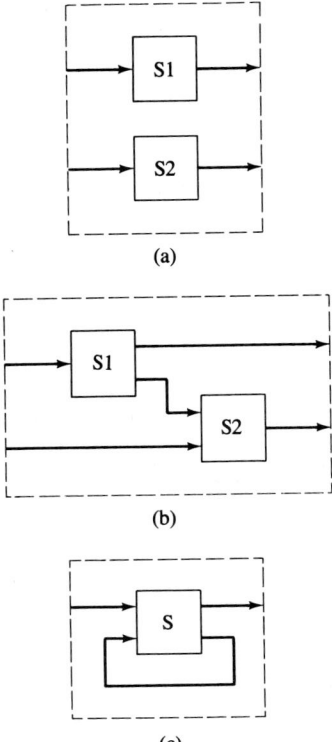

Figure 1.2 Fundamental block diagram composition rules. (a) Direct product. (b) Cascade. (c) Feedback loop.

realizable. The n.s.s. realizability of a feedback system, on the other hand, is not guaranteed in general. A feedback system will be n.s.s.-realizable, however, if the output that is fed back is *state-determined* [45]. A system is said to have state-determined output at y_i if observation function G_i has the form $G_i(s_1(n)\ldots s_N(n))$, so that it is independent of inputs $x_1(n)\ldots x_M(n)$. A simple example of a system with state-determined output is the unit delay, which has the n.s.s. realization

$$s(n + 1) = x(n)$$
$$y(n) = s(n) \qquad (1.17)$$
$$s(0) = y^0$$

Informally, if an output is state-determined it will take at least one time unit for the output to be affected by any particular input sample. The requirement of state-determined feedback is analogous to the condition that a computable digital filter structure contain no delay-free loops [43].

1.3.2 A Simple Stream Abstraction

The concept of a *stream* or *first-in-first-out queue* (FIFO) is familiar to most programmers, although perhaps not as an explicit data type. Streams are commonly used to provide sequential access to the elements of a data set. For example, streams

```
(make-stream)              ;create new (empty) stream
(stream-put val x)         ;append element to end
(stream-get x)             ;remove element from front
(stream-put-eos x)         ;append end-of-stream-token
(stream-is-empty? x)       ;test for elements in stream
```

Figure 1.3 Operations of a simple stream abstraction.

are used in many contemporary programming languages as the basic file input-output facility [52]. Similarly, streams are often used for communication between concurrent processes, as illustrated by the UNIX *pipe* construct [53].

Figure 1.3 shows a partial set of operations for a simple stream abstraction. The operation `make-stream` creates and returns a new stream object. The operation `stream-put` appends an element to the end of a stream, and `stream-get` removes an element from the front of a stream. The operation `stream-put-eos` appends a special *end-of-stream* token to the end of a stream. When returned by `stream-get`, the end-of-stream token signals that the stream will contain no additional elements. Finally, the operation `stream-is-empty?` is a predicate that returns `t` or `nil` according to whether or not there are elements waiting to be read. A complete specification for a stream abstraction would also define the response of a stream to exceptional conditions such as invoking `stream-get` on an empty stream or invoking `stream-put` after `stream-put-eos`. For the purpose of discussing signal representation, however, the operations identified in Figure 1.3 are sufficient.

1.3.3 The Block Diagram Language BDL

The form of (1.16) suggests a simple algorithm for computing the outputs of an n.s.s.-realizable system, using a stream representation for the input, output, and state sequences. The output samples are computed in order of increasing sample index (i.e., $n = 0\ldots$) during successive passes through the body of a main loop. During each pass, samples $s_1(n)\ldots s_N(n)$ and $x_1(n)\ldots x_M(n)$ are removed from their respective state or input streams. Next, procedures that implement $F_1\ldots F_N$ and $G_1\ldots G_P$ are applied to these values to obtain $s_1(n+1)\ldots s_N(n+1)$ and $y_1(n)\ldots y_P(n)$. Finally, the computed values are appended to the state and output streams. This cycle is repeated for each vector of input samples. Before the iteration begins the initial values of the state variables are appended to the state streams.

Block diagram programming languages are based on the implementation of n.s.s.-realizable systems by algorithms with the above general structure. A block diagram language typically provides a set of primitive systems, whose n.s.s. realizations are known, plus facilities for describing networks of these primitives and deriving their realizations. Some block diagram languages are embedded in general-purpose programming languages such as C and allow the user to create new types of primitive operators by implementing them in the base language [12, 20, 16].

Figure 1.4 shows the implementation of three simple block types—adder,

```
(defprimtype adder
    :inputs (x₁ x₂)
    :outputs (y)
    :fire (stream-put ( + (stream-get x₁)
                          (stream-get x₂))
                      y))
```

(a)

```
(defprimtype scalor
    :inputs (x)
    :outputs (y)
    :parameters (a)
    :fire (stream-put ( × a (stream-get x))
                      y))
```

(b)

```
(defprimtype unit-delay
    :inputs (x)
    :outputs (y)
    :parameters (initial-value)
    :state (s)
    :init (stream-put initial-value s)
    :fire (progn
              (stream-put (stream-get s) y)
              (stream-put (stream-get x) s)))
```

(c)

Figure 1.4 Primitive BDL block types for first-order filter. (a) Adder. (b) Scalor. (c) Unit delay.

scalor, and unit-delay—using defprimtype, the primitive block definition form of BDL. The definition of a two-input, single-output adder corresponding to (1.15) is given in Figure 1.4(a). The :inputs and :outputs clauses of the definition specify that the input and output "ports" of an adder are named x_1, x_2, and y, respectively. These ports are connected to streams when an instance of an adder is created. The :fire clause specifies the body of a procedure (a *fire* method) which is invoked during each pass through the n.s.s. computation cycle. The fire method implements the state-update and observation functions and performs input/output operations on the streams connected to the block ports. The fire method of the adder retrieves a pair of elements from input streams x_1 and x_2 and appends the sum to output stream y.

Figure 1.4(b) shows an implementation of a single-input, single-output

Sec. 1.3 Streams and Block Diagram Languages

scalor defined by (1.9). This example illustrates the concept of block *parameters*. The :parameters option indicates that the class of scalor blocks is parameterized by *a*, the applied gain. A value for this parameter is supplied whenever an instance of type scalor is created.

The adder and scalor are examples of memoryless systems implemented without the use of a state sequence. Figure 1.4(c) illustrates the use of a state stream to implement the memory of the unit-delay defined by (1.17). The unit-delay fire method removes $s(n)$ and $x(n)$ from the state and input streams and appends them to the output and state streams, respectively. The initial value s^0 is appended to the state stream by the unit-delay *init* method, a function that is called whenever a new instance of a unit-delay is created.

As Figure 1.4 illustrates, in many block diagram languages the state-update and observation functions are combined into a single fire method rather than implemented as separate procedures. In addition, it is common for the state vector to be maintained by the fire method internally, rather than implemented as an explicit state stream. Using a fire method rather than separate state-update and observation procedures simplifies the programming of many block types. However, when fire methods are used, it is no longer possible to construct a feedback loop around every system with state-determined output [26].

The examples in Figure 1.4 illustrate the definition of primitive block types, defined by explicitly supplying a fire method written in a general-purpose programming language. The final BDL example illustrates the definition of a block diagram *network*, an interconnected collection of block operator instances. In a typical block diagram language, a fire method for a network is generated automatically by a compiler which sorts the component blocks according to the data dependencies implied by the network topology.

Figure 1.5 shows the definition of first-order-filter, a specification for the network shown in Figure 1.1. A first-order-filter network contains three "parts," which are instances of types adder, scalor, and unit-delay, respectively. Each component of the :parts clause is a list describing one embed-

```
(defnettype first-order-filter
    :inputs (x)
    :outputs (y)
    :parameters (gain)
    :parts ((adder :type adder)
            (scalor :type scalor :parameters ((a gain)))
            (delay :type unit-delay :parameters ((initial-value 0))))
    :connections ((x (adder x₁))
                  (y (adder y) (delay x))
                  ((delay y) (scalor x))
                  ((scalor y) (adder x₂))))
```

Figure 1.5 BDL definition of first-order filter network type.

ded block instance. Each description consists of a name to be assigned to the block instance, the type of the block, and the values of the block parameters.

The :connections clause specifies the sets of block ports that are connected together. For example, "((delay y) (scalor x))" specifies that output *y* of the unit-delay block is connected to input *x* of the scalor. Each set of connected ports consists of exactly one output port and one or more input ports. A separate stream is associated with each input port; the block diagram compiler arranges for the samples produced at the output port to be copied to each of the streams.

1.3.4 Discussion of Block Diagram Languages

The criterion of uniform reference requires a signal representation to support dimensions and fetch, defined by (1.12), as generic inquiry operations which can be applied to any signal. The stream does not support a dimensions operation since the "length" of a stream is not a well-defined concept. Given a stream object, the only way to determine the length of the represented signal is by repeatedly invoking stream-get until the end-of-stream token is encountered. This strategy will fail if stream-put-eos was not previously invoked.

The strictly sequential access provided by stream-get does not satisfy the requirement for a "random access" fetch operation. The lack of random access makes streams inconvenient for implementing noncausal systems such as the discrete Fourier transform (DFT) [43]

$$y(k) = \sum_{n=0}^{N-1} x(n) e^{-j2\pi kn/N}, \qquad k = 0 \ldots N-1 \qquad (1.18)$$

which requires all of the input samples in order to compute each of the output samples. Similarly, sequential access complicates the implementation of "multirate" systems whose inputs and outputs have different lengths, such as time-domain decimators.

Difficulties with noncausal and multirate systems reflect the fact that the class of n.s.s.-realizable systems, while very large, excludes many systems of practical importance. Some recent block diagram languages are based on a more general model of computation in which independent concurrent processes communicate via streams [12, 20, 16]. These languages support the implementation of a wider class of systems than do n.s.s.-based languages. However, the greater generality appears to reflect the fact that these languages are more like general-purpose programming languages for parallel processing than specialized signal processing languages.

The immutability criterion requires that the observable properties of a signal (i.e., its dimensions and sample values) remain invariant after the signal is created. Since successive invocations of stream-get may return different values, it is clear that stream objects per se are mutable. However, the values returned by `stream-get` during successive invocations correspond to different signal samples. Since each sample is returned only once, it is vacuously true that its value will be the same every time it is observed. Thus it may be argued that streams satisfy the signal immutability requirement.

The criterion of dimensional extensibility requires a signal abstraction to support both one- and multidimensional signal processing. Streams do not satisfy the criterion of dimensional extensibility. Fundamental to the use of streams is the assumption that all systems will access the samples of a signal in the same order. A significant subset of one-dimensional signal processing algorithms may be structured to access signal samples in the order of increasing "time" index. However, sequential streams are inappropriate as a general representation for multidimensional signals since there is no single access order that is suitable for all multidimensional systems. The order in which samples of a multidimensional signal are accessed greatly affects important performance attributes of a system, such as the dimension of its state space, its quantization noise and coefficient sensitivity properties, and the possible arithmetic parallelism [47].

The criterion of deferred evaluation requires the ability to create the representation of a discrete-time signal without computing the values of its samples. Invocations of the stream operations `stream-get` and `stream-put` may be safely interleaved. The stream mechanism ensures that each sample returned by `stream-get` was previously set by `stream-put`. Thus, the samples of a signal represented as a stream may be incrementally calculated while they are being used. Since this computation occurs after the stream object has been created, the stream abstraction supports deferred evaluation.

The criterion of manifest typing requires a signal representation to record information about the specific type of each signal. The manifest typing criterion is not satisfied by the stream. The discrete-time signal represented by a particular stream object is determined by specifying the numerical value of each sample using `stream-put`. There is no indication in the resulting representation of how those values are related to each other or how they were computed from the samples of other signals.

Although streams do not provide manifest *signal* typing, the block diagram language BDL can support manifest typing of n.s.s.-realizable *systems*. The BDL `defprimtype` form explicitly describes the n.s.s. realization of a class of primitive systems. Similarly, `defnettype` explicitly describes the network structure of a class of composite systems. Instances of a given block type in a BDL network share a common fire method and are distinguished by the values of their parameters and the streams to which they are connected. Although the previous discussion of BDL did not describe block diagram systems as explicit data abstractions, it is clear that such a representation is possible and that `defprimtype` and `defnettype` provide a syntactic framework for manifest system typing. The definition and instantiation of system objects form the basis for the use of block diagram languages in code generation for specialized signal processors [19, 16, 14, 15, 17].

1.4 ARRAY PROCESSING

The term *array processing* is usually associated with high-speed arithmetic processors that are architecturally optimized for performing computations on collections (arrays, blocks, or vectors) of signal samples [1, 2]. More generally, the term array processing may be applied to any computation in which a discrete-time signal is represented by an array containing the values of its samples. With this extended definition, array processing languages include many that are not based on specialized processors, such as FANLZ [10], ILS [23], and SIG [24]. In addition, the broader definition encompasses many of the signal processing implementation techniques commonly used with general-purpose languages such as Fortran and C [21, 22].

This section considers array processing in the extended sense. A simple array abstraction and a model for systems realizable by array processing operators are described. These form the basis for the Lisp-based array processing language AOL.

1.4.1 A Simple Array Abstraction

The concept of an array is supported as a data type in nearly all modern programming languages. Figure 1.6 shows a set of operations for a simple multidimensional array abstraction. The operation allocate-array creates and returns a new array object of specified dimensions. The elements of an array with dimensions $N_1 \ldots N_m$ have vector indices $(i_1 \ldots i_m)$, where $i_k \in [0, N_k)$. The operation array-dimensions returns a list containing the dimensions of its array argument. The operation array-fetch returns the value of an array element, and array-store changes the value of an element. As with streams, a complete array abstraction would include other operations as well as mechanisms for detecting and handling exceptional conditions.

1.4.2 The Array Operator Model

Informally, a system may be implemented by array processing if each of its input and output signals can be represented as an array containing the signal samples. This notion is formalized in the following definition. An M-input, P-output discrete-time system is *a.o.m.*-(array operator model) *realizable* if it can be represented by a set of equations of the form

$$y_p = T_p(x_1 \ldots x_M), \, p = 1 \ldots P \qquad (1.19)$$

```
(allocate-array N₁...Nₘ)      ;create new array
(array-dimensions x)           ;return list of dimensions
(array-fetch x i₁...iₘ)        ;fetch specified element
(array-store val x i₁...iₘ)    ;store specified value
```

Figure 1.6 Operations of a simple array abstraction.

Sec. 1.4   Array Processing

where $x_1 \ldots x_M$ and $y_1 \ldots y_P$ are vectors whose components are the samples of the input and output signals.

It is clear that any discrete-time system whose inputs and outputs are signals of the form (1.1) is a.o.m.-realizable. In particular, a.o.m. realizability is preserved, in principle, by the direct product, cascade, and feedback productions of Figure 2.1. A more important question is whether an a.o.m. realization of a composite system can be mechanically derived from realizations of its components. The affirmative answer to the analogous question about n.s.s. realizability is the basis for the use of next-state simulation in block diagram programming.

Constructing the a.o.m. realization of a direct product or cascade system is straightforward. On the other hand, the a.o.m. realization of a feedback loop may be formally expressed in terms of a realization of the system inside the loop only when the equated input and output signals are totally independent [26]. For example, for the system described by (1.19), y_1 may be equated to x_1 only when

$$T_1(x_1 \ldots x_M) \equiv F_1(x_2 \ldots x_M) \tag{1.20}$$

i.e., when x_1 is not required by T_1.

The independence condition necessary for automatic generation of the a.o.m. realization of a feedback loop is a much stronger constraint than the state-determinedness required for n.s.s. realizability. As a result, the role of feedback is severely limited in array processing. For example, an a.o.m. realization of the filter in Figure 1.1 cannot be constructed from a.o.m. realizations of the adder, scalor, and unit delay.

1.4.3 Array Processing Languages

The form of (1.19) suggests a simple approach to implementing a.o.m.-realizable systems. Each input and output signal is represented by an array that contains the values of its samples. The dimensions of a signal array are the same as those of the corresponding signal. A system is represented by a set of procedures that implement $T_1 \ldots T_P$; the arguments to such a *system procedure* are the arrays of input samples, and the value returned is an array containing one of the output signals.

The above general approach is the basis for the hypothetical array processing language AOL (array operator language). AOL differs slightly from the model presented previously in that, for each a.o.m.-realizable system, all of the mappings $T_1 \ldots T_P$ are implemented by a single system procedure which returns $y_1 \ldots y_P$. This difference is analogous to the use of a single `:fire` method, rather than separate state-update and observation procedures, in BDL.

Figure 1.7 contains several examples of AOL array operators. Figure 1.7(a) shows `array-2d-add`, an implementation of the two-dimensional signal adder defined by (1.15). An important feature of `array-2d-add`, which illustrates a general characteristic of AOL operators, is that it allocates the array y in which the output samples are returned. In many array processing systems, y would be among the arguments supplied to `array-2d-add` when it is invoked. By convention,

```
(defun array-2d-add (x₁ x₂)
    (loop with (n₁ n₂) = (array-dimensions x₁)
          with y = (allocate-array n₁ n₂)
          for i from 0 below n₁
          do (loop for j from 0 below n₂
                   for v₁ = (array-fetch x₁ i j)
                   for v₂ = (array-fetch x₂ i j)
                   do (array-store ( + v₁ v₂) y i j))
          finally (return y)))
```

(a)

```
(defun array-even-and-odd samples (x)
    (loop with nₓ = (first (array-dimensions x))
          with n_y = (/ nₓ 2)
          with y_e = (array-allocate n_y)
          with y_o = (array-allocate n_y)
          for i from 0 below nₓ by 2
          for j from 0 below n_y
          for v_e = (array-fetch x i)
          for v_o = (array-fetch x ( + i 1))
          do (array-store v_e y_e j)
          do (array-store v_o y_o j)
          finally (return (values y_e y_o)))))
```

(b)

Figure 1.7 Example of AOL array operators. (a) Two-dimensional signal adder. (b) Time-domain decimator.

supplying output arrays as arguments is prohibited in AOL because it introduces two serious issues—*array aliasing* and *length computation* [26].

Allowing caller-supplied output arrays raises the possibility that the same array will be used to represent both an input signal and an output signal. This situation is called *array aliasing* because the array involved becomes accessible through two distinct argument names. Array aliasing is the basis for *in-place computation* [43], an implementation technique frequently used to reduce the storage requirements of a program. In addition, array aliasing can be used to broaden the class of feedback systems that can be implemented using array operators [26]. However, array aliasing can modify the input/output mapping realized by a system procedure in unforeseen ways. Furthermore, the notion of using the same array to represent two signals violates the immutability requirement for signal representation. For these reasons, array aliasing is avoided in AOL by requiring each system procedure to allocate the arrays in which its output samples are returned.

A second problem with supplying output arrays as arguments is that the calling

Sec. 1.4 Array Processing

procedure must predict the length of the output signals. This is illustrated by `array-even-and-odd samples` in Figure 1.7(b), an implementation of the simple 2-to-1 time domain decimators defined by

$$y_e(n) = x(2 \cdot n)$$
$$y_o(n) = x(2 \cdot n + 1), n = 0 \ldots N/2 \quad (1.21)$$

where N is the length of x. The decimator is typical of the class of multiple sample rate systems whose inputs and outputs have different lengths. If arrays y_e and y_o were supplied as arguments, the procedure invoking `array-even-and-odd-samples` would need to anticipate the exact lengths of the output signals. In the case of the decimator, the output length is a relatively simple function of the input length. More generally, the output length may depend on the values of the input samples as well. An example of such a system is one that computes the histogram of a signal whose samples are nonnegative integers (e.g., digitized image data), where the length of the histogram is one greater than the maximum value of the signal. In the extreme, it is even possible that an output length cannot be computed without duplicating the computation of the entire output signal.

1.4.4 Discussion of Array Processing

The array operations `array-dimensions` and `array-fetch` satisfy the uniform reference requirement. The array operation `array-store` modifies its array argument and thus violates the immutability criterion. Although immutability may be adopted as a programming convention, by prescribing against the use of `array-store` on arrays that represent signals, the array abstraction provides no enforcement mechanism for such a convention. The array signal representation clearly provides dimensional extensibility, as `array-2d-add` in Figure 1.7(a) illustrates.

The criterion of deferred evaluation is not satisfied by the array abstraction. In order for an invocation of `array-fetch` to return the correct sample value, `array-store` must have been previously invoked to set that value. However, the array abstraction places no constraints on the relative order of `array-fetch` and `array-store` invocations. Thus, the only reliable way of ensuring that each element of an array will be properly defined when it is fetched is to compute and store all of the elements when the array is first created. The lack of deferred evaluation is the fundamental reason why arrays are generally unsuitable for representing feedback signals.

The criterion of manifest typing requires a signal representation to record information about the specific type of each signal. The manifest typing criterion is not satisfied by the array. The discrete-time signal represented by a particular array is determined by specifying the numerical value of each sample using `array-store`. There is no indication in the resulting representation of how those values are related to each other or how they were computed from the samples of other signals.

1.5 ABSTRACT SIGNAL OBJECTS

Historically, the concept of discrete-time signal as a formal data abstraction appears to have evolved in three steps. The general notion that data abstractions might be useful in signal processing was first explored in connection with the specification of specialized signal processing structures such as FIR filters [39, 40, 41].

Subsequently, the desirability of a specific type of data abstraction for signals was argued on the basis of software methodology and the elementary mathematics of discrete-time signals [26]. This work culminated in a formal computational model for discrete-time signals, called the *closure model* [26, 42], and the design and implementation of the Lisp-based signal processing languages SRL [27] and SPLICE [29], which attempt to adhere rigorously to this model. In addition, several other signal processing languages, including ISPUD [31], QuickSig [34], and the commercial signal processing package N!Power [32], have been motivated by the closure model but trade off rigorous adherence to the mathematical requirements for improved program efficiency.

Finally, interest in combining symbol manipulation techniques from the field of artificial intelligence with traditional numerical signal processing methods has recently motivated the application of knowledge representation and automated reasoning concepts to signal representation and processing [29, 35, 36, 37]. This has led to significant progress in the design and implementation of signal representations for both numerical and symbolic signal processing [29, 35].

This section reviews the basic concepts of abstract signal representation in the context of a brief review of the closure model and the signal representation language SRL.

1.5.1 The Closure Model for Signals

The closure model for signals is motivated by the use of mathematical expressions to define classes of discrete-time signals, as discussed previously. A signal expression usually contains a number of "free" parameters which define the space of represented signals; each signal in the class corresponds to a set of specific values for the parameters.

The notion of representing a signal as a parameterized procedure plus a set of values for the parameters is termed the *closure model* [26, 42] to acknowledge its similarity to the concept of a *function closure* [54]. A function closure is a representation of a function in terms of a syntactic expression for its values and a set of *bindings* for some of the symbols in the expression. Closures have been used as abstract program representations in programming language theory [54] and as concrete program representations in languages such as LISP [50].

Formally, the closure model for a discrete-time signal x of the form (1.1) is a triple (P, B, D) consisting of three components:

Sec. 1.5 Abstract Signal Objects

- A *procedure P* for computing $x(n)$ as a function of n. Formally, P is a *lambda expression* [51] of the form $P = (\text{lambda }(n)E)$, where E is an expression that may contain *free variables*.
- An *environment B* containing bindings for some of the free variables of E.
- A *dimensions* specification $D = (N_1 \ldots N_m)$ which identifies the domain of x.

For example, the closure model for a length-256 complex exponential with $\omega = \pi/5$ is

$$P = (\text{lambda }(n)e^{j\omega n})$$
$$B = [\omega : \pi/5] \qquad (1.22)$$
$$D = (256)$$

Figure 1.8 shows a graphical representation of the closure model for output y of the first-order filter of Figure 1.1. Comparing this graph with Figure 1.1 reveals that the arcs in Figure 1.1 (which represent signals) correspond to the boxes in Figure 1.8. This correspondence is similar to the duality of signal flow graphs and block diagrams in classical network theory.

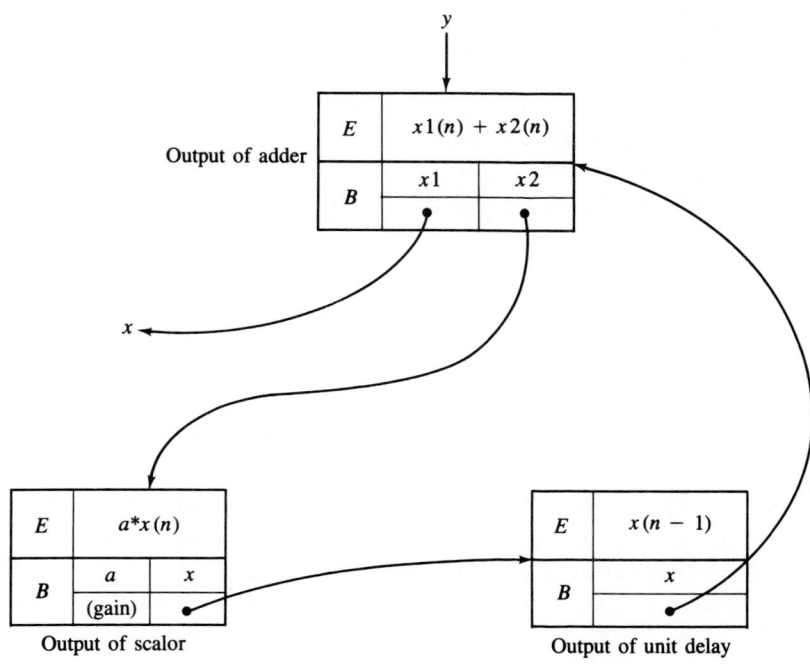

Figure 1.8 Closure model for output of first-order filter.

1.5.2 The Signal Representation Language (SRL)

The concepts and facilities of SRL are based on the closure model and are explicitly designed to satisfy the signal representation criteria of section 1.2. The fundamental activity in SRL programming is the implementation of *signal types*. A signal type is a representation for a class of signals that share a common parameterized procedure for computing the sample values. Instances of a signal type are created by binding the parameters of this procedure. The basic observable properties of a signal are the dimensions of its domain and the values of its samples. Furthermore, signal objects are immutable, so that these properties remain fixed after a signal is created.

The following is a brief overview of SRL signal inquiry operations and facilities for defining signal types.

Numerical Signal Inquiry Operations

Figure 1.9 lists the basic inquiry operations on SRL signal objects. These operations are divided into two groups: Figure 1.9(a) identifies "numerical" operations which are motivated by the mathematical definition of signal equality, and Figure 1.9(b) contains examples of "symbolic" operations which return history-related information about signal objects. The numerical operations are described immediately below; the symbolic operations will be described after the discussion of signal type definitions.

The signal inquiry operations in Figure 1.9(a) are similar to the array inquiry

```
(signal-dimensions x)              ;return list of dimensions
(signal-fetch x i₁ ... i_N)        ;return value of single sample
(with-signal-values ( (var₁ s₁)    ;bind var₁ ... var_N to
                      .             ;arrays containing
                      .             ;samples of s₁ ... s_N
                      .
                     (var_N s_N))
    ...⟨body⟩...                    ;read-only access to arrays
)
```

(a)

```
(≫ name x)                         ;return value of parameter
(signal-what x)                    ;return list of finder and parameters
(signal-ancestors x)               ;return list of direct ancestor signals
(signal-dependents x)              ;return list of direct dependent signals
```

(b)

Figure 1.9 SRL signal inquiry operations. (a) Numerical operations. (b) Symbolic operations.

operations of Figure 1.6. The operation `signal-dimensions` returns a list containing the dimensions of its argument. The operation `signal-fetch` returns the value of a specified signal sample.

In principle, the single-element access provided by `signal-fetch` is an adequate means of obtaining the samples of any signal. In practice, however, a problem arises because such access is not directly compatible with array-oriented algorithms, such as the fast Fourier transform (FFT) [43], which compute the values of many samples simultaneously.

Although a basic approach to eliminating the mismatch between `signal-fetch` and array-oriented algorithms is easily identified (i.e., buffering), formulating a detailed strategy requires addressing a number of issues [27, 29]. For example, storage limitations may make it impractical to allocate a separate array for each signal that requires one for buffering. On the other hand, the desired immutability of signals may be compromised if the same array is used to buffer the samples of more than one signal. An important practical consideration in the design of signal representation languages such as SRL is developing an array access discipline that allows storage to be shared while preserving signal immutability.

The SRL `with-signal-values` form is an inquiry operation for obtaining an array containing the samples of a signal. The body of the form is a collection of expressions requiring access to the samples of signals $s_1 \ldots s_N$. SRL executes the body after binding local variables $var_1 \ldots var_N$ to arrays containing the requested samples. The dimensions of each array are the same as those of the corresponding signal; the arrays contain *all* of the signal samples.

User access to a signal values array within the body of a `with-signal-values` form is subject to the following two restrictions:

1. A signal values array should be regarded as a temporary object that "disappears" when evaluation of the body terminates. The array should not be made part of any data structure that is accessible outside the dynamic scope of the `with-signal-values` form.
2. A signal values array should be regarded as "read-only" and must not be modified in any way. In particular, operations that change the values of array elements or the array dimensions are prohibited.

Adherence to these restrictions allows the formulation of a simple and efficient array storage management strategy that is consistent with the requirement of signal immutability [27]. Alternative strategies for signal values array management are also possible. For example, SPLICE [29, 30] implements a sophisticated copying mechanism that allows signal arrays to be read/write.

Signal Type Definitions

The primary activity in SRL programming is the implementation of signal types. Fundamentally, implementing a type involves supplying mechanisms for computing the dimensions and sample values of any signal of the type. In concept, a signal type is similar to a Smalltalk *class* [55] or CLOS *class* [56].

Figure 1.10 illustrates the use of `defsigtype`, the SRL signal type definition form. Figure 1.10(a) shows a definition of `2d-signal-sum`, a signal type for the output of the two-dimensional signal adder defined by (1.15). The `:parameters` clause of a signal type definition identifies the parameters by which individual signals of the type are distinguished; the parameters of a `2d-signal-sum` are x_1 and x_2, the two signals whose sum is represented. The `:finder` clause specifies the name to be given to an automatically generated function (the signal *finder*) which returns

```
(defsigtype 2d-signal-sum
    :parameters (x₁ x₂)
    :finder signal-2d-sum
    :init (setq-my dimensions (signal-dimensions x₁))
    :fetch ((i₁ i₂)
           ( + (signal-fetch x₁ i₁ i₂)
               (signal-fetch x₂ i₁ i₂))))
```

(a)

```
(defsigtype 2d-signal-sum
    :parameters (x₁ x₂)
    :finder signal-2d-sum
    :init (setq-my dimensions (signal-dimensions x₁))
    :values (with-signal-values ((x₁-vals x₁)
                                 (x₂-vals x₂))
              (array-2d-add x₁-vals x₂-vals)))
```

(b)

```
(defsigtype first-order-filter-output
    :parameters (x gain)
    :finder signal-first-order-filtered
    :init (setq-my dimensions (signal-dimensions x))
    :fetch ((n)
           (if ( < n 0)
               0.0
               ( + (signal-fetch x n)
                   (* gain
                      (signal-fetch self (− n 1)))))))
```

(c)

Figure 1.10 Examples of SRL signal type definitions. (a) Sum of two-dimensional signals. (b) Array-based two-dimensional signal sum. (c) First-order recursive filter.

Sec. 1.5 Abstract Signal Objects

instances of the signal type. The arguments to a finder are the values of the signal parameters. For example, a `2d-signal-sum` representation for the sum of signals s_1 and s_2 may be obtained by the invocation

$$(\texttt{signal-2d-sum}\ s_1 s_2) \qquad (1.23)$$

The `:fetch` clause specifies the argument list and body of a parameterized procedure (the *fetch method*) which returns a sample of any signal of the class. The fetch method is invoked by the generic inquiry operation `signal-fetch`. The parameters of the signal are accessed as free variables within the body of the fetch method. The fetch method of `2d-signal-sum` is a straightforward implementation of (1.15).

Figure 1.10(b) shows an alternate implementation of `2d-signal-sum` that is based on the use of arrays. The `:values` clause specifies the body of a procedure (the *values method*) which generates the signal values array returned by the inquiry operation `with-signal-values`. The values method in Figure 1.10(b) obtains the values arrays for x_1 and x_2 using `with-signal-values` and generates the output array using the AOL function `array-2d-add` from Figure 1.7(a).

Finally, Figure 1.10(c) shows `first-order-filter` output, a simple SRL representation of output y of the system shown in Figure 1.1. The fetch method computes $y(n)$ using (1.3) by recursively invoking itself to compute $y(n)$ if $n \geq 0$.[3]

Although the fetch method in Figure 1.10(c) is functionally correct, it is based on an inefficient algorithm that computes $y(0)\ldots y(n-1)$ every time $y(n)$ is requested. As a result, if all of the samples of an N-point `first-order-filter-output` signal are fetched, each sample will be computed, on the average, $(N+1)/2$ times. A variety of approaches to eliminating this general type of inefficiency are possible, based on the use of a *cache buffer* to store computed sample values for subsequent retrieval [26, 29].

Figure 1.10(c) demonstrates that SRL supports the self-reference capability that is necessary to implement feedback signals as primitive signal types. It is also possible to implement feedback by interconnecting a collection of signal objects into a closed loop of parameter bindings [26]. For example, output y of the first-order filter in Figure 1.1 may be represented by constructing a network of adder, scalor, and unit delay output signals corresponding to Figure 1.8.

The fetch and values methods illustrated in Figure 1.10 represent two common ways of specifying the computation of signal sample values—by pointwise operations and by array operators. Other algorithmic structures for sample calculation may also be supported in a signal representation language. For example, SPLICE includes a *state-machine* model which allows n.s.s.-realizable systems to be described in a natural way and a *composition* model for building more complicated signal definitions out of simpler ones [29].

---

[3] SRL binds the variable *self* to a signal object when the fetch method for the object is activated.

Symbolic Signal Inquiry Operations

Figure 1.9(b) identifies several signal inquiry operations that return "nonmathematical" information about how a signal was created and the values of its parameters. The operation $\gg$ is an inspection function which retrieves the value of a specified parameter of a signal. The operation `signal-what` returns a brief description of a signal in the form of a list containing the name of the signal finder and the values of the signal parameters. This list can be evaluated as a Lisp expression to obtain the signal it describes. The operation `signal-ancestors` returns a list containing those parameters of a signal that are themselves signals. Conversely, the operation `signal-dependents` returns a list of all signals for which the specified signal is a parameter.

1.5.3 Discussion of Abstract Signal Objects

The SRL operations `signal-dimensions` and `signal-fetch` clearly satisfy the uniform reference requirement. The SRL inquiry operations `signal-dimensions` and `signal-fetch` cannot be used to modify the observable properties of a signal. On the other hand, immutability may be violated using `with-signal-values` if the restrictions on array access are not obeyed. However, while there are no mechanisms to enforce these restrictions, the existence of a simple and syntactically supported convention contributes to voluntary compliance.

The SRL signal representation clearly provides dimensional extensibility, as `2d-signal-sum` in Figure 1.10(a) illustrates. SRL supports deferred evaluation for signals whose samples are computed using fetch methods. The samples of such a signal are calculated only when they are explicitly requested. Thus, for example, it is possible to implement feedback structures, as Figure 1.10(c) illustrates. A signal type implemented using a values method does not support deferred evaluation, since all samples are computed simultaneously; such signals are generally ill-suited for feedback applications.

The signal representation language SRL provides manifest signal typing. The inquiry operations in Figure 1.9(b) illustrate that information related to the history of a signal object is explicitly recorded and visible.

1.6 CONCLUSIONS

Five signal representation criteria—uniform reference, immutability, dimensional extensibility, deferred evaluation, and manifest typing—are motivated by elementary observations about the mathematics of discrete-time signals. These criteria are used to assess the signal representations used in block diagram languages, array processing languages, and the signal representation language SRL. Table 1.1 evaluates the strengths and weaknesses of each representation in terms of the five criteria.

TABLE 1.1 SUMMARY APPRAISAL OF SIGNAL REPRESENTATIONS.

Requirement	Stream	Array	SRL
Uniform reference	No	Yes	Yes
Immutability	Yes	No	Yes
Dimensional extensibility	No	Yes	Yes
Deferred evaluation	Yes	No	Yes
Manifest typing	No	No	Yes

REFERENCES

[1] J. Allen, "Computer Architecture for Signal Processing," *Proc. IEEE*, 63, no. 4 (April 1975).

[2] A. Salazar, ed., *Digital Signal Computers and Processors* (New York: IEEE Press, 1977).

[3] L. Morris, "Automatic Generation of Time-Efficient Digital Signal Processing Software," *IEEE Trans. Acoustics, Speech, and Signal Processing*, ASSP-25 (February 1977).

[4] L. Morris and C. Mudge, "Speed Enhancement of Digital Signal Processing Software via Microprogramming a General Purpose Minicomputer," *IEEE Trans. Acoustics, Speech, and Signal Processing*, ASSP-26 (April 1978).

[5] J. Kelley, C. Lochbaum, and V. Vyssotsky, "A Block Diagram Compiler," *BSTJ*, 40, no. 3 (May 1961), 669–76.

[6] B. Karafin, "A Sampled Data System Simulation Language," in *System Analysis by Digital Computer*, eds. F. Kuo and J. Kaiser (New York: Wiley, 1966).

[7] B. Gold and C. Rader, *Digital Processing of Signals* (New York: McGraw-Hill, 1969).

[8] M. Dertouzous, M. Kaliske, and K. Polzen, "On-Line Simulation of Block Diagram Systems," *IEEE Trans. Comput.*, C-18, no. 4 (April 1969).

[9] G. Korn, "High-Speed Block-Diagram Languages for Microprocessors and Minicomputers in Instrumentation, Control, and Simulation," *Computers in Elec. Eng.*, 4 (1977), 143–59.

[10] W. Henke, "MITSYN—An Interactive Dialog Language for Time Signal Processing," RLE TM-1, Cambridge, Mass.: MIT (February 1975).

[11] T. Crystal and L. Kulsrud, "Circus," Institute for Defense Analysis, Princeton, N.J., CRD Working Paper 435 (1974).

[12] D. Messerschmitt, "A Tool for Structured Functional Simulation," *IEEE Journal on Selected Areas in Communications*, SAC-2, no. 1 (January 1984), 137–47.

[13] D. Hait, "The BLOSIM Simulation Program" (Master's thesis, University of California, Berkeley, November 1985).

[14] C. Covington, G. Carter, and D. Summers, "Graphic Oriented Signal Processing Language—GOSPL," in *Proc. 1987 IEEE Int. Conf. Acoustics, Speech, and Signal Processing*, Dallas, Tex. (1987).

[15] M. Zissman, G. O'Leary, and D. Johnson, "A Block Diagram Compiler for a Digital Signal Processing MIMD Computer," in *Proc. 1987 IEEE Int. Conf. Acoustics, Speech, and Signal Processing*, Dallas, Tex. (1987).

[16] E. Lee, W.-H. Ho, E. Goei, J. Bier, and S. Bhattacharyya, "GABRIEL: A Design Environment for DSP," *IEEE Trans. Acoust., Speech, Signal Processing*, ASSP-37, no. 11 (November 1989).

[17] M. Karjalainen and S. Helle, "Block Diagram Compilation and Graphical Editing of DSP Algorithms in the QuickSig System," *Proc. IEEE Int. Symp. Circuits and Systems*, Espoo, Finland (1988).

[18] D. Bursky, "Software Tool Kit Eases Design of DSP Algorithms," *Electronic Design* (June 25, 1987).

[19] J. Kuusama, "A Prototype of an Integrated Signal Processing Environment" (Master's thesis, Tampere University of Technology, Tampere, Finland, 1988).

[20] D. Johnson and R. Vaughan, "A Software Environment for Digital Signal Processing Simulations," *Circuits, Syst., Signal Processing*, 6 (January 1987), 31–44.

[21] Digital Signal Processing Technical Committee, IEEE Acoust., Speech, Signal Processing Society, *Programs for Digital Signal Processing* (New York: IEEE Press, 1979).

[22] W. H. Press, B. P. Flannery, S. A. Teukolsky, and W. T. Vetterling, *Numerical Recipes in C* (Cambridge: Cambridge University Press, 1988).

[23] *The Interactive Laboratory System* (*ILS*), Signal Technology Inc. (Goleta, Calif., 1989).

[24] D. Lager and S. Azevedo, "SIG—A General-Purpose Signal Processing Program," *Proc. IEEE*, 75, no. 9 (September 1987), 1322–32.

[25] *Matlab™ for Macintosh Computers*, The Mathworks, Inc. (South Natick, Mass., 1989).

[26] G. Kopec, "The Representation of Discrete-Time Signals and Systems in Programs" (Ph.D. dissertation, MIT, Cambridge, Mass., 1980).

[27] G. Kopec, "The Signal Representation Language SRL," *IEEE Trans. Acoustics, Speech, and Signal Processing*, ASSP-33, no. 4 (August 1985).

[28] G. Kopec, "An Overview of Signal Representations in Programs," in *VLSI and Modern Signal Processing*, eds. S.-Y. Kung, H. Whitehouse, and T. Kailath (Englewood Cliffs, N.J.: Prentice Hall, 1985).

[29] C. Myers, "Signal Representation for Symbolic and Numerical Processing," RLE Technical Report 521, Cambridge, Mass.: MIT (August 1986).

[30] W. Dove, C. Myers, and E. Milios, "An Object-Oriented Signal Processing Environment: The Knowledge-Based Signal Processing Package," RLE Technical Report 502, Cambridge, Mass.: MIT (October 1984).

[31] P. Peterson and J. Frisbie, "An Interactive Environment for Signal Processing on a VAX Computer," in *Proc. 1987 IEEE Int. Conf. Acoust., Speech, Signal Processing*, Dallas, Tex. (1987).

[32] *N!Power™*, Signal Technology Inc. (Goleta, Calif., 1990).

[33] M. Karjalainen, T. Altosaar, and P. Alku, "Quicksig—An Object-Oriented Signal Processing Environment," in *Proc. 1988 IEEE Int. Conf. Acoustics, Speech, and Signal Processing*, New York, N.Y. (1988), pp. 1682–85.

[34] M. Karjalainen, "Object-Oriented Programming: A Case Study of Quicksig," *IEEE ASSP Magazine* (April 1990), 21–31.

[35] M. Covell, "An Algorithm Design Environment for Signal Processing," RLE Technical Report 549, Cambridge, Mass.: MIT (December 1989).

[36] W. Dove, "Knowledge-Based Pitch Detection," RLE Technical Report 513, Cambridge, Mass.: MIT (June 1986).

[37] E. Milios, "Signal Processing and Interpretation Using Multilevel Signal Abstractions," RLE Technical Report 516, Cambridge, Mass.: MIT (June 1986).

[38] G. Kopec, "The Integrated Signal Processing System ISP," *IEEE Trans. Acoustics, Speech, and Signal Processing*, ASSP-32, no. 4 (August 1984).

[39] H. Gethoffer, A. Lacroix, and R. Reis, "A Unique Hardware and Software Approach for Digital Signal Processing," in *Proc. 1977 IEEE Int. Conf. Acoustics, Speech, and Signal Processing*, Hartford, Conn. (1977).

[40] H. Gethoffer, K. Hoffmann, A. Lenzer, N. Roeth, and H. Waldschmitt, "A Design and Computing System for Signal Processing Applications," in *Proc. 1979 IEEE Int. Conf. Acoustics, Speech, and Signal Processing*, Washington, D.C. (1979).

[41] H. Gethoffer, "SIPROL: A High Level Language for Digital Signal Processing," in *Proc. 1980 IEEE Int. Conf. Acoustics, Speech, and Signal Processing*, Denver, Colo. (1979).

[42] G. Kopec, "A High Level Block Diagram Signal Processing Language," in *Proc. 1979 IEEE Int. Conf. Acoustics, Speech, and Signal Processing*, Washington, D.C. (1979).

[43] A. Oppenheim and R. Schafer, *Digital Signal Processing* (Englewood Cliffs, N.J.: Prentice Hall, 1975).

[44] D. Dudgeon and R. Mersereau, *Multidimensional Digital Signal Processing* (Englewood Cliffs, N.J.: Prentice Hall, 1984).

[45] L. Zadeh and C. Desoer, *Linear System Theory: The State-Space Approach* (New York: McGraw-Hill, 1963).

[46] A. Willsky, *Digital Signal Processing and Control and Estimation Theory: Points of Tangency Areas of Intersection, and Parallel Directions* (Cambridge, Mass.: MIT Press, 1979).

[47] D. Chan, "Theory and Implementation of Multidimensional Discrete Systems for Signal Processing" (Ph.D. dissertation, MIT, 1978).

[48] B. Liskov and V. Berzins, "An Appraisal of Program Specifications," in *Research Directions in Software Technology* ed. (Cambridge, Mass.: MIT Press, 1979).

[49] B. Liskov, A. Snyder, R. Atkinson, and C. Schaffert, "Abstraction Mechanisms in CLU," *CACM*, 20, no. 8 (August 1977), 564–76.

[50] G. L. Steele, *Common Lisp*, 2nd ed. (Bedford, Mass.: Digital Press, 1990).

[51] H. Abelson and G. Sussman, *Structure and Interpretation of Computer Programs* (Cambridge, Mass.: MIT Press, 1985).

[52] B. Kernighan and D. Ritchie, *The C Programming Language* (Englewood Cliffs, N.J.: Prentice Hall, 1978).

[53] G. Anderson and P. Anderson, *The UNIX C Shell Field Guide* (Englewood Cliffs, N.J.: Prentice Hall, 1986).

[54] J. Stoy, *Denotational Semantics* (Cambridge, Mass.: MIT Press, 1977).

[55] A. Goldberg and D. Robson, *Smalltalk-80: The Language and Its Implementation* (Reading, Mass.: Addison-Wesley, 1983).

[56] S. Keene, *Object-Oriented Programming in Common Lisp* (Reading, Mass.: Addison-Wesley, 1989).

2

COMPUTER-AIDED ALGORITHM DESIGN AND REARRANGEMENT

- Michele M. Covell
  SRI International
- Cory S. Myers
  Lockheed Sanders, Inc.
- Alan V. Oppenheim
  Massachusetts Institute of Technology

2.1 INTRODUCTION

The design of signal processing systems typically involves a high-level specification of the requirements of the system, development of an appropriate algorithm or set of algorithms to accomplish these requirements, and the implementation of the algorithms in an appropriate technology. Often these stages are not independent, and in particular, the detailed structure of the algorithms for implementing the system needs to take into account a variety of cost measures and options relating to system requirements, such as speed, modularity, and so on, and a variety of architectural constraints or technologies available for the implementation. For these reasons, it is important that algorithm design explore a wide variety of implementations which are found by exploiting the underlying mathematics of signal processing and taking into account a variety of cost measures. This is usually done by a design engineer with a detailed knowledge of and insight into a variety of transformations. From an input/output point of view, these transformations result in equivalent signal processing operations, but may have very different implications with regard to various cost measures.

In carrying out signal processing system design, there are a few design tools currently available, but these primarily provide a convenient environment for writing programs or for testing algorithms on data. There are no systems available presently that remap signal processing algorithms, specified at a high level of abstraction, to algorithmic descriptions that are more efficient in the context of particular implementational or architectural constraints. Such a design environment, if successful, might generate designs that improve on those that would be generated by experienced systems designers. A more likely and also highly desirable outcome, at least in the short run, is an environment that, with modest human intervention, achieves fast designs of signal processing algorithms that are reasonably efficient in relation to hand designs by sophisticated systems designers. The potential ability to do this in a signal processing context stems from the fact that, for signal processing, algorithmic transformations tend to follow a clearly defined set of mathematical rules. Based on these rules, the space of equivalent algorithms can, in principle, be explored to determine those that are most efficient using appropriate cost measures.

This chapter attempts to demonstrate both the feasibility and the advantages of a signal processing design environment which incorporates symbolic algorithm manipulation along with numerical processing. In particular, we focus on the problem of developing automated tools to support algorithm manipulation. As viewed in this chapter, algorithm manipulation includes property and transform analyses and algorithm rearrangement. Property and transform analyses provide information about an algorithm or its output signal. For example, determination that the output of an FFT will be conjugate symmetric because its input is real, is property analysis. Some examples of properties that are widely used in signal processing are computational cost, stability, causality, symmetry, linearity, and time-invariance. The use of a z-domain representation of a linear, time-invariant system to determine stability is an example of transform analysis. An example of algorithm rearrangement is shown in Figure 2.1. Figure 2.1(a) describes one implementation of a noninteger sampling rate conversion: upsample by five, low-pass filter, and downsample by four. This simple description of a noninteger sampling rate conversion is easily designed and implemented but is less computationally efficient than the implementations shown in Figures 2.1(b) and (c), which also provide 4 : 5 noninteger sampling rate conversions. The desired capabilities of an environment that integrates algorithm definition, manipulation and analysis, along with numerical processing, are discussed in section 2.3, along with the constraints that these capabilities impose.

In order to demonstrate the advantages of a design environment that combines symbolic algorithm manipulation with numerical processing, two application areas are considered: noninteger sampling rate conversion and code-division sonar imaging. Both of these application areas are discussed in section 2.2. Algorithms for these applications are defined and manipulated in ADE [1]. ADE (A Design Environment), described in more detail in section 2.4, is an experimental environment that shows the feasibility of a signal processing workstation with integrated tools for the specification of algorithms, computer-aided analysis, and manipulation of those algorithms, as well as application of the algorithms to numerical sequences.

ADE is based on E-SPLICE [2], the first system to demonstrate automated property analysis and algorithm manipulation for digital signal processing. ADE and

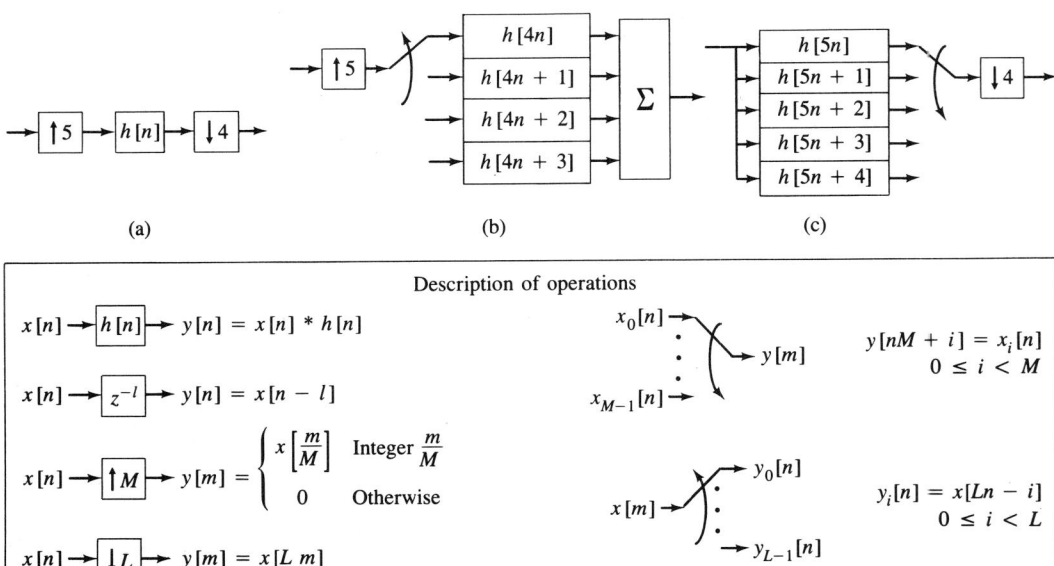

Figure 2.1 Alternate implementations of a 4:5 sampling rate conversion.

E-SPLICE are the results of a sequence of research efforts into integrated signal processing environments. These research efforts started in the late 1970s with Kopec's development of SRL [Chapter 1 this volume; 3]. SRL provides data abstractions that both reflect the basic characteristics of signals and support numeric manipulations. SPLICE [2, 4] resulted from an effort to improve the computer representation of signals beyond the work that had already been done by Kopec. Subsequently, with E-SPLICE, Myers [2] expanded the scope of the software environment to include symbolic manipulation of algorithms in addition to the previously supported numerical manipulation of signals. ADE [this chapter; 1] refines the symbolic rearrangement capabilities of E-SPLICE, making possible the manipulation of large signal processing expressions.

2.2 SIGNAL PROCESSING TOPICS USED IN EXAMPLES

In this chapter, two examples, noninteger sampling rate conversion and sonar FSK-code detection, are used to illustrate the potential of an integrated signal processing environment that combines algorithm definition, manipulation, and analysis with numerical processing. These applications are described in this section.

2.2.1 Noninteger Sampling Rate Conversion

Discrete-time sequences are often used to represent a bandlimited, continuous-time signal. Often, it is desirable to change the sampling rate that defines the conversion between discrete and continuous time. For example, film is shot at 24 frames/sec.

Sec. 2.2   Signal Processing Topics Used in Examples

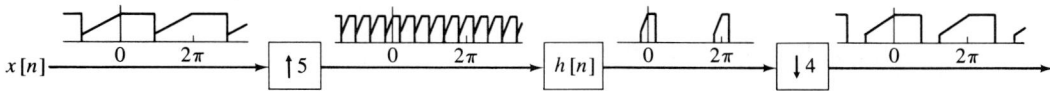

Figure 2.2 Effect of a 4:5 sampling rate conversion in the Fourier domain.

To display the film at the correct rate on American television, the temporal sampling rate must be changed to 30 frames/sec.[1] This 4:5 sampling rate conversion can be implemented using the structure shown in Figure 2.1(a). The effect of these operations in the frequency domain is shown in Figure 2.2. The low-pass filter between the upsampling and downsampling operations is necessary to prevent temporal aliasing. Unfortunately, this filter operates at a high data rate: for each four input points and five output points of the overall system, twenty input and output points pass through the low-pass filter. Figures 2.1(b) and (c) show two alternate implementations in which the filter runs at lower rates. Figure 2.1(b) shows a polyphase implementation of the filter/downsample part of the sampling rate conversion: the four, shorter filters each operate at one fourth the rate of the filter in Figure 2.1(a) and, if the convolutions are implemented directly, the computational requirements are reduced by a factor of four from the original implementation. Figure 2.1(c) shows a polyphase implementation of the upsample/filter part of the sampling rate conversion: the five, shorter filters each operate at one fifth the rate of the original filter and, again assuming direct implementation of the convolutions, the computational requirements are reduced by a factor of five.

The two polyphase implementations shown in Figure 2.1(b) and (c) are well documented in multirate filtering literature [5]. Another polyphase implementation for noninteger sampling rate conversion, which has only recently been documented in signal processing literature, was generated by E-SPLICE [2]: this alternate implementation will be introduced and discussed in section 2.5.

2.2.2 Modulated Filter Banks and Short-Time Fourier Transforms

Conventional sonar imaging systems achieve spatial resolution either through the use of a single, swept beam or through the use of multi-element arrays. These techniques, while highly successful, present some inherent difficulties. In the case of the single swept beam, the time required to scan through the desired aperture can result in the failure to detect transients. When multi-element arrays are used instead, the hardware requirements necessary to achieve high resolution can result in a large, costly system. Jaffe and Richardson [6] propose an alternative to these two techniques using the simultaneous transmission of a set of coded waveforms.

The transmitter in the proposed system is a set of N transducers, each illuminating a different direction and each transmitting a distinct signal, S_i for

---

[1] Each frame in American television is made up of two vertically interlaced fields. Sixty fields are displayed per second.

$i = 0, \ldots, N - 1$. One wide-beam hydrophone is used as a receiver. Multiple-hypothesis testing is then used to detect and discriminate the returns from the separate beams. In order to achieve good spatial resolution, the set of signals $\{S_0, \ldots, S_{N-1}\}$ must have good signal-to-signal rejection for all possible time delays. In addition, to achieve good range resolution, each signal should have a sharply peaked autocorrelation function. Jaffe and Richardson [6] propose a specific set of FSK codes with these properties:

$$S_i(t) = \sum_{k=0}^{N-1} Re\{P_{i,k}(t)e^{-j2\pi f_c t}\} \quad \text{for } i = 0, \ldots, N - 1$$

$$P_{i,k}(t) = C_l(t - kT)$$

where $l = p_i(k)$ provides, for each i, a different permutation of the numbers $0, \ldots, N - 1$ versus k and

$$C_l(t) = w(t)e^{-j(2\pi/T)lt}$$

where $w(t) = 0$ for $t < 0$ and for $t \geq T$

Specifically, each signal is made up of a sum of N individual, uniformly spaced frequency bursts (commonly referred to as frequency chips). When $N + 1$ is prime, $p_i(k)$ can be selected such that the signals and all their circular shifts achieve maximal Hamming distance separation.[2] The window $w(t)$ allows the frequency chips to be shaped to adjust their side-lobe characteristics.

The received signal can be modeled as a superposition of the reflected energy from each of the illuminated scattering centers:

$$r(t) = \sum_{i=0}^{N-1} \sum_{m=1}^{M_i} \rho_{i,m} \sum_{k=0}^{N-1} Re\{e^{j\varphi_{i,m,k}} P_{i,k}(t - \tau_{i,m})e^{-j2\pi f_c(t - \tau_{i,m})}\}$$

The summation over i represents the superposition of the returns from different transmitter beams, and the summation over m represents the superposition of the returns from the M_i scattering centers within the ith beam. $\rho_{i,m}$ is a positive number representing the strength of the return from the mth scattering center in the ith transmitter beam: it is determined by the scattering cross section and the distance of the target. $\tau_{i,m}$ is the propagation delay for the combined forward and return paths to and from the scattering center. $\varphi_{i,m,k}$ represents a nonuniform phase distortion on the kth chip of the FSK code introduced by the scattering characteristics of the target and by the fluctuations in the propagating medium. The possibility of a Doppler frequency shift is ignored in this model.

From this model, the values of $\rho_{i,m}$ and $\tau_{i,m}$ as a function of i and m provide the desired sonar "image." For any given time delay, $\rho_{i,m}$ can be estimated using a detector that minimizes the mean square error. Estimation of $\tau_{i,m}$ can be avoided by simply estimating $\rho_{i,m}$ for all resolvable time delays.[3] Using this approach, the

---

[2] The Hamming separation distance is the minimum number of elements that differ between a code word and any of the circular shifted or unshifted versions of another code word.

[3] Since the total bandwidth of the transmitted signal is N/T Hz, the resolvable time delay is approximately T/N seconds.

discrete-time approximation to the detectors is shown in Figure 2.3. In-phase and quadrature samples are taken of the received signal after demodulation by the carrier frequency, f_c. Matched filters are used to detect the individual frequency chips. Since the model allows for an unknown, nonuniform phase distortion between frequency chips, incoherent summation is used between frequency chips: this incoherent combination is completed in the last box in Figure 2.3. Finally, to avoid à priori estimation of $\tau_{i,m}$, the output from these detectors is computed at each point in time.

The digitized frequency chips will be $w(nT/N)e^{-j(2\pi/N)kn}$ for $k = 0, \ldots, N-1$. Thus, the outputs from the frequency-chip matched filters are:

$$y_k[t] = x[t] * \left(w\left(-t\frac{T}{N}\right)e^{j(2\pi/N)kt}\right)$$

$$= \sum_{n=-(N-1)}^{0} x[t-n]w\left(-n\frac{T}{N}\right)e^{j(2\pi/N)kn} \qquad (2.1)$$

$$= \sum_{n=0}^{N-1} x[t+n]w\left(n\frac{T}{N}\right)e^{-j(2\pi/N)kn} \qquad (2.2)$$

From (2.1), the frequency-chip matched filters can be implemented as a modulated filter bank.

A well-known implementation of a modulated filter bank is the short-time Fourier transform (STFT). By defining $v_t[n] = x[t+n]w(nT/N)$, (2.2) can be seen to be the N-point discrete Fourier transform (DFT) of $v_t[n]$. Thus, the matched filters can be implemented using a STFT. Formulating the matched filters as a STFT allows the use of well-known, computationally efficient implementations of the STFT. The most obvious of these is the FFT: an N-point FFT of $v_t[n]$ can be separately computed at each time sample t. This approach requires $O(N \log N)$ computations per time sample t and allows temporally sparse computation of $y_k[t]$ without any increase in complexity per output point.

A third implementation of the frequency-chip matched filters can be derived by again considering (2.2). Defining

$$x_t[n] = \begin{cases} x[t+n] & 0 \le n < N \\ 0 & \text{otherwise} \end{cases}$$

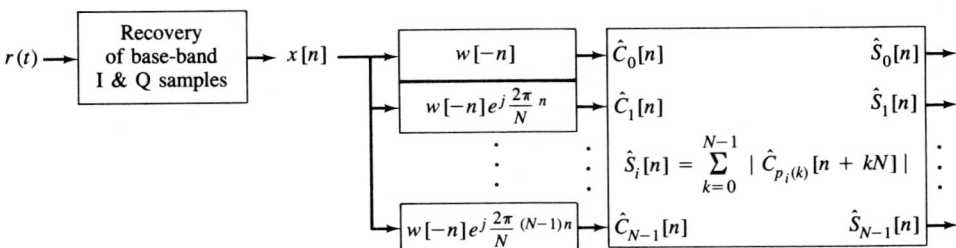

Figure 2.3 The discrete-time approximation to the optimal detectors for N FSK-coded sonar signal beams.

with $X_t[k]$ and $W[k]$ as the N-point DFT's of $x_t[n]$ and $w(nT/N)$, $y_k[t] = (1/N)X_t[k] \circledS W[k]$ where $\circledS$ is an N-point circular convolution. This approach imposes the shaping provided by $w(t)$ through frequency-domain convolution. The advantage of shaping $x_t[n]$ after computing the DFT is that the Goertzel algorithm can then be used. In particular

$$X_{t+1}[k] = e^{j(2\pi/N)k}(X_t[k] + x[t+N] - x[t])$$

Using this recursive approach to find $X_t[k]$ requires $O(N)$ computations per time sample t. The total computational cost of computing $Y_k[t]$ depends on the form of $W[k]$: if $W[k]$ has only a few nonzero samples, the additional computational cost imposed by shaping may be much lower than would be required by a general convolution of two sequences. A disadvantage of this approach is a problem with stability: the recursive computation depends on pole/zero cancellation on the z-domain unit circle.

As will be described in section 2.4, ADE generated a fourth, innovative implementation of the frequency-chip matched filters [1]. This new approach, which will be referred to as the pruned FFT, could not be found in either modulated filter bank or short-time Fourier transform literature. Like the recursive computation of the STFT, the pruned FFT imposes the shaping provided by $w(t)$ through frequency-domain convolution and, like the recursive STFT, only $O(N)$ computations per time sample t are required to compute $X_t[k]$. The pruned FFT has the advantage over the recursive STFT of unconditional stability.

2.3 CAPABILITIES AND REQUIREMENTS FOR AN INTEGRATED SIGNAL PROCESSING ENVIRONMENT

Section 2.1 asserted that a signal processing workstation that provides an integrated environment for numerical processing, algorithm definition, and algorithm manipulation could potentially simplify the design of new signal processing algorithms, as well as improve the reliability of the design process. Supporting both algorithm definition and numeric processing allows the engineer to test the behavior of algorithms as soon as they are defined. Similarly, supporting both algorithm definition and property/transform analysis would allow analytical tools to be easily applied by computer, such as studying the z-domain representations of linear time-invariant systems. By including algorithm rearrangement in the workstation's capabilities, a high-level signal processing "compiler" can be used to apply well-known global transformations to an initial description of an algorithm. One example of this high-level compilation would be the transformation from the noninteger sampling rate conversion shown in Figure 2.1(a) to one of the polyphase structures shown in Figures 2.1(b) and (c). Another possible advantage to a high-level signal processing compiler is the derivation of new, computationally efficient implementations for the given signal processing operation. Two examples of new, innovative implementations found by high-level signal processing compilers will be discussed later in this chapter.

Given that our goal is to expedite signal processing algorithm design, some effort should be made to determine what capabilities are desired and what constraints are imposed by these desired capabilities. A general statement of the desired capabilities has already been made: to support numeric processing of signals, algorithm definition, signal and system property analysis, and algorithmic rearrangement, all in a consistent, well-integrated manner. Each of these requirements in turn imposes constraints on the signal representation that is chosen. Some of these constraints have been pointed out by Kopec in Chapter 1. In this section, we review the requirements described by Kopec and we introduce other requirements, to allow for algorithm manipulation and analysis.

2.3.1 Support for Numerical Signal Processing

Providing consistent, well-formulated support for numeric processing of signals and for algorithm definition imposes some basic constraints on the representation of signals, some of which were discussed in Chapter 1. The signal representation should be explicit and unique, with uniform external behavior; it should distinguish between the domain and the nonzero support of the signal; and it should be externally immutable. The same signal representation should be used for both numerical processing and for the signal analysis and manipulation operations. In more detail:

- **Explicit, unique signal identity:** In signal processing, signals are not just an ordered collection of sample values, but instead have a unique identity and inherent properties of their own [Chapter 1]. Many of their properties, such as nonzero support, domain, and symmetry, are closely tied to the sample values, but others, like the algorithm that generated the signal or its computational cost, cannot be derived from the sample values. Therefore, an explicit signal representation, distinct from a simple ordered set of sample values, is necessary.
- **Uniform external interface to signals:** To simplify the interface of signals and systems, signal representations should all have the same external behavior, independent of the internal computational method used to generate the sample values of the signal [2, 4]. For example, the retrieval of a sample value from a point operator, such as a multiplication block, should externally appear the same as the retrieval of an array operator, such as an FFT block, and the same as the retrieval from a state-machine operator, such as an IIR filter.
- **Distinct signal domain and nonzero support:** The signal representation should distinguish between the domain and the nonzero support of the signal [2, 4]. The domain of a signal determines where the signal is defined: discrete-time sequences are defined on all integer time indices and undefined elsewhere; discrete-time Fourier transform signals are defined on all real frequency indices; and z-transform signals are defined on the annulus of complex indices inside its region of convergence and undefined elsewhere. Any sample value within the defined domain of the signal should be accessible. Accessing a signal inside its domain but outside its nonzero support should return the sample value of the signal at that point, namely zero.

- **Immutable signal representation:** An additional constraint on a signal representation for numerical processing is that signals should be immutable, i.e., there is no operation that can change the properties of a signal once it is defined. Mathematically, signals are immutable objects: their identity and their properties are fixed and unchanging. For example, the sample values, symmetry, and nonzero support of the 256-point Hanning window are completely defined and immutable. Using this sequence as input to a system does not alter the sequence, but instead produces a new sequence. As pointed out in Chapter 1, this immutability in signals also simplifies and clarifies the signal processing algorithms that use them: immutability makes signals referentially transparent.

2.3.2 Support for Algorithm Definition

Supporting algorithm definition constrains the internal representation of signals and systems, just as the support of numerical processing of signals constrains the external behavior of signals. Besides the constraints already discussed, algorithm definition is simplified if there are multiple computational models that are supported by the signal processing environment. Four computational methods that are widely used in signal processing are point operations, array operations, state-machine models, and composition operations.

- **Array operators**, such as FFTs, compute multiple sample values simultaneously.
- **State-machine models** generate sample values sequentially using an internal state vector: an IIR filter, for example, could be easily implemented using a state-machine model.
- **Point operations**, in contrast to array operators and state-machine models, generate one sample value at a time in any random order. Two examples of point operations are addition or multiplication.
- **Composition operations** are implicitly defined through the cascade or "composition" of other, previously defined systems. An example would be the definition of a cosine sequence as the sum of two complex exponentials.

Supporting all four of these computational methods simplifies the programmer's task, since algorithm definitions do not have to be forced into a computational form that is ill-suited for the operation at hand. Providing multiple computational methods while maintaining a uniform external interface decouples the internal and external characteristics of the signal [2, 4]. Thus, the user of a signal need not be concerned with the computational method that the programmer of the signal chose.

2.3.3 Support for Signal Property and Transform Analysis

The premise of this chapter is that a well-integrated signal processing design environment should support property and transform analysis, as well as the more standard algorithm definition and numeric processing steps. Property analysis allows specific

Sec. 2.3 Capabilities and Requirements

questions about the characteristics of a signal or system (e.g., symmetry or linearity) to be answered. Transform analysis uses an alternate representation of a signal or system (e.g., a z-domain representation) to emphasize some aspect of the signal or system that is not apparent in the time-domain representation. The automation of property and transform analysis would allow these analytical tools to be easily and reliably applied by computer. Although property and transform analysis could theoretically be completed using "first principles," this approach to property and transform analysis would be slow and unwieldy at best. Instead, the approach that is envisioned relies on explicitly including information in the signal and system definitions about properties and transforms. This approach to the automation of property and transform analysis would require:

- **Explicit descriptions of signal properties:** For example, the definition of cosine sequences would explicitly include the fact that these sequences are real and symmetric. With this type of information, the environment could provide the values of signal properties, such as those listed in Table 2.1. Furthermore, by providing the user with the appropriate tools, additional properties (e.g., cyclostationarity) could be added by simply adding this extra information to the signal definitions.

- **Explicit descriptions of the effects of systems on signal properties:** For example, by indicating the effect of a shift operator on the symmetry of a sequence (i.e., that it shifts the point of symmetry) and on the sample value type of a sequence (i.e., no effect), the symmetry and sample value type of a shifted cosine signal could be determined by the signal processing environment. This information about a system would describe its effect on the signal properties of its inputs.

- **Explicit identification of signal transforms:** For example, the definition of rectangular-window sequences would include the fact that their discrete-time Fourier transform is an aliased sinc signal. With this information, the environment could provide closed-form expressions for signal transforms, such as discrete-time Fourier transforms (DTFTs) and z-transforms.

- **Explicit descriptions of the effects of systems in the signal transform space:** For example, by indicating the effect of a shift operator on the Fourier-domain representation of its input (i.e., modulation by a complex exponential signal), the Fourier-domain representation of a shifted rectangular window could be determined by the signal processing environment. This information about a system would describe its effects on the DTFT and z-domain representations of its inputs.

TABLE 2.1 SOME USEFUL SIGNAL AND SYSTEM PROPERTIES

INVERTIBLE-P, INVERSE-SYSTEM: Whether or not the system is invertible and if so, its inverse.
SAMPLES-COMPUTABLE-P, COMPUTABLE-P: The computability of individual sample values of the signal and of all the sample values of the signal.
SAMPLE-TYPE: The data type of the sample values of the signal.
RANGE: The range of the sample values of the signal.
NON-ZERO-SUPPORT: The indices on which the sample values of the signal may be nonzero.
PERIODICITY: The repetition period of the signal.
SYMMETRY: The description of the symmetry characteristics of the signal.

2.3.4 Automation of Algorithm Rearrangement

One of the most desirable capabilities for a signal processing environment is the automation of algorithm rearrangement. This would in effect provide the engineer with a high-level signal processing "compiler." This compiler could be used both to apply well-known global transformations to an algorithm and to derive customized, computationally efficient implementations for unusual algorithms. High-level compilation of algorithms requires both the ability to enumerate mathematically equivalent implementations and the ability to compare these alternate implementations to determine their relative merit, based on computational efficiency [1, 2].

Enumeration of Mathematically Equivalent Implementations

Since high-level compilation of an algorithm relies on enumeration of mathematically equivalent implementations, the distinction between mathematical equivalence and computational equivalence is important in algorithm manipulation. *Mathematical equivalence*, or equality, between signals implies that the domain and all of the sample values of the signals are equal, even though the signals may have used different computations to arrive at those sample values. An example of mathematical equivalence is provided by the 256-point DFT of the 256-point Hanning window[4] and the sequence $\frac{1}{2}\delta[k] - \frac{1}{4}\delta[k-1] - \frac{1}{4}\delta[k-255]$. Assuming infinite computational precision, the domain and the sample values of the sequences are equal, even though they were arrived at through very different paths. *Computational equivalence* between signals implies that *all* the signal's properties are identical. This includes the sequence of computations used to arrive at the signal sample values (i.e., the generating system or algorithm). Thus, the only way to get to computationally equivalent signals is to use the same input signals into the same sequence of operations: the output of an FFT operator applied to a discrete-time impulse is computationally equivalent only to the output from the same FFT operator applied to an identical discrete-time impulse. Computational equivalence may seem like a tautology, involving statements like "$\sin x = \sin x$," but it is an equivalence that is often lost in computer programming languages. For example, in most programming languages, the mathematical equality of two copies of the output from an FFT operator would be difficult to determine, requiring a sample-by-sample comparison of the two output sequences, and computational equality would not be determinable based on these output samples.

In algorithm rearrangement, mathematical equivalence must be maintained

---

[4] An M-point Hanning window, $w[n]$, is [7]

$$w[n] = \begin{cases} \frac{1}{2} - \frac{1}{2}\cos(2\pi n/M) & 0 \leq n < M \\ 0 & \text{otherwise} \end{cases}$$

Sec. 2.3 Capabilities and Requirements

even though computational equivalence is deliberately lost. To provide these lists of mathematically equivalent (but computationally distinct) signals requires:

- **Explicit identification of equivalent signals:** For example, by noting the mathematical equivalence between the Hanning window sequence and a raised, windowed cosine sequence, a list of two mathematically equivalent signals could be collected.

- **Recursive identification of equivalent signals and identification and substitution of equivalent subexpressions:** This requirement is most easily explained by example. Using the previous example of the Hanning window, assume that another mathematical equivalence was noted between the cosine sequence ($\cos \omega n$) and the sum of the conjugate pair of complex exponentials (($e^{j\omega n} + e^{-j\omega n}$)/2). The environment must be able to put these two pieces of information together and increase the list from the two equivalent sequences (the Hanning window and the raised, windowed cosine) to three (the Hanning window; the raised, windowed cosine; and the raised, windowed sum of the conjugate pair of complex exponentials). This requires that the environment search recursively for equivalent signals using newly discovered signals (in this case, the raised, windowed cosine sequence) and that the environment search for signals that are equivalent to the subexpressions of the complete signal description (in this case, the cosine sequence is a subexpression of the raised, windowed cosine). This search strategy is described in detail in sections 2.5 and 2.6.

To provide these lists of mathematically equivalent systems requires:

- **Explicit identification of equivalent output signals:** For algorithm manipulation, lists of mathematically equivalent systems must be derived from information included in the system definitions. Often, the most straightforward way to provide this information is to describe the signals that are mathematically equivalent to the output signal when the system has been applied to some general or partially specified input. Without this shift in focus, most algorithmic transformations are difficult to specify, since the input signals and parameters of the system are unbound until the system is applied. For example, the description of the Goertzel algorithm as being mathematically equivalent to one channel from the rectangularly windowed STFT is not possible without being able to refer to the input signal, $x[n]$. Thus, this approach to finding equivalent algorithms places two requirements on the environment: that the system definitions include information about signals that are equivalent to their output signal and, as discussed next, that general or partially specified signals can be represented and manipulated.

- **Representation of general or partially specified signals:** As mentioned previously, one straightforward method of finding mathematically equivalent implementations of an algorithm is to use general or partially specified signals as the inputs to the algorithm and to then find signals that are mathematically equivalent to the output signal from the algorithm: the equivalent algorithms are then given by the composition of systems used to generate the equivalent output signals. This approach to algorithm manipulation is highly reminiscent of the algebraic manipulation that engineers commonly do on signal processing equations. In algebraic manipulation of signal processing expressions, the inputs are represented by an algebraic variable, such as $x[n]$, and the output signal generated by processing this input is then manipulated. For example, the derivation of the Goertzel algorithm from the direct formulation of the STFT would start with the input to the STFT being represented by $x[n]$. The output signal from the DFT would then be

$$y_k[t] = \sum_{n=0}^{N-1} x[t+n] e^{-j(2\pi/N)kn}$$

The Goertzel algorithm could then be derived by manipulating $y_k[t]$ without making any assumptions about the sample values or properties of $x[n]$:

$$y_k[t+1] = \sum_{n=0}^{N-1} x[t+1+n] e^{-j(2\pi/N)kn} \tag{2.3}$$

$$= e^{j(2\pi/N)k} \sum_{m=1}^{N} x[t+m] e^{-j(2\pi/N)km} \tag{2.4}$$

$$= e^{j(2\pi/N)k} \sum_{m=0}^{N-1} x[t+m] e^{-j(2\pi/N)km} + e^{j(2\pi/N)k} x[t+N] - e^{j(2\pi/N)k} x[t] \tag{2.5}$$

$$= e^{j(2\pi/N)k}(y_k[t+1] + x[t+N] - x[t]) \tag{2.6}$$

The representation of a general or partially specified signal requires a mechanism for representing signals whose sample values and properties are not completely known.

Comparison of Computational Cost

To provide a high-level compiler of signal processing algorithms, the environment must include some method for ranking the mathematically equivalent implementations of a given algorithm and selecting the best one. The metric that is generally used by compilers is the computational cost of the alternative algorithms. Thus, the signal processing environment must be able to determine the relative costs of equivalent algorithms, in order to select the most computationally efficient. Myers [2] and Covell [1] discuss computational cost metrics in more detail.

2.3.5 Summary of Requirements for an Integrated Signal Processing Environment

An integrated signal processing environment should support the numeric processing of signals, the definition of new signal processing algorithms, the analysis of the properties of signals and systems, and the rearrangement or "compilation" of signal processing algorithms. To provide support for these capabilities, certain constraints have been placed on the signal processing environment.

- It must provide an explicit representation for signals.
- The sample values of signals must be accessible at random from anywhere in

the domain of the signal: this requirement implies both that the external access of the sample values is unaffected by the internal computational model and that sample values must be accessible outside the nonzero support of signals.
- The signal must be immutable in its external characteristics: its property values and sample values must be unchanging from the time they are first referenced.
- Analyses of signal and system properties and transforms must be provided by the environment, using information included in the definitions of signal and system classes.
- Finally, algorithm rearrangement and cost analysis must also be provided by the computer, to allow for high-level compilation of signal processing algorithms.

Figure 2.4 reviews the capabilities of some currently available signal processing environments. This figure does not attempt to exhaustively list the currently available software. Instead an attempt is made to examine the range of signal processing environments currently in use. The first set of signal processing environments [17, 18] listed in Figure 2.4 were developed to only support the definition and numerical application of completely specified, numeric algorithms. As mentioned in section 2.1 and described in Chapter 1, SRL is the result of research by Kopec into data abstractions both to reflect the basic characteristics of signals and to support numeric manipulations. Kopec [3] advocated the immutability of signals and the explicit availability of their nonzero supports as being essential for simplifying and clarifying signal processing programs. nthPOWER (Signal Technology, Inc.) is a commercial version of SRL and one of its descendants, ISPUD [8]. SPLICE [2, 4] was also mentioned in section 2.1. In SPLICE, as in SRL, sequences are immutable data objects with an explicit nonzero support. Unlike SRL, sample values outside the nonzero support can be accessed by the same operations that access the sample values inside the nonzero support. Sequences, defined by the generating system and its inputs, behave uniformly independent of the signal processing model used to define the system: for example, sample values of a sequence defined using a state-machine model can be fetched at any index without explicitly determining the

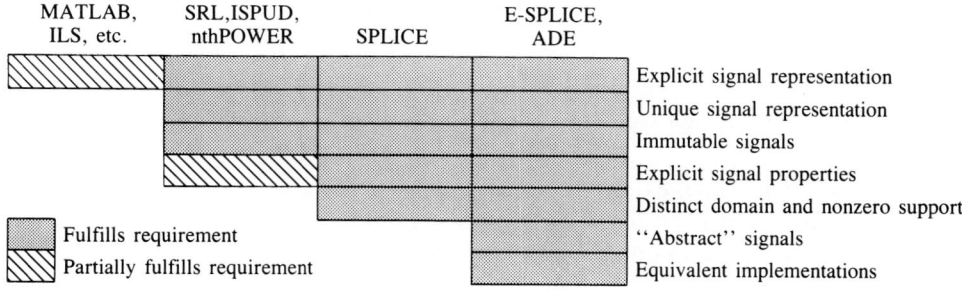

Figure 2.4 The capabilities of some currently available signal processing environments.

previous states. E-SPLICE [2] and its descendent, ADE [1], were designed to support not only numerical processing and algorithm definition but also automated property analysis and algorithm rearrangement. Hence, these are only environments that provide signal representations which meet most of the constraints developed in this section. In particular, the signal representations are unique and immutable with a distinct domain and nonzero support and with explicit signal properties, such as symmetry and computational cost. Furthermore, these environments automate the analysis of properties and signal transforms and can represent and manipulate "general" signals. Finally, both E-SPLICE and ADE provide a tool that approaches the desired high-level signal processing compiler discussed earlier: both environments will provide an enumeration of alternate implementations of an algorithm, partially ranked on the basis of the computational cost vectors.

The remainder of this chapter focuses on the capabilities and characteristics of ADE. Section 2.4 includes an example of the use of ADE to define and analyze the properties of the FSK-code detector, shown in Figure 2.3. The remaining sections of the chapter explore the process through which alternate implementations of an algorithm are found by ADE. Some results from using ADE to find alternate implementations are discussed in section 2.4.

2.4 ADE: A FEASIBILITY PROOF FOR AN INTEGRATED SIGNAL PROCESSING ENVIRONMENT

The previous section pointed out some of the requirements for a complete and well-integrated signal processing environment. As shown in Figure 2.4, ADE attempts to meet each of these requirements. Therefore, ADE will be used within this chapter to illustrate the potential benefits of a well-integrated signal processing environment. The illustration of these potential benefits is simplified by considering a specific signal processing design problem: the problem that has been chosen for this illustration is the design of a detector for sonar FSK-code reflections.

This section provides a brief description of ADE. It is guided by a discussion of two short sessions in ADE, one illustrating the programming of the environment and the other, its interactive use.

2.4.1 An Overview of ADE

ADE provides an integrated environment for numerical processing, algorithm definition, signal and system property analysis, and algorithm rearrangement. The underlying signal representation is an object-oriented signal representation, satisfying all the desirable properties discussed in section 2.3. Information about signal processing, used in property analysis and in algorithm rearrangement, is represented in a rule-based system. Although the capacity for forward chaining[5] is included in ADE, the majority of the environments resources are devoted to backward chaining from specific inquiries. Tools are provided within ADE for extending this rule base, both through the introduction of new signal and system classes and through the introduction of new properties and signal transforms.

ADE is a descendant of the SPLICE and E-SPLICE environments. ADE inherits its basic approach to signal definition and representation from SPLICE. The influences of E-SPLICE and to a lesser extent PDA [9] are reflected in the structure of some parts of the rule base. In particular, as in E-SPLICE, ADE uses backward-chaining rules to describe the properties of signals. ADE, like E-SPLICE, supports multilevel matching within the patterns of these rules. The approach used in ADE for matching forward-chaining rules was introduced by Dove [9]. ADE makes use of a subset of QM [10] and a limited number of functions from MACSYMA [11]. QM is the product of research into qualitative mathematics. It represents, manipulates, and describes piecewise-continuous functions. A subset of QM is used to record and propagate constraints on symbolic numbers. ADE includes an extension to QM to support limited reasoning about symbolic integers as well as the continuously variable numbers. ADE also makes limited use of MACSYMA to simplify and factor the polynomials used in the characterization of z-transform signals. ADE is written in Symbolics Common Lisp [12]. This choice of language provides both the flexibility of a LISP dialect and support for object-oriented programming.

The remainder of this section provides examples of the use of ADE in the context of the FSK-code problem introduced earlier. Examples are given of programming (algorithm definition), property analysis, and algorithm rearrangement in ADE.

2.4.2 Examples of Algorithm Definition and Manipulation in ADE

In the sonar imaging problem, an important problem is to find a way to achieve good spatial resolution without requiring a large, costly array and without missing transients through the use of a swept beam. The first step in solving this problem is to select a method by which it will be solved. Although this selection draws heavily on signal processing experience and creativity, the selection process can be accelerated by providing a support environment in which the signal and system representations closely match the internal models used by the system designer. These representations must change according to the problem at hand, since different problems give

---

[5] Forward chaining in a blackboard/rule-based system is the use of current information to determine additional information, without requiring an inquiry to explicitly trigger that line of reasoning. In general, this approach triggers any rule whose preconditions are satisfied based on the current state of the blackboard. This contrasts with backward chaining, which starts from a question and then, based on the preconditions of rules that could answer that question, asks additional questions until a question is asked that can be answered.

rise to different signal models. To provide this range of representations, ADE allows the system designer to introduce his own signal and system definitions. For example, in the FSK-code detector shown in Figure 2.3, the incoherent combination of the matched filter outputs is modeled as a single processing block which follows, but is separate from, the matched filtering itself. To support this model of the detector, a new system class, INCOHERENT-COMBINATION, is defined in Figure 2.5. The definition relies on the composition of other, previously defined signal processing systems to provide the output signals with their observable characteristics: lines 8–12 of Figure 2.5 describe this composition.

To simplify the programming task, signal and system definitions closely mimic the notational conventions used in signal processing. As illustrated by lines 1–4 of Figure 2.5, signal and system definitions form new "classes" of signals and systems. Hierarchies of classes are used to make similarities explicit and to reduce the amount of coding required. Signals are formed by one of two paths: either as independent entities that are inherently defined, like an impulse or a complex-exponential sequence, or as the output from a system that has been applied to some inputs. Some of the 43 inherent signal classes and the 169 system classes currently defined in ADE are listed in Table 2.2.

Once all the appropriate signal and system classes have been defined, the process of creating and analyzing the signals and systems involved in the design problem is simplified. Figure 2.6 illustrates the design sequence. Line *I-1* of Figure 2.6 shows the definition of a partially specified discrete-time signal, $x[n]$, which is periodic and complex valued with both its real and imaginary parts in the range from -1 to $+1$. This description is only a partial description of the input since there are a large number of discrete-time sequences that satisfy all parts of this description. The resulting object, printed on line *O-1*, is an "abstract" signal.

The FSK-code detector is compactly described in lines *I-2* and *I-3* of Figure 2.6. As can be seen by comparing these lines with the model for the detector shown in Figure 2.3, the computer representation and the designer's representation are closely matched. As is shown in the remainder of this figure, ADE provides information about the properties of the output signals from this detector and about alternate implementations of the matched filters used in the detector.

In more detail, lines *I-2* and *I-3* of Figure 2.6 create an incoherent detector for the set of 16 FSK codes, using a 16-point rectangular window to shape each frequency chip. The input to this detector is $x[n]$. The output from the modulated filter bank is a two-dimensional signal, YC. YC is used as the input to the incoherent combiner and the final output signal is Y. Lines *I-4* and *I-5* of Figure 2.6 request the range and periodicity of the output from the detector, Y. The range and periodicity of a signal are among the properties that can be explicitly requested. Signal properties, such as symmetry, sample type, and nonzero support, are explicitly available characteristics of every signal. Similarly, system properties, such as equivalent systems and invertibility, are explicitly available characteristics of every system. ADE determines the values of signal and system properties by using the property information explic-

ADE Code	Description
1 (DEFINE-SYSTEM-CLASS-ALIAS	1
2 (INCOHERENT-COMBINATION N@INTEGER)	2 accept an integer (N) as a system parameter
3 (INPUT@2D-SEQUENCE)	3 accept a 2-D sequence (INPUT) as an input
4 (SHIFT-INVARIANT-SYSTEM HOMOGENEOUS-SYSTEM 2D-SYSTEM)	4 a subclass of these classes
5 ("an incoherent combiner of shifted versions of the sequences in INPUT"	5
6 SELF	6 the systems "alias" to themselves
7 'the output sequences from the incoherent combination'	7 the output signals "alias" to the composition of operations:
8 (MAP-OVER BANK-OF-SEQUENCES I 0 N	8 (BANK-OF-SEQUENCES $S_0,...S_{N-1}$)
9 (MAP-OVER SEQUENCE-ADD K 0 N	9 where S_i = (SEQUENCE-ADD $p_i(0)...p_i(N-1)$)
10 (OUTPUT-OF(SEQUENCE-SHIFT (* K N))	10 where $p_i(k)$ = (OUTPUT-OF (SHIFT (* $k\ N$))
11 SEQUENCE-MAGNITUDE	11 (MAGNITUDE
12 (FETCH-SEQUENCE INPUT (MOD(− (*(1 + I)(1 + K))1)(1 + N)))))))	12 INPUT$_{((i + 1)*(k + 1) - 1)\bmod (N+1)}$))

Figure 2.5 An example of the programming of ADE

This example defines an incoherent combiner for the FSK system of Figure 2.3 in the special case when the permutation function, $p_i(k) = ((i + 1)*(k + 1) - 1) \bmod (N + 1)$

TABLE 2.2 SOME OF THE SIGNAL AND SYSTEM CLASSES CURRENTLY DEFINED IN ADE

Inherent signal classes (43 hierarchical classes)	
DISCRETE-TIME-SEQUENCE	COMPLEX-EXPONENTIAL
FOURIER-DOMAIN-SIGNAL	CAUSAL-RECTANGULAR-WINDOW
Z-DOMAIN-SIGNAL	SINC
2D-SEQUENCE	COSINE-SEQUENCE
RATIONAL-ZT	SINE-SEQUENCE
CONSTANT	FIR-SEQUENCE
POWER-SEQUENCE	IIR-SEQUENCE
IMPULSE	CAUSAL-IIR-SEQUENCE
GENERAL-EXPONENTIAL	ANTICAUSAL-IIR-SEQUENCE
UNIT-STEP-SEQUENCE	STABLE-IIR-SEQUENCE
CAUSAL-HAMMING-WINDOW-SEQUENCE	

System classes (169 hierarchical classes)		
DISCRETE-TIME-SYSTEM	ADD	BANK-OF-SEQUENCES
2D-SYSTEM	SUBTRACT	ROTATED-BANK-OF-SEQUENCES
FOURIER-DOMAIN-SYSTEM	MULTIPLY	SHORT-TIME-WINDOW
Z-DOMAIN-SYSTEM	CONVOLVE	SHORT-TIME-FT
SHIFT-INVARIANT-SYSTEM	SHIFT	MAPOVER-SYSTEM
GENERALIZED-SHIFT-INVARIANT-SYSTEM	SCALE	FIR-FILTER
MEMORYLESS-SYSTEM	RECIPROCAL	CAUSAL-IIR-FILTER
ASSOCIATIVE-SYSTEM	DIVIDE	ANTICAUSAL-IIR-FILTER
ADDITIVE-SYSTEM	REAL-PART	SIGNAL-ALIAS-IN-2PI
HOMOGENEOUS-SYSTEM	IMAG-PART	FOURIER-TRANSFORM
GENERALIZED-HOMOGENEOUS-SYSTEM	MAGNITUDE	INVERSE-FOURIER-TRANSFORM
LINEAR-SYSTEM	INPUT-PHASE	Z-TRANSFORM
GENERALIZED-LINEAR-SYSTEM	ABSOLUTE-VALUE	INVERSE-Z-TRANSFORM
SEQUENCE-CIRCULAR-SHIFT	SCALE-INDEX	INVERSE-TRANSFORM
SEQUENCE-CIRCULAR-REVERSE	UPSAMPLE	DISCRETE-FOURIER-TRANSFORM
SEQUENCE-CIRCULAR-CONVOLVE	DOWNSAMPLE	COMPLEX-CONJUGATE
SEQUENCE-CONVOLVE-OVERLAP-SAVE	INTERLEAVE	

Interactive sessions	Description

I-1 ADE: (NAMED-SETQ X
  (A-MEMBER-OF 'DISCRETE-TIME-SEQUENCE
  &PROPERTIES :PERIODICITY 256
  :RANGE (CREATE-RANGE {−1 1} {−1 1})))
	I-1 ADE: create X, an "abstract" discrete-time sequence with a periodicity of 256 samples and with real and imaginary sample values above −1 and below 1

O-1 ⇒ X	O-1 ⇒ the abstract discrete-time sequence

I-2 ADE: (NAMED-SETQ YC
  (OUTPUT-OF (MODULATED-FILTER-BANK
  (REVERSE (RECTANGULAR-WINDOW 16))
  16)
  X))
	I-2 ADE: create YC, the output of a 16-channel modulated filter bank with a 16-point anticausal rectangular window as the base impulse response and with X as the input to the filter bank

O-2 ⇒ YC	O-2 ⇒ the output from the modulated filter bank
I-3 ADE: (NAMED-SETQ Y (OUTPUT-OF (INCOHERENT-COMBINATION 16) YC))	I-3 ADE: create Y, the output of the incoherent combiner with YC as the input
O-3 ⇒ Y	O-3 ⇒ the output from the incoherent combiner
I-4 ADE: (RANGE Y)	I-4 ADE: what is the range of Y?
O-4 ⇒ (RANGE {0 32} {0 0})	O-4 ⇒ Y is real with values between 0 and 32
I-5 ADE: (PERIODICITY Y)	I-5 ADE: what is the periodicity of Y?
O-5 ⇒ 256	O-5 ⇒ the periodicity of Y is 256 samples
I-6 ADE: (EFFICIENT-IMPLEMENTATIONS YC)	I-6 ADE: what are the efficient ways to compute?

O-6 ⇒ ((MAP-OVER 'BANK-OF-SEQUENCES K 0 16
  (MAP-OVER (IF (< K 8) 'ADD 'SUBTRACT) N8 0 2
  (OUTPUT-OF
  (SCALE (EXP (COMPLEX 0 (* −2 PI K N8))))
  …)))
  (BANK-OF-SEQUENCES (SEQUENCE-ADD …) …) …)
	O-6 ⇒ the list of efficient implementations includes the classic FFT implementation pruned FFT implementation shown in Figure 2.7 and other, partially pruned FFT implementations

Figure 2.6 A sample of an interactive session in ADE The I-lines with the token "ADE:" designate the user's inputs and the O-lines with the token "⇒" designate the outputs. See Covell (1989) for a more detailed discussion of the use of ADE.

itly included within the definitions of signal and system classes or, if this information is missing, by using the default value for the property. Some of the signal and system properties that are currently included in ADE were listed in Table 2.1. Tools are also available within ADE for adding other signal properties that could be useful to the particular problem under consideration (e.g., stationarity).

Line *I-6* of Figure 2.6 requests a list of all the computationally efficient implementations that can be found for the matched filters, used in the detector. Computational efficiency is determined in ADE on the basis of memory requirements and the required additions, multiplications, and memory references. Line *O-6* shows some of the efficient implementations generated by ADE. The given form of the modulated filter bank is not included in this list of implementations, since there are other implementations that are more computationally efficient. Instead, the list includes the classic FFT-based structure, described in section 2.2, and a variety of "pruned" FFT-based structures: one of the pruned FFT-based structures is shown in Figure 2.7.[6] The pruned FFT implementations have the same underlying structure as the classic FFT implementation. The difference lies in the *number* of butterflies that are computed at each stage. For example, the pruned FFT structure shown in Figure 2.7 has only one butterfly in the first stage, two in the second, four in the third, eight in the fourth, and so on, while the classic FFT structure has $N/2$ butterflies in each stage. As can be seen from the comparison of costs shown in Figure 2.7, a trade-off exists between the minimum number of memory locations, achieved by the classic FFT structure, and the minimum number of operation counts, achieved by the pruned FFT structure.[7]

Since the selection of the frequency-chip window affects both the range resolution and the signal-to-signal rejection of the sonar system, another two frequency-chip windows were considered to explore the trade-offs between resolution and signal-to-signal rejection: the 16-point Hanning window and the 32-point Hanning window with overlapping frequency chips within the FSK code.[8] Figure 2.8 shows the pruned FFT structure generated by ADE for the 16-point Hanning window. The resulting structure consists of the pruned FFT structure of Figure 2.7 followed by frequency-domain convolution. Figure 2.9(a) shows the pruned FFT structure gen-

---

[6] Figure 2.7 shows operations of the form z^i for positive values of i. These operations are (anticausal) sequence advance operations and arise from the use of the anticausal window within the modulated filter bank (see line *I-2* of Figure 2.6). The use of the anticausal window within the modulated filter bank is the result of using a causal window for the frequency chips in the FSK codes [see (2.1)].

[7] The memory counts do not include the registers necessary for storing the intermediate sequence values. If these additional memory locations were included in the cost structures, the amount of memory for the classic FFT structure using the method given by Singleton [13] and the pruned FFT structure would be identical.

[8] These requests for efficient implementations of the two Hanning-window modulated filter banks are not shown in Figure 2.6 since they are not substantially different in form from the request already shown in that figure.

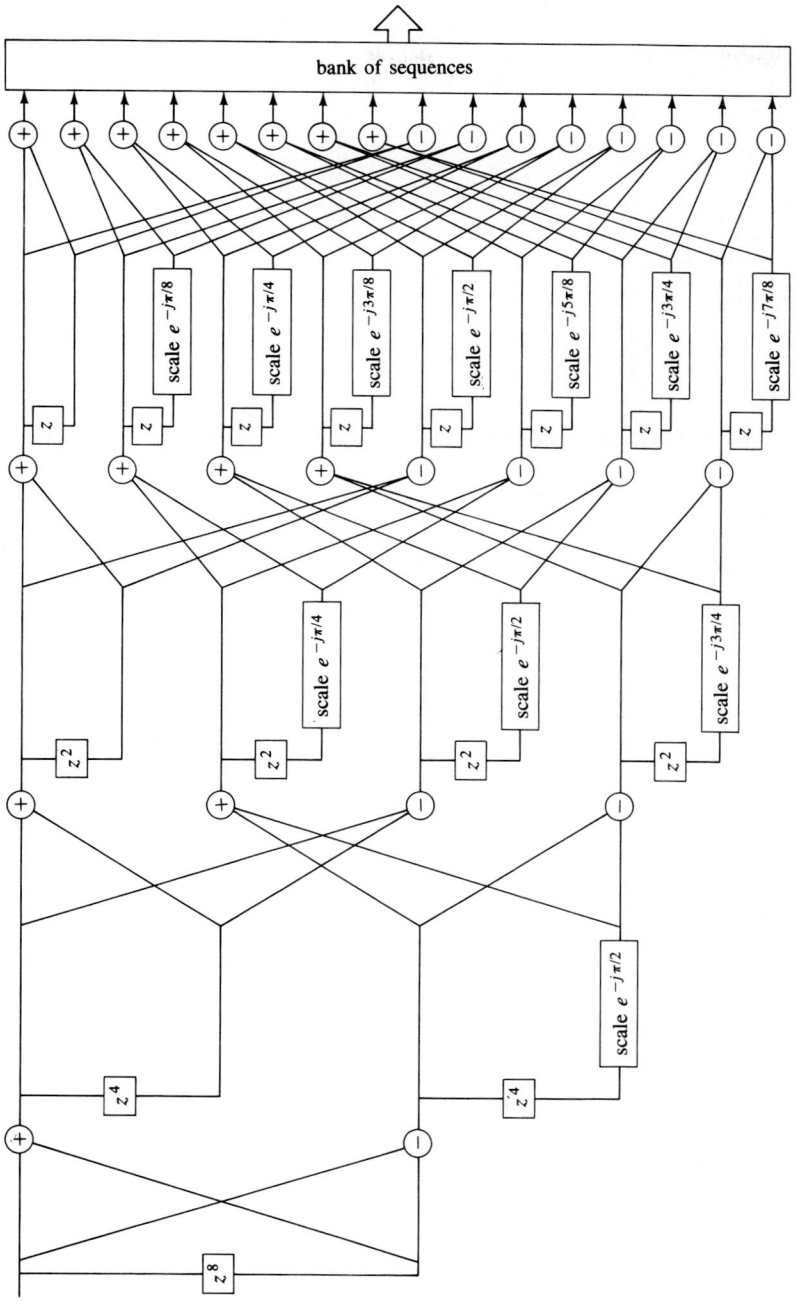

Figure 2.7 The pruned FFT implementation of the modulated filter bank for a 16-point rectangular window.

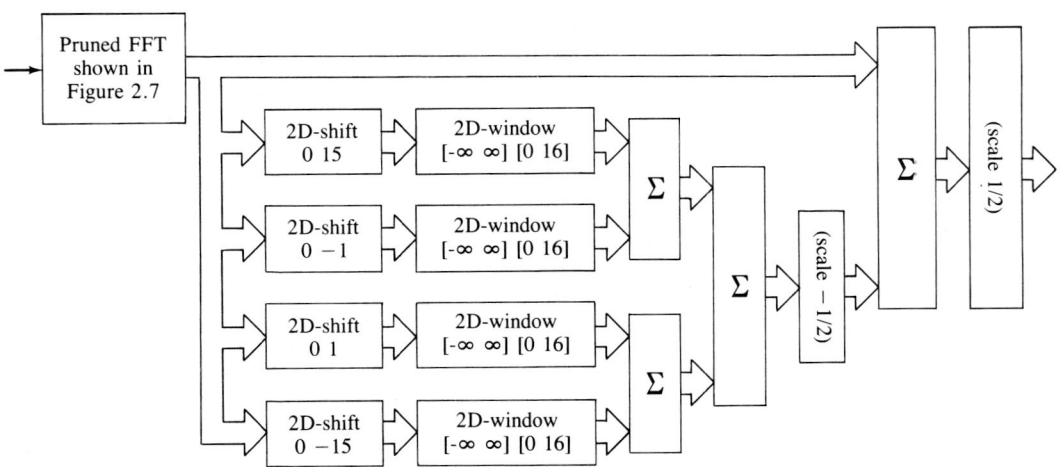

Figure 2.8 The pruned FFT implementation of the modulated filter bank for a 16-point Hanning window.

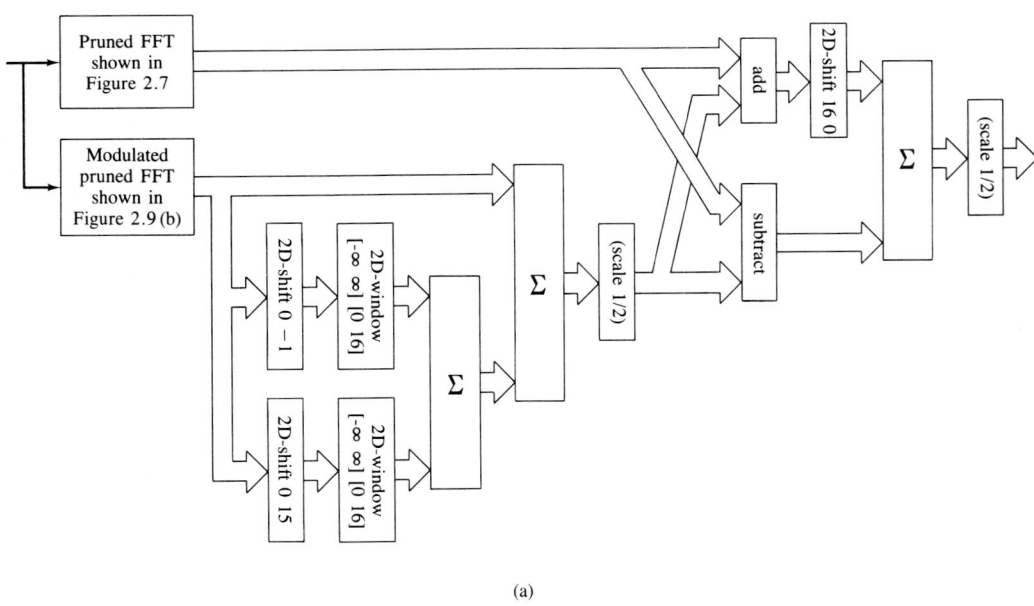

(a)

Figure 2.9 The pruned FFT implementation of the 16-channel modulated filter bank using a 32-point Hanning window.

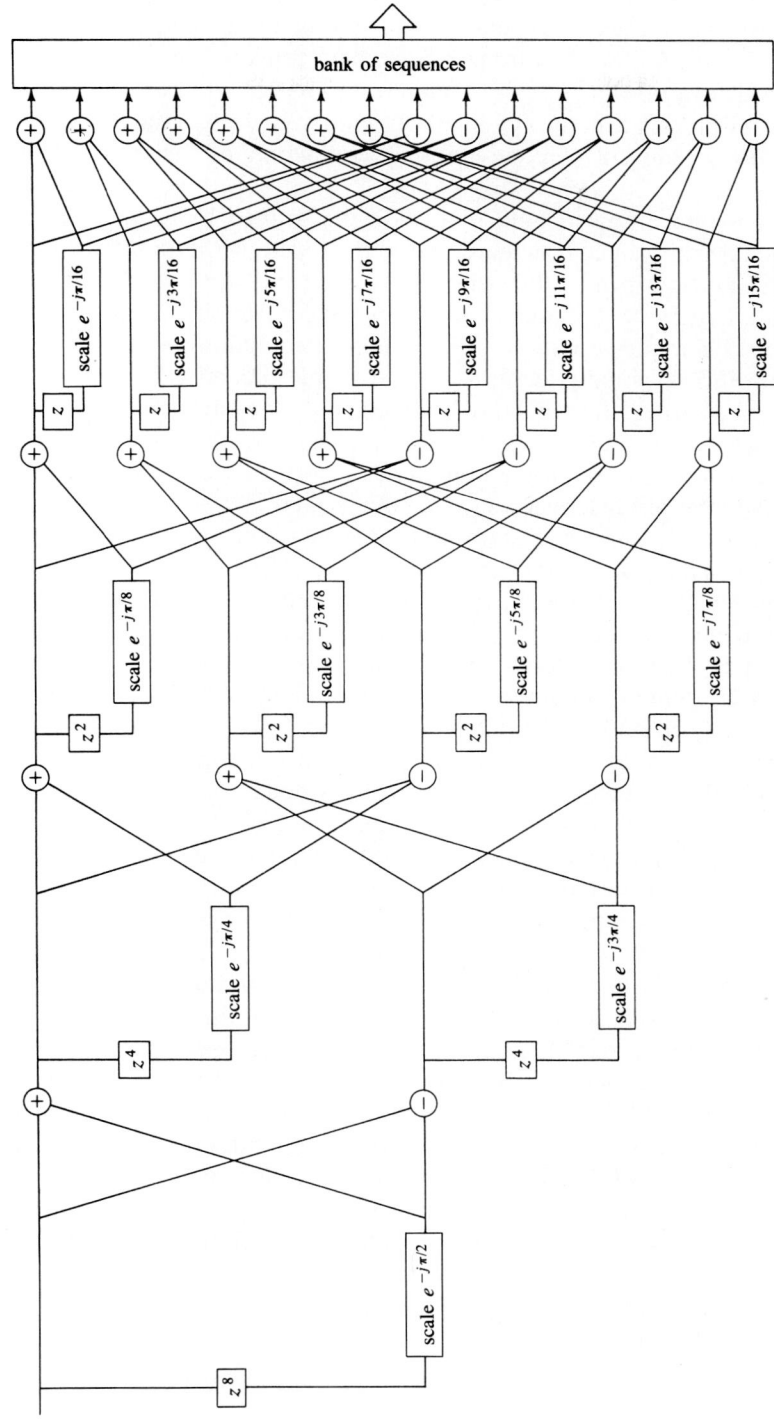

(b) The modulated pruned FFT structure.

erated by ADE for the overlapping, 32-point Hanning window. The resulting structure uses two pruned FFTs, one taken from Figure 2.7 and the other illustrated in Figure 2.9(b), followed by frequency-domain convolution. The modulated pruned FFT structure of Figure 2.9(b) computes a 16-point pruned FFT, offset by half a frequency bin. As with the rectangularly windowed design, the implementations that were found reduced the computational complexity of the matched filter bank to $O(N)$.

It is interesting to note that, with the pruned FFT, the *order* of the computational complexity is actually reduced as well as the number of computations themselves. Specifically, the order is reduced from $O(N^2)$ for the direct-form implementation or from $O(N \log N)$ for the classic FFT implementation to $O(N)$ for the pruned FFT implementation. The amount of computation that is required for the pruned FFT is actually identical to that of the recursive implementation and, as mentioned earlier, the pruned FFT structure has the advantage of being numerically stable while the recursive formulation is marginally unstable due to its reliance on pole/zero cancellation on the unit circle.

Another interesting aside is that the pruned FFT structure has not been found in the currently published literature. Although other pruned FFT structures have been published [14, 15], these structures depend on the characteristics of the inputs as opposed to the characteristics of the desired outputs. Thus, in addition to the standard implementations, ADE generated a new structure for efficiently computing the outputs from a modulated filter bank.

ADE generates equivalent implementations of an algorithm by testing and applying algorithm transformation rules. These rules are included within the definitions of signal and system classes. Detailed derivations of the pruned FFT implementations of the modulated filter banks are provided in Appendix A. In the derivation of the pruned FFT implementation of the rectangularly windowed modulated filter bank, the actual transformation rule, which is crucial, is relatively straightforward. The crucial transformation rule simply pulls common shifts through a generalized shift-invariant system. With a generalized shift-invariant system, $H\{\ \}$, if $y(t) = H\{x_1(t), \ldots, x_N(t)\}$ then $y(t - T) = H\{x_1(t - T), \ldots, x_N(t - T)\}$. Lines 13–30 of Figure 2.10 show how the shift-invariant property of the generalized shift-invariant system is used in ADE to generate an equivalent form. By pulling all the common shift operations through the butterfly and twiddle stages of the classic FFT structure, the classic FFT structure collapses into the pruned FFT structure. The derivations of the pruned FFT implementations of the other two windows are longer and more involved but, again, rely only on relatively straightforward transformations.

As illustrated with the rule describing the shift-invariant property of the generalized shift-invariant system, the transformation rules that generate the equivalent implementations are relatively straightforward. However, as illustrated in section 2.3.4, the search process that must be used to discover and combine all the appropriate transformations is comparatively complex. The remaining sections of this chapter focus on this search process.

	ADE code		Description
1	(DEFINE-ABSTRACT-SYSTEM-CLASS		
2	(GENERALIZED-SHIFT-INVARIANT-SYSTEM *)*	2	accept any parameters or inputs
3	()	3	no superclasses
4	("h{ } s.t. y[n] = h{x_1[n]...x_L[n]} ⇒ y[n − N] = h{x_1[n − N]...x_L[n-N]}")	4	
5	nil ()	5	generate a new output signal class
6	("the output from a generalized shift-invariant system")	6	
7	(GOAL SIMPLIFICATION	7	a simplification rule: if all the inputs are shifted identically,
8	:NAME SHIFTED-INPUT	8	pull shift system outside.
9	:OBJECT (OUTPUT-OF ?SYSTEM@(NOT SHIFT-SYSTEM) &REST	9	the inputs are all outputs from a single shift system.
10	?INPUTS$((OUTPUT-OF ?SHIFT ?[SHIFT-INPUTS]))	10	?SHIFT-INPUTS will be bound to the list of shift inputs
11	:ANSWER (OUTPUT-OF SHIFT	11	
12	(APPLY 'OUTPUT-OF SYSTEM SHIFT-INPUTS)))	12	
13	(GOAL EQUIVALENT-FORM	13	an equivalent form rule: if all the inputs are shifted,
14	:NAME UNEQUALLY-SHIFTED-INPUT	14	pull one of the shifts outside.
15	:OBJECT	15	
16	(OUTPUT-OF ?SYSTEM@(NOT SHIFT-SYSTEM) &REST	16	the inputs are all outputs from shift systems.
17	?INPUTS$((OUTPUT-OF (SPECIFIC-MEMBER SHIFT-SYSTEM	17	?SHIFT-FACTORS will be bound to the list of shift amounts.
18	&REST ?[SHIFT-FACTORS])	18	?SHIFT-INPUTS will be bound to the list of shift inputs.
19	?[SHIFT-INPUTS]))	19	
20	:ANSWER	20	
21	(LET ((COMMON-SHIFT (FIRST SHIFT-FACTORS)))	21	pull first shift outside
22	(OUTPUT-OF (APPLY 'SHIFT COMMON-SHIFT	22	
23	(APPLY 'OUTPUT-OF SYSTEM	23	
24	(MAPCAR	24	
25	'(LAMBDA (SHIFT-FACTOR SHIFT-INPUT)	25	
26	(LET ((REMAINING-SHIFT	26	compensate for the shift that was pulled outside
27	(MAPCAR '$- SHIFT-FACTOR COMMON-SHIFT)))	27	
28	(OUTPUT-OF (APPLY 'SHIFT REMAINING-SHIFT)	28	
29	SHIFT-INPUT)))	29	
30	SHIFT-FACTORS SHIFT-INPUTS))))	30	
31	...))	31	additional information about generalized shift-invariant system outputs

Figure 2.10 The definition for the system class, GENERALIZED-SHIFT-INVARIANT-SYSTEM

2.5 UNCONSTRAINED DERIVATION AND RANKING OF EQUIVALENT ALGORITHMS

To simplify the development of efficient algorithms, an integrated signal processing environment should provide a high-level, signal processing "compiler." This compiler must make use of rules, such as the one shown in Figure 2.10, to explore the space of alternative implementations. This section and the next describe the structure of the search space that is explored in finding the equivalent implementations of an algorithm.

2.5.1 Unconstrained Search for Equivalent Algorithms

The task of finding alternate implementations of a signal processing expression is the same as finding all the algorithmic transformations that are applicable to the signal processing expression; to its input/output equivalent expressions; and to the subexpressions used by these expressions. For example, to find the equivalent implementations of the filter bank used in the FSK-code detector, all the applicable algorithmic transformations for the filter bank should be completed, as should the transformations on the modulated window sequences and the input sequence. In addition, once an alternate implementation is generated, all of the algorithmic transformations that are applicable to this new expression or to one of its subexpressions must also be applied. Thus, equivalent implementations of a signal processing expression can be obtained in any of a variety of ways: a transformation can be applied to the original signal processing expression itself; a subexpression of the original expression can be replaced by an equivalent implementation of the subexpression; or either of these approaches can be applied to one of the newly generated equivalent implementations of the signal processing expression.

To simplify this discussion, a graphical representation of the search process is presented in Figure 2.11. The problem of finding the equivalent forms of a signal processing expression, without consideration of its subexpressions, can be represented graphically as a planar net, as shown in Figure 2.11(a). The nodes of the net represent the signal processing expression and its equivalent forms. The directed arcs connecting the nodes within the planar net represent the application of simple transformation rules. For example, in Figure 2.11(a), the modulated filter bank (node A-1) is replaced by a short-time Fourier transform (node A-2) using a rule included in the definition of the system class MODULATED-FILTER-BANK: this rule is shown below the transformation arc.

The new nodes that result from algorithm transformations can themselves be used as the starting point for other transformations. An example of this recursive transformation is also shown in Figure 2.11(a): the modulated filter bank is first replaced by a short-time Fourier transform, which is then expanded into the basic addition, subtraction, time shifts, and multiplications (node A-3) that make up the short-time Fourier transform.

Each of the nodes of the planar net can also be viewed as a combination of

Sec. 2.5 Unconstrained Derivation and Ranking of Equivalent Algorithms 57

subexpressions: the subexpressions are the inputs to the generating system. For example, as shown in Figure 2.11(b), the four inputs to the BANK-OF-SEQUENCES system in node A-3 can each be manipulated independently. In particular, each of these four expressions can be replaced by any of their equivalent implementations, without changing the input/output mapping of the overall algorithm. Thus, this replacement provides additional equivalent implementation to the original signal processing expression.

Graphically, requesting the equivalent forms of the inputs to the generating system of a signal drops the problem down to another set of nets and again tries to

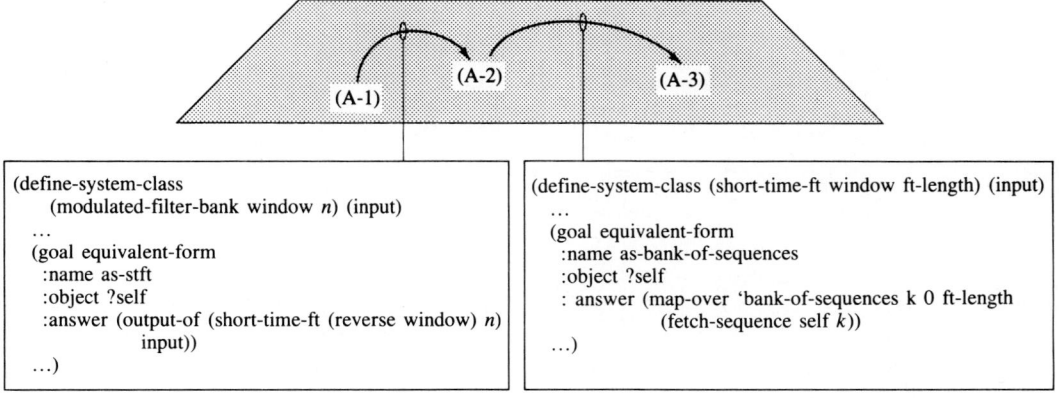

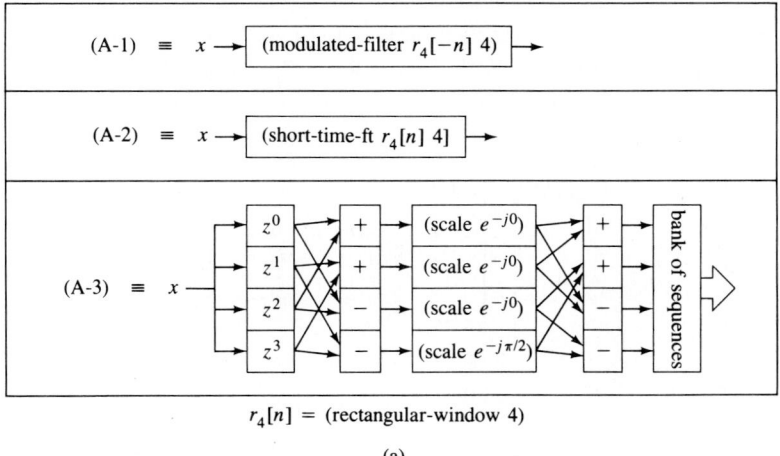

(a)

Figure 2.11 A net representation of the search for equivalent forms. This figure shows an example of a search for equivalent forms. Each node (e.g., A-1 or D-4) represents an expression. The name for each node consists of a letter (A through E) and a number. The letter indicates which expressions are equivalent (e.g., D-4 is equivalent to D-5) and the number indicates the order within the sequence of manipulations (e.g., D-5 is modified to create D-6).

58　　　Computer-Aided Algorithm Design and Rearrangement　　Chap. 2

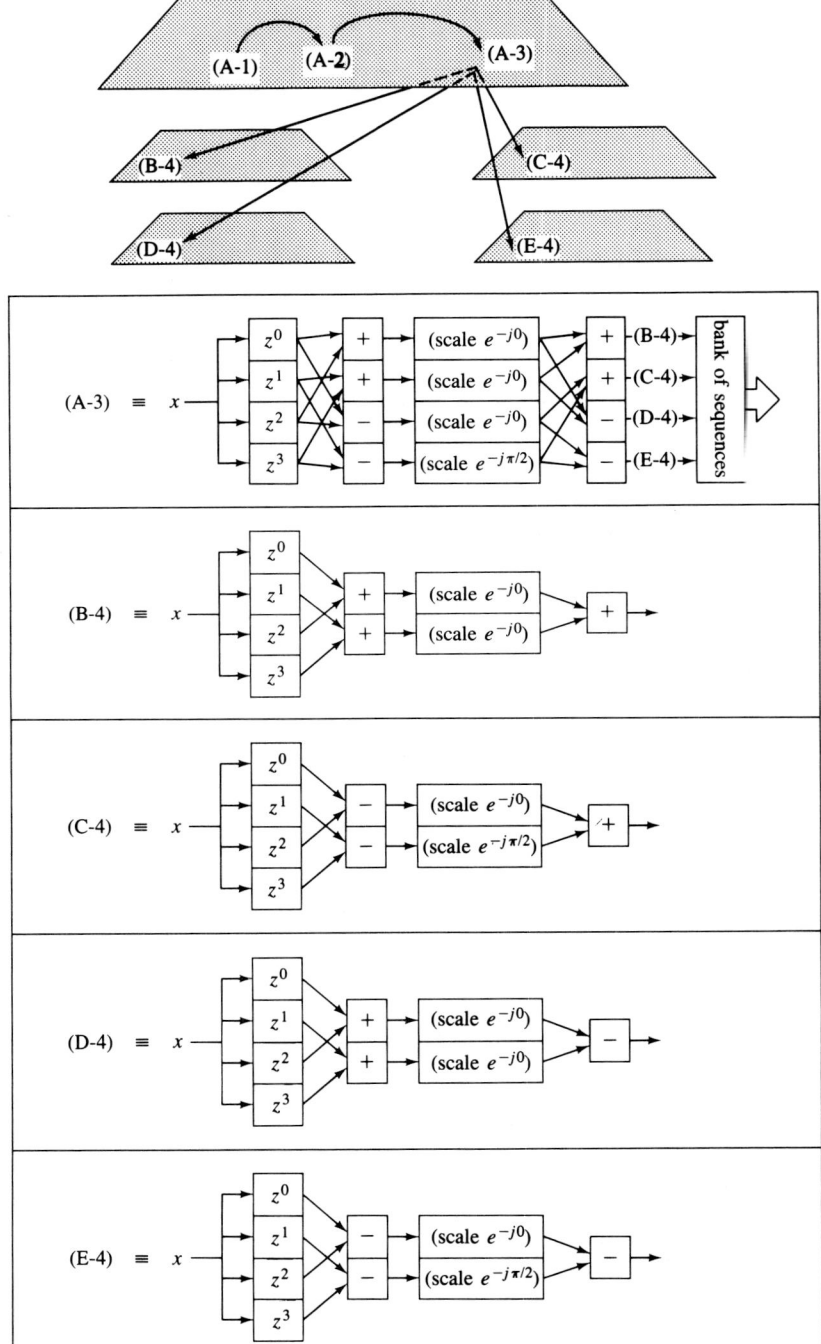

(b)

Figure 2.11 (continued)

Sec. 2.5 Unconstrained Derivation and Ranking of Equivalent Algorithms

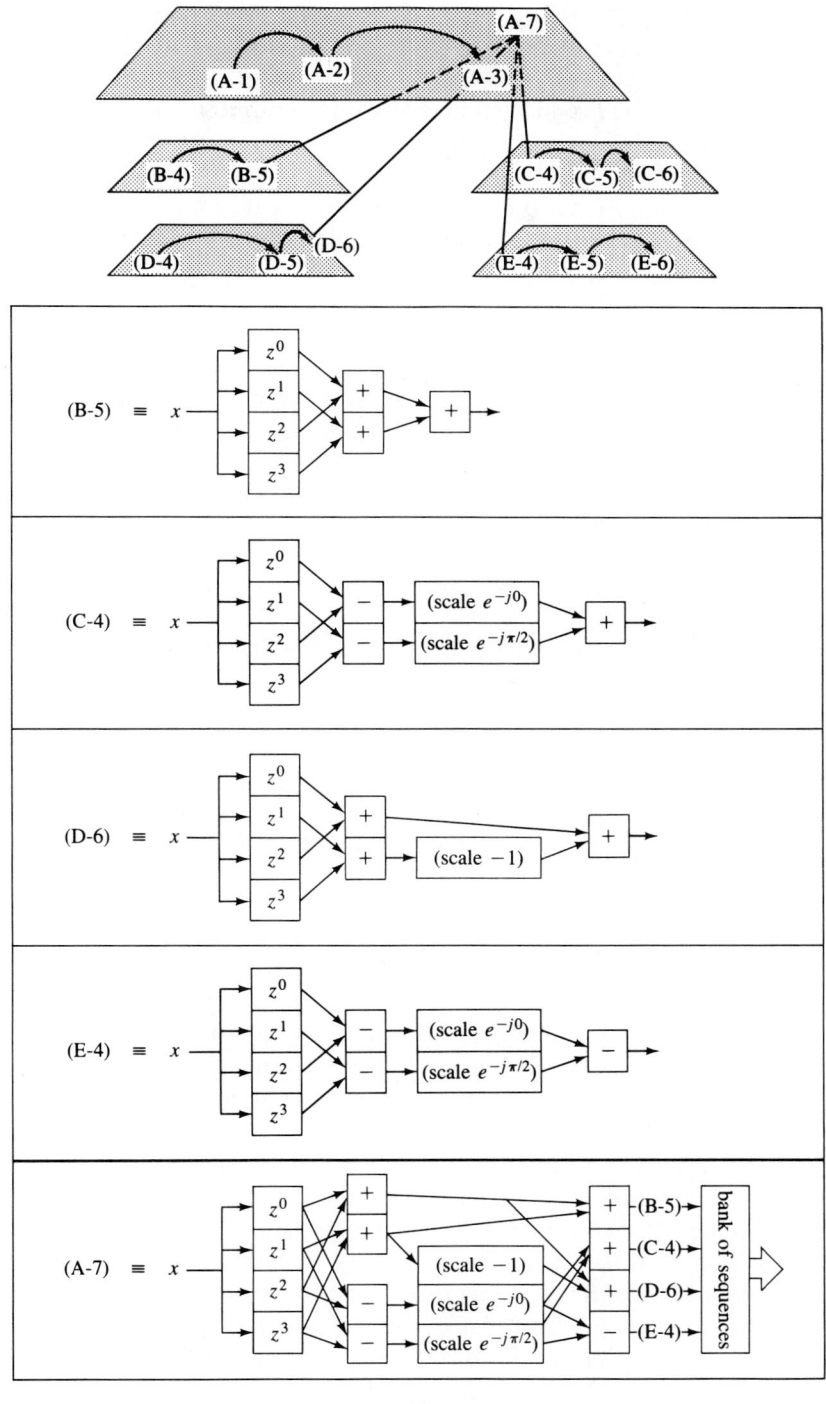

(c)

Figure 2.11 (continued)

find connected nodes. In our example, this process generates four new subsearches for the equivalent forms of the four inputs to the BANK-OF-SEQUENCES system in node A-3: Nodes B-4, C-4, D-4, and E-4 represent those inputs in Figure 2.11(b). The subsearches must also find all equivalent implementations of their given algorithm using simple transformations, repeated transformations, and subexpression transformations.

Once all the equivalent forms of the inputs are found, these equivalent forms can be used to replace the original input expressions. This replacement process is shown in Figure 2.11(c) as a projection of the nodes on the lower planar nets back into the original net: nodes B-5, C-4, D-6, and E-4 are used as inputs into the BANK-OF-SEQUENCES system, resulting in the new expression, node A-7. As with this example, the projection upward often generates new nodes in the original net (node A-7). A new node in the original net is generated whenever the input replacement results in an expression that has not already been generated through some other transformation path. These new nodes are equivalent forms of the original expression: thus, the new nodes can also be the starting point for further transformations.

This process of repeated transformation and subexpression transformation continues until no new equivalent expressions (nodes) can be found.

2.5.2 Infinite Expansion fo the Search Space for Equivalent Algorithms

Two major difficulties with the search process described above are apparent after careful consideration: the possibility of the infinite expansion of the search space and the finite but exponential growth of the space due to the separate manipulation of subexpressions. The search space will expand indefinitely, if a simple transformation or a combination of simple transformations repeatedly introduce operators that have no net effect (e.g., a delay operator followed by an advance operator). This difficulty is considered briefly in this subsection. The problem of limiting the exponential growth is the subject of the next section of this chapter.

The transformations used in generating equivalent implementations often result in signal processing algorithms whose complexity is greater than the manipulated algorithm. For example, consider the problem of finding equivalent implementations of the matched filters for the frequency chips in the rectangularly windowed FSK-code detector. One of the rearrangements that is found is shown in Figure 2.12(a). The subexpressions of this implementation will be manipulated, due to the combination of recursive search and subexpression manipulation. One of the implementations discovered by this process is shown in Figure 2.12(b): this structure results from the application of a rule shown on lines 13–30 of Figure 2.10. Applying repeated transformation and subexpression manipulation to the structure shown in Figure 2.12(b) will result in, among others, the structure shown in Figure 2.12(c). In fact, the recursive transformations and the increasing complexity of the algorithmic description would result in the infinite expansion of the search space.

Sec. 2.5 Unconstrained Derivation and Ranking of Equivalent Algorithms

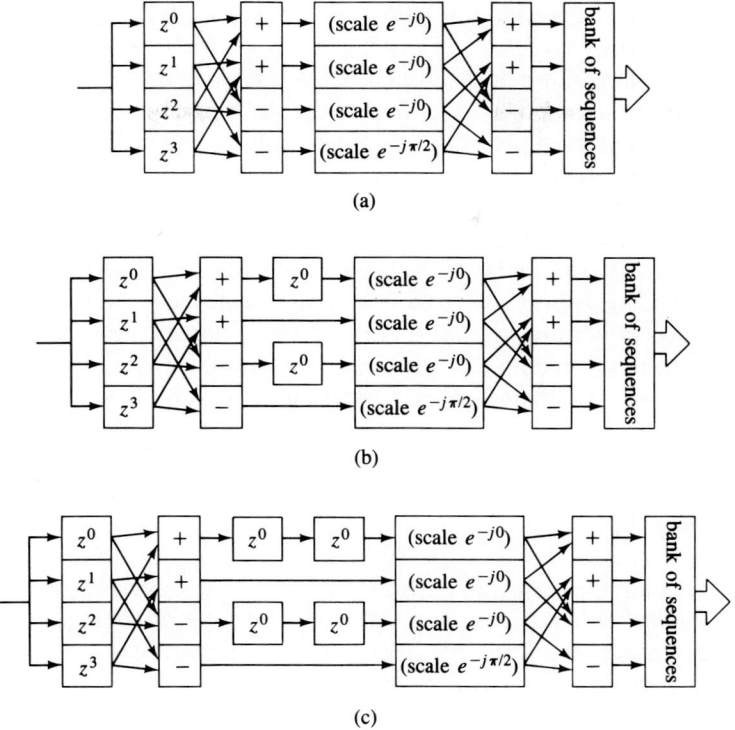

Figure 2.12 An example of increasing complexity resulting from equivalent-form manipulations.

As can be seen from this example, care must be taken to limit the complexity of the algorithms before they are used as starting points for further transformations. In ADE, simplifications are used to control the complexity of the signal processing expression. SIMPLIFICATION, when applied to a signal processing expression, returns the simplest direct description of the expression that the environment can find. The simplest description of a signal processing expression is obtained both by simplifying its subexpressions and by repeatedly simplifying the modified description. The actual simplifying transformations are encoded in ADE using over 300 of the 850 rules currently included.

The simplification process in ADE has two potential shortcomings. First, a good simplification can be missed because the steps required to generate the simplified form include steps that would cause subexpressions to not be in their simplified form. This restriction is imposed in order to prevent an unlimited potential growth in the number of expressions that must be considered. Second, and perhaps more fundamentally, ADE defines simplifications by a set of rules. It does not have any hooks for the user to define in what ways an expression is simpler than another one.

Hence, it is possible for ADE to "simplify" an expression to a form that is not appropriate.

Although we have pointed out that unconstrained algorithm manipulation has limitations, we must also point out that in many cases it is still a valuable tool. One such example is provided by the noninteger sampling rate conversion problem introduced in section 2.2. By applying unconstrained manipulations and simplifications to the straightforward implementation of a noninteger sampling rate conversion [shown in Figure 2.1(a)], ADE can automatically derive the computationally efficient implementation shown in Figure 2.13.[9] While the structure shown in Figure 2.13 can also be derived using the constrained manipulations, described in the next section, there are cases when efficient implementations cannot be derived using constrained manipulation. One such example is the efficient implementation of maximally decimated, octave band filters (Figure 2.14).

2.6 CONSTRAINED DERIVATION AND RANKING OF EQUIVALENT ALGORITHMS

The search for equivalent implementations of a signal processing expression must consider the equivalent implementations of the subexpressions as well as the complete expression itself. Since each of the subexpressions is independently manipulated and their equivalent forms are independently recombined to form new equivalent expressions, the size of the search space under consideration grows exponentially with the number of subexpressions. To illustrate, consider the problem of implementing the full FSK-code detector for 16 channels. Five independent descriptions of a simple, finite-length convolution are included in ADE: direct-form convolution, overlap-save convolution, the Fourier-domain representation of convolution, the z-domain representation of convolution, and the representation of convolution as the sum of scaled, shifted versions of the input. Thus, using these alternate forms as inputs into the incoherent summation, there will be $5^{16} \approx 10^{11}$ equivalent forms to consider. None of these implementations exploit the special structure of the modulated filter bank: the actual number of equivalent implementations that have to be considered is more than 10^{19}. Each of these implementations would then be reconsidered to see if any additional equivalent forms could be found, due to interactions between the implementations of the matched filters and the implementations of the incoherent processing. As illustrated by the projected size of the design space, some set of constraints must be imposed on the search process to avoid this exponential growth.

---

[9] Unconstrained manipulation was first shown to be effective by Myers [2] using a 2 : 3 noninteger sampling rate conversion in E-SPLICE. E-SPLICE generated a multirate structure of the same form as the one shown in Figure 2.13 for a 2 : 3 rate conversion. This demonstration of the potential of high-level signal processing compilation was made even more convincing by the subsequent publication of an independent article presenting this new type of polyphase structure [16]: E-SPLICE actually anticipated results from research in the area of noninteger sampling rate conversion algorithms.

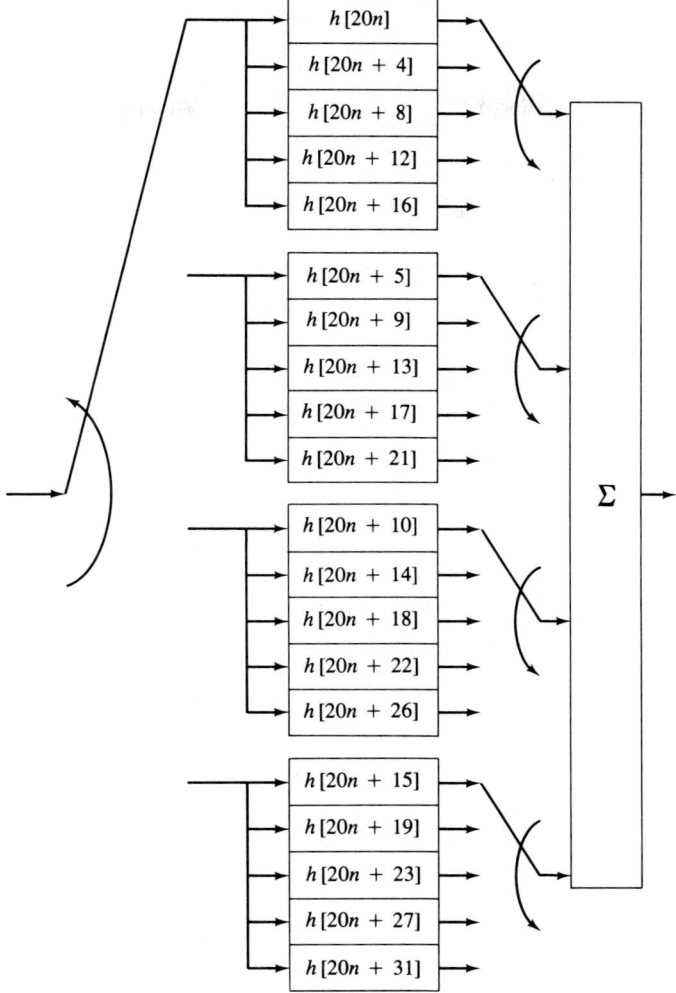

Figure 2.13 An efficient implementation of the 4:5 noninteger sampling rate conversion.

2.6.1 Approaches for Avoiding Exponential Growth of the Algorithm Design Space

One possible strategy for limiting the exponential growth in searches for efficient implementations relies on the cost measure of each subexpression to heuristically prune the space. Instead of enumerating all the equivalent implementations of a signal processing expression and then filtering out the inefficient and uncomputable structures using the overall cost measure, this strategy would immediately prune the

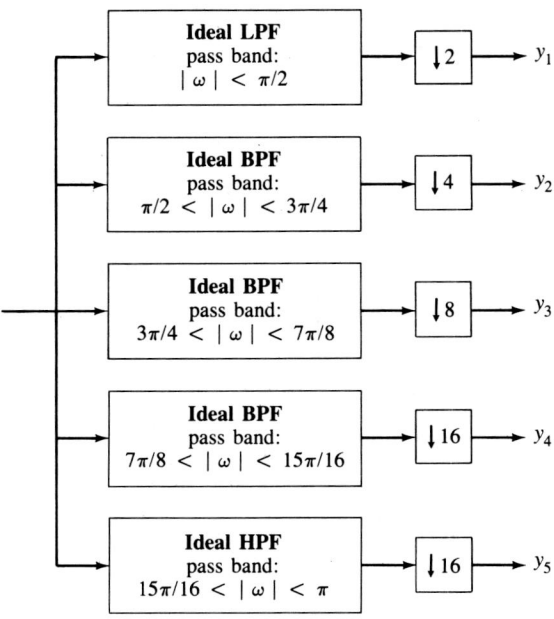

(a) Direct implementation of maximally decimated octave-band filters

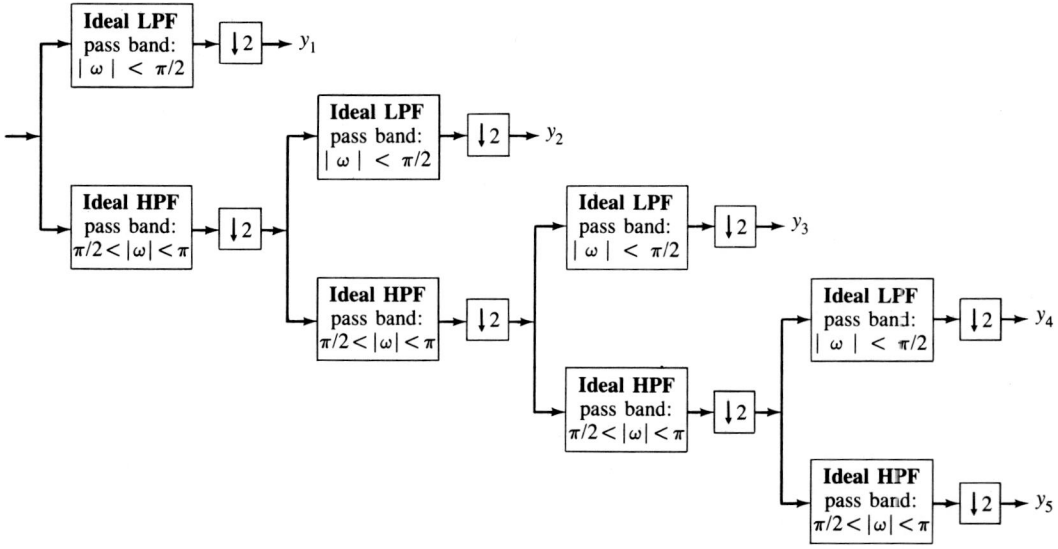

(b) A more efficient implementation of maximally decimated octave band filters

Sec. 2.6 Constrained Derivation and Ranking of Equivalent Algorithms

number of subexpression implementations, prior to their upward propagation, based on their relative costs. This approach relies on the assumption that, when propagating two alternate implementations upward, the more expensive implementation will not be incorporated into any of the efficient implementations of the enclosing expression. Unfortunately, this pruning strategy suffers from the interaction of subexpression costs: the cost of using one implementation of a subexpression is often ameliorated by reusing part or all of the subexpression in some other part of the enclosing expression. For example, the computational savings of the FFT-based STFT results from the interaction of the computational cost of the subexpressions: it is the reuse of the partial summations which reduces its computational cost to $O(N \log N)$. Thus, the contribution of a subexpression to the overall cost of an enclosing expression is not independent of the other parts of the enclosing expression.

The approach that ADE takes to limiting the search space is to attempt to exploit the internal regularity of signal processing algorithms. Signal processing algorithms are often described at different levels of detail. For example, the four-point, rectangularly windowed, short-time Fourier transform of a sequence can be described by either of the structures shown in Figure 2.15. The structure in Figure 2.15(a) is by definition the STFT. Figure 2.15(b) shows a fully expanded short-time FFT structure, i.e., one in which common subexpressions are not combined. Starting from the high-level description of an algorithm, the regularity in the low-level computational structure can often be asserted. For example, when the SHORT-TIME-FT system is expanded into the structure in Figure 2.15(b), the underlying regularity inherent in Figure 2.15(b) can be noted. By enforcing these internal correspondences in the low-level descriptions, the space of equivalent forms that is explored can be drastically reduced. This approach to pruning the search is heuristic. However, the regularity of the computation suggests that the efficient implementations will reflect the same regularity: if separate sections of an algorithm are very similar, then the efficient implementations of these separate sections are likely to coincide.

To illustrate what is meant by internal regularity within an algorithm, consider the description of the short-time Fourier transform given in Figure 2.15(b). This

Figure 2.14 Implementation of maximally decimated, ideal, octave-band filters. Maximally decimated octave-band filters are useful in speech and image coding as well as perceptual modeling. A simple and direct implementation of maximally decimated, octave-band filters, using ideal low- and band-pass filters, is shown in part (a) of this figure. An implementation that is more computationally efficient is shown in part (b): here the band-pass filters in the octaves above base-band have been divided into cascades of high-pass filters, followed by decimation, followed by another stage of high- and low-pass filters. This implementation is one example of a computationally efficient implementation which would not be found using constrained manipulation as it is described in section 2.6 of this chapter. The reason that this implementation would be missed using constrained manipulation is that each branch of the algorithm is treated differently: the branch that computes y_1 is unchanged, the branch that computes y_2 is separated into two stages of filtering/decimation, the branch that computes y_3 is separated into three stages of filtering/decimation, and so on.

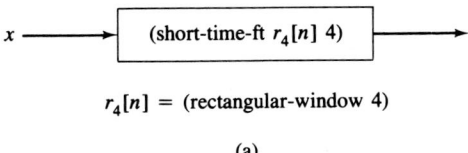

(a)

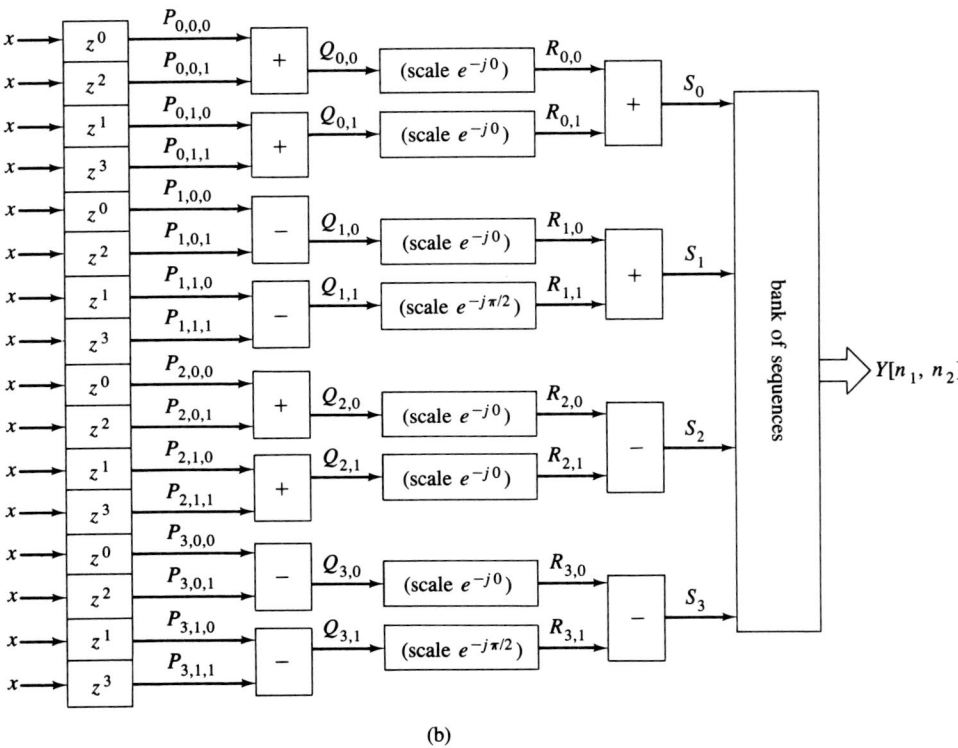

(b)

Figure 2.15 Two alternate descriptions of a four-point, rectangularly windowed short-time Fourier transform.

implementation of the short-time Fourier transform is provided explicitly by one of the transformation rules included in the definition of SHORT-TIME-FT. The regularity of this implementation is also explicitly noted by the rule. In particular, the similarity of the sequences feeding into the BANK-OF-SEQUENCES is pointed out using a "correspondence constraint": by placing a correspondence constraint on the subexpressions feeding into the BANK-OF-SEQUENCES, the similarity of these subexpressions is recorded. As will be discussed later in this section, a correspondence constraint forces the similar subexpressions to be subject to a single series of transformation rules, so that, if a transformation is applied to one of a set of similar expressions, that transformation must also be applied to all the other members of the set.

Sec. 2.6 Constrained Derivation and Ranking of Equivalent Algorithms

In addition to the point of similarity at the BANK-OF-SEQUENCES, the structure shown in Figure 2.15(b) has two other levels of similarity at the inputs to the addition/subtraction systems: at each of the two butterfly stages, the first addend into the kth butterfly is similar to the second addend into the same kth butterfly. Thus, more correspondence constraints are placed on the inputs into each of these systems. Through these correspondence constraints, the similarity of the corresponding subexpressions is explicitly noted, resulting conceptually in the manipulation of $Y[n_1, n_2]$, $S_k[n]$, $R_{k,l}[n]$, $Q_{k,l}[n]$, and $P_{k,l,m}[n]$ where

$$Y[n_1, n_2] = S_{n_2[n_1]}$$
$$S_k[n] = R_{k,0}[n] \pm R_{k,1}[n]$$
$$R_{k,l}[n] = e^{-j(2\pi/4)lk} Q_{k,l}[n]$$
$$Q_{k,l}[n] = P_{k,l,0}[n] \pm P_{k,l,1}[n]$$
$$P_{k,l,m}[n] = x[n + l + 2m]$$

By enforcing these constraints, the rearranged algorithms will also have a regular internal structure and the number of independently manipulated subexpressions is reduced, in this case from $O(N^2)$ to $O(\log N)$.

2.6.2 Algorithmic Transformations in the Presence of Correspondence Constraints

This subsection examines the process of finding equivalent forms of an algorithm in the presence of correspondence constraints.

When a correspondence constraint is imposed, it is imposed on the inputs to a system. For example, a correspondence constraint was imposed on the inputs to the BANK-OF-SEQUENCES in Figure 2.15(b), creating nine sets of similar sequences: S_k, $R_{k,0}$, $R_{k,1}$, $Q_{k,0}$, $Q_{k,1}$, $P_{k,0,0}$, $P_{k,0,1}$, $P_{k,1,0}$, and $P_{k,1,1}$. Additional correspondence constraints were also imposed on the inputs for S_k, reducing the number of separately manipulated sets of signals to five (S_k, $R_{k,l}$, $Q_{k,l}$, $P_{k,l,0}$, and $P_{k,l,1}$) and on the inputs for $Q_{k,l}$, further reducing the number of separately manipulated sets to four (S_k, $R_{k,l}$, $Q_{k,l}$, and $P_{k,l,m}$). As mentioned above, imposing these constraints forces the similar signals to be manipulated as a set: the transformation rules are applied to the set of similar signals, instead of just to the individual signals. To continue the STFT example, if the rule shown on lines 13–30 of Figure 2.10 is applied to one of the sequences labeled $Q_{k,l}$, then this same rule must be applied to all the other sequences labeled $Q_{k,l}$. The remainder of this subsection considers the changes required in the recursive search and subexpression manipulation, described in section 2.5, to accommodate these requirements.

The approach to finding equivalent forms described in section 2.5 involved both recursive search and subexpression manipulation. The recursive search, in which newly generated equivalent forms act as the starting point for further transformations, is used without modification in searches for constrained equivalent forms. The subexpression manipulation must be modified for constrained equivalent-form searches. In particular, subexpression manipulation generates equivalent forms of an expression by replacing its subexpressions by their equivalent forms. Naturally, if no correspondence constraint has been imposed on the inputs to a system, the strategy for finding the constrained equivalent forms of the subexpressions is similar to that for finding unconstrained equivalent forms. In this case, the only difference is whether the constrained or unconstrained equivalent forms of the subexpressions are used. This distinction will make a difference when there are correspondence constraints imposed within one or more of the subexpressions.

When a correspondence constraint has been imposed on the inputs to a system, the constrained equivalent forms of the inputs cannot be generated separately. Instead, the constrained equivalent forms of the inputs are found by manipulating these subexpressions as a set. The set of alternate implementations that result from these manipulations of the subexpressions are then used to replace the subexpressions in the generating expression. In terms of the search process illustrated in Figure 2.11, correspondence constraints imply that the four subsearches in Figures 2.11(b) and (c) apply the same ordered sequence of transformation rules to the four input expressions.

To illustrate this process, consider the task of finding the constrained equivalent forms of the four-point, short-time Fourier transform structure (node A-1 in Figure 2.16). Some of these equivalent forms will be found by manipulating the input sequences to the system BANK-OF-SEQUENCES as a set. Thus, the sequences B-2 are considered simultaneously. Under this set manipulation, each of the equivalent-form transformations that could be applicable to the individual expressions is considered. Only the equivalent-form transformations that are applicable to *all* of the sequences in the set are used. The point of manipulation progresses inward (or downward, in the graph representation shown in Figure 2.16) by considering the inputs to the constrained expressions. Thus, the sequences C-3 are examined as a set; then the sequences D-4 are examined as a set; and, finally, the sequences E in Figure 2.16 are examined as a set.

Each equivalent implementation of a set of subexpressions propagates upward by replacing the original subexpressions in the containing expression with the set of new implementations. Thus, any equivalent forms found for node E will propagate upward by replacing the subexpressions E in the expressions D-4 with the new equivalent forms. This replacement process will create alternate implementations of the parallel expressions D-4. Each new set of equivalent expressions is simplified using CONSTRAINED-SIMPLIFICATION and used as a starting point to find other constrained equivalent forms. This recursive search continues until no new constrained equivalent forms are found. The sets of implementations equivalent to node D-4 then propagate upward by replacing the subexpressions D-4 in expressions C-3 with these new implementations. The cycle of constrained simplification and recursive searches is repeated. The upward progression of constrained equivalent forms continues until all the equivalent forms of the original short-time Fourier transform are found. Figure 2.16 sketches one line of manipulation. Here, the original expression, A-1, is recursively decomposed into the structures B-2, C-3, and

Sec. 2.6 Constrained Derivation and Ranking of Equivalent Algorithms

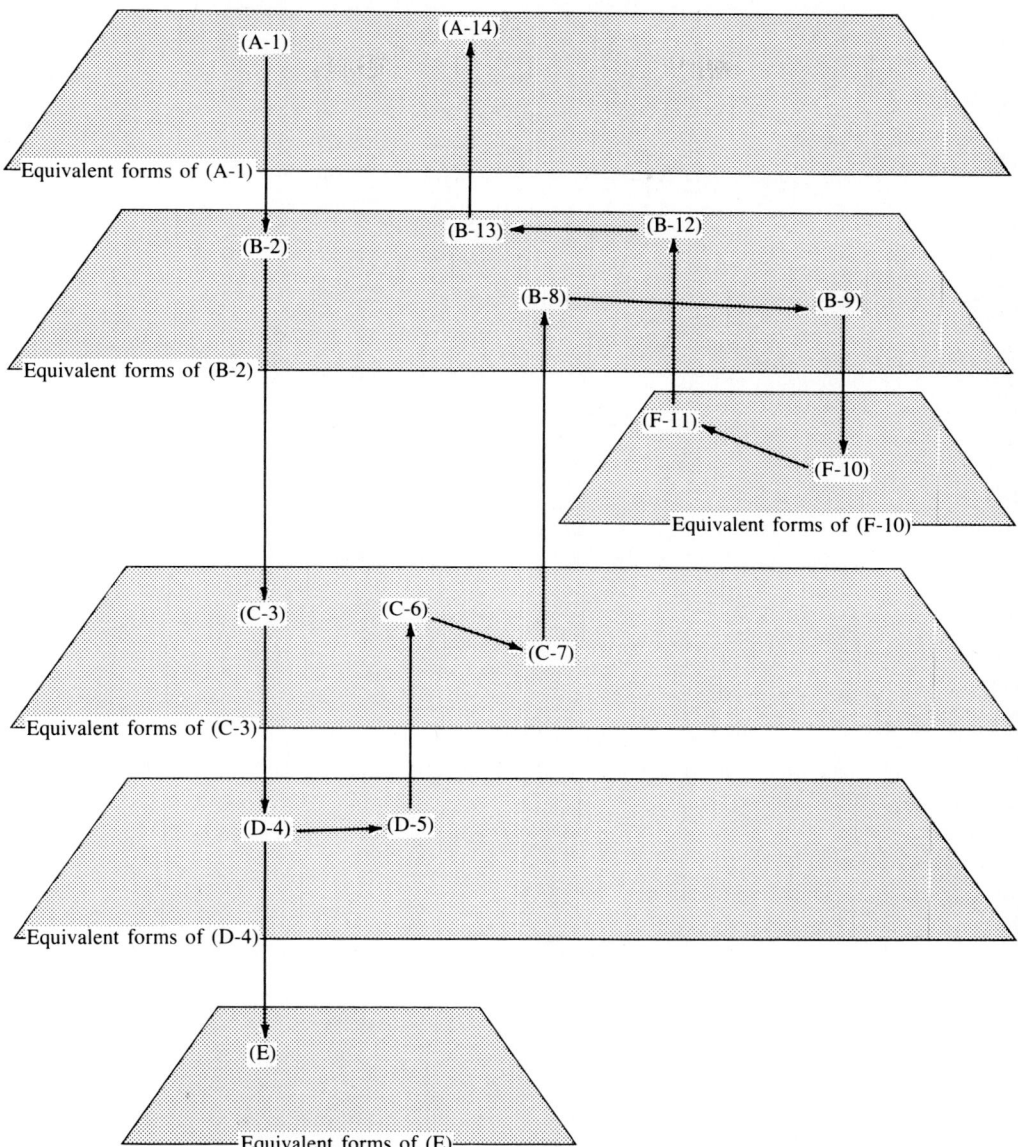

Figure 2.16 An example of parallel expressions and the effects of imposing regularity constraints. This figure shows one line of manipulation that occurs in the search for the constrained equivalent forms of the 4-point STFT labelled (A-1). Each node (e.g., A-1 or D-4) represents an expression or a set of parallel expressions. The name for each node consists of a letter (A through F) and a number. The letter indicates which expressions are equivalent (e.g., D-4 is equivalent to D-5) and the number indicates the order within the sequence of manipulations (e.g., D-5 is used to create C-6). The only node without a number, node E, occurs as a "side branch" to this line of manipulations: that is, although the equivalent forms of node E are considered, none of these equivalent forms are used in this line of manipulation. The remainder of the figure shows the expressions that correspond to the nodes.

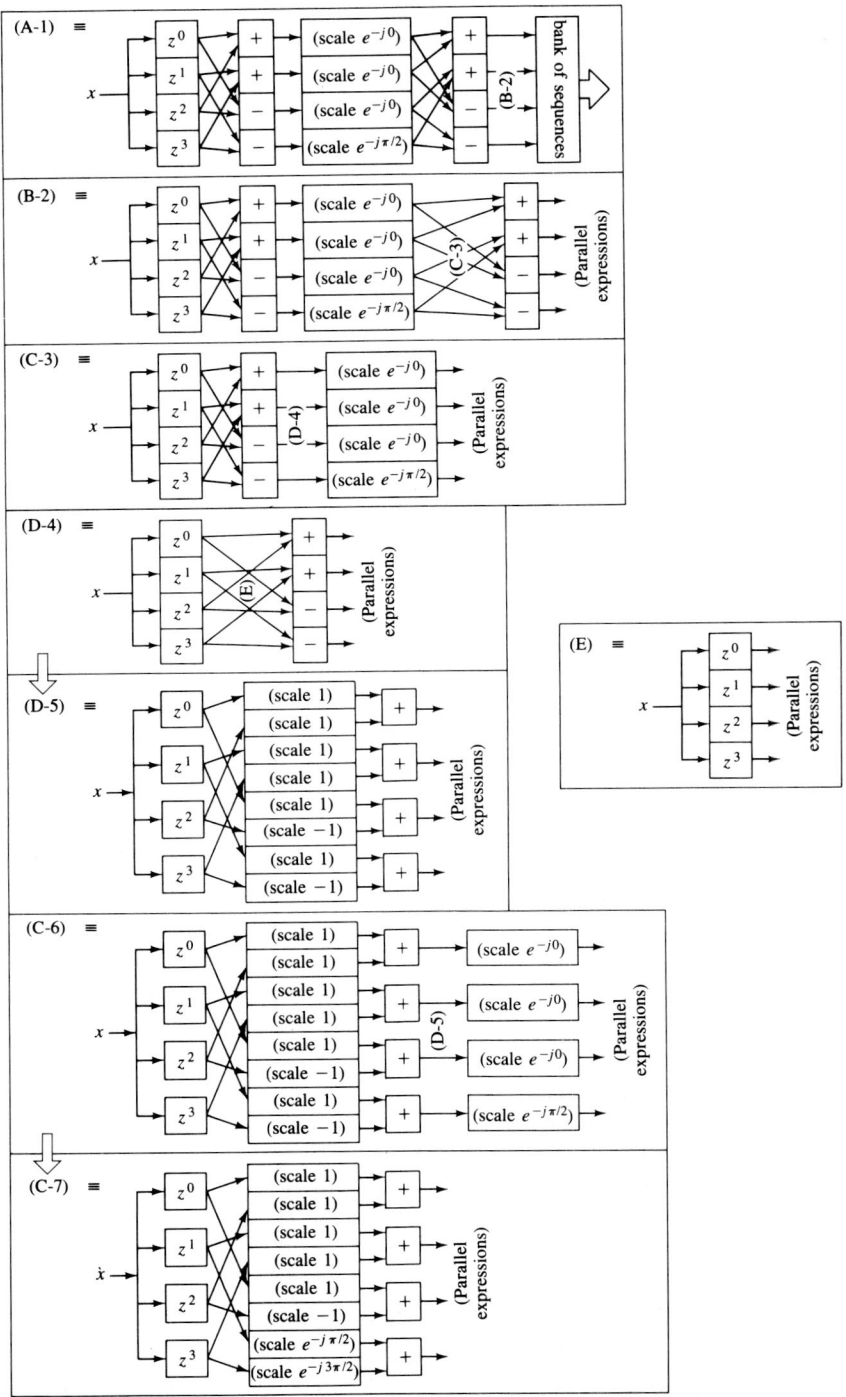

Figure 2.16 (continued)

Sec. 2.6 Constrained Derivation and Ranking of Equivalent Algorithms

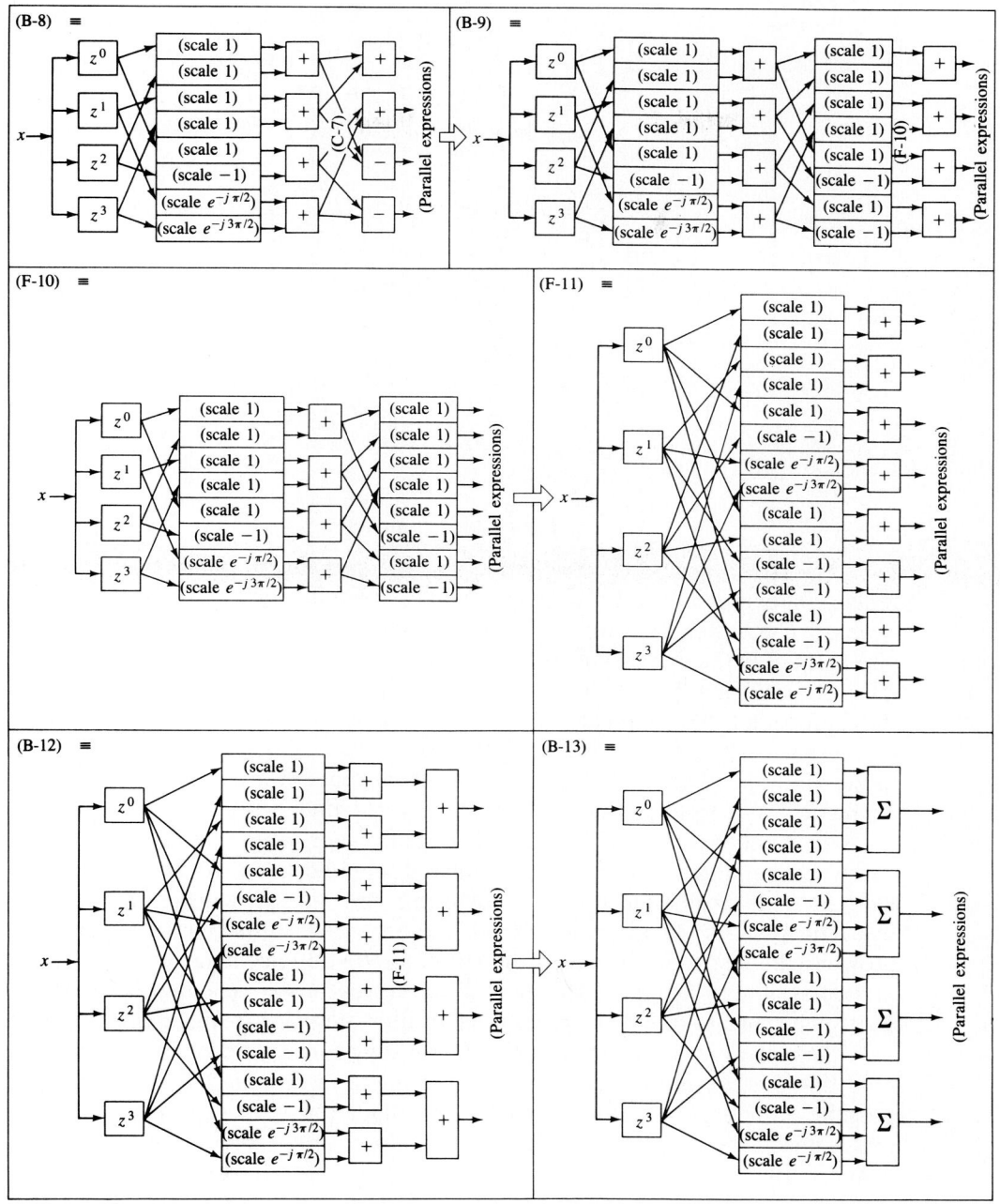

Figure 2.16 (continued)

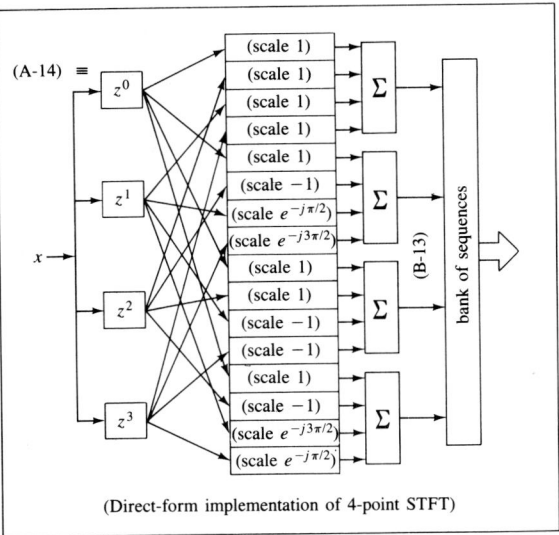

Figure 2.16 (continued)

D-4. The structure D-4 is transformed into the structure D-5 and this is used to replace subexpression D-4 in the expression C-3, resulting in the structure C-6. The structure C-6 is then transformed into the structure C-7, which then replaces subexpression C-3 in the expression B-2, resulting in the structure B-8. The structure B-8 is transformed into the structure B-9. The subexpression F-10 is then extracted from the expression B-9, manipulated into subexpression F-11, and replaced back into the expression B-9, resulting in the structure B-12. Next, the structure B-12 is transformed into the structure B-13 and this structure replaces subexpression B-2 in the expression A-1, yielding the structure A-14 as the final result.

As mentioned earlier, the actual transformation rules that are used to find constrained equivalent forms and constrained simplifications are the same as those used to find unconstrained equivalent forms and unconstrained simplifications, respectively. It is the manner in which these transformations are combined that provides the distinction between the constrained and unconstrained searches. In particular, in unconstrained searches, all subexpressions are manipulated independently while in constrained searches, similar subexpressions are manipulated as sets.

As a measure of the importance of imposing correspondence constraints, we note that for the modulated filter bank example of section 2.4, the number of possible structures using unconstrained manipulation is more than 10^{19} for the rectangular-window matched filters and it is more than 10^{58} for the Hanning-window matched filters. Using constrained manipulations, the numbers are 13 and 20, respectively.[10] Thus, the use of regularity constraints is essential for solving these problems.

---

[10] The large number of structures generated by unconstrained manipulation is a result of considering all possible combinations of the alternate implementations of the subexpressions. Since there are 16 branches with 16 inputs each, the number of possible combinations grows very quickly.

2.6.3 Propagating Correspondence Constraints to a Modified Structure

When constrained expressions are manipulated, new expressions are often generated on which the same correspondence constraints should be imposed. To illustrate, consider the manipulations shown in Figure 2.16. The result of these manipulations is a direct-form implementation of the short-time Fourier transform (node A-14). In order to reflect the correspondence constraints of the original structure, this new bank of sequences should also include two correspondence constraints: the inputs to the bank of sequences should be constrained to coincide as should the inputs to the addition systems. Unless these constraints are imposed, this new form will introduce unconstrained structures into the constrained manipulations. ADE propagates structural constraints to new expressions automatically. When a constrained expression is manipulated, the inputs that are constrained to be parallel are noted prior to manipulation. After each manipulation of a constrained structure, ADE attempts to impose the analogous correspondence constraints on the modified structure.

2.7 CONTRIBUTIONS AND LIMITATIONS

Our goals in this chapter were to describe the mechanisms by which an integrated signal processing environment could manipulate signal processing algorithms and to demonstrate the potential of these manipulations. ADE has been used within this chapter to demonstrate this potential.

ADE makes extensive use of general signals, rules, and regularity constraints. General or abstract signals provide the variables in the manipulation of algorithms while the rules in ADE about signal processing provide the information required to manipulate this "signal processing algebra." Regularity constraints limit the exponential expansion that manipulation of subexpressions introduces by exploiting the internal regularity of signal processing algorithms to limit the size of the explored search space. This regularity in the low-level signal processing descriptions is noted using information provided by the higher level description of the same operation. Without these constraints, many FFT-based and polyphase-based algorithms would be beyond the scope of consideration, due to exponential expansion of these design spaces.

ADE demonstrates the potential of integrated signal processing environments. Its algorithm rearrangement capabilities have been used to generate innovative, computationally efficient implementations in two well-developed areas of signal processing. However, much work remains to be done to transform the concepts embodied in ADE into practical signal processing workstations:

- The user interface should be made more accessible and pleasant.
- A fast algorithm-rearrangement facility, providing only well-known algorithm implementations, should be included as an alternative to the current rearrange-

ment facilities which do constrained searches of the full design space. This facility would provide quick compilations by ignoring the possibility of efficient nonconventional implementations. An example of this facility would be one that would provide the classic implementations of a noninteger sampling rate conversion [e.g., Figures 2.1(b) and (c)] without exploring the full design space. This facility would miss the more nonconventional implementations (e.g., Figure 2.13).
- The derivation of explicitly recursive algorithms, such as the one shown in (2.3)–(2.6), is an unexplored area of research.
- Regularity within expressions should be automatically detected. In ADE, the regularity of a signal processing algorithm must be explicitly pointed out. Although propagation of the regularity constraints, both within the algorithm and to modified expressions, is supported by the environment, the initial description of the constraints must be done manually.

As can be seen by the breadth of this partial list of "things to do," much work still remains to be done to achieve a signal processing environment that facilitates the algorithm manipulation as well as numeric processing of signals and algorithm definition. However, as asserted previously, the rewards for research in this field have thus far been high.

ACKNOWLEDGMENTS

We would like to acknowledge the contributions to this work by Professor Hal Abelson, Professor Randy Davis, Dr. Webster Dove, Dr. Robert Kahn, Dr. Evangelos Milios, Dr. Douglas Mook, Dr. Bruce Musicus and Professor Victor Zue. We would also like to thank Dr. Gary Kopec, Dr. Evangelos Milios and Dr. Malcolm Slaney for their comments and help with this chapter. This work was supported in part by the Advanced Research Projects Agency, Lockheed Sanders, Inc., the AMOCO Foundation, and Schlumberger-Doll Research.

REFERENCES

[1] M. M. Covell, "An Algorithm Design Environment for Signal Processing," RLE Technical Report 549, Cambridge, Mass.: MIT (1989).

[2] C. S. Myers, "Signal Representation for Symbolic and Numerical Processing," RLE Technical Report 521, Cambridge, Mass.: MIT (1986).

[3] G. E. Kopec, "The Signal Representation Language SRL," *IEEE Trans. Acoustics, Speech, and Signal Processing*, 33 (August 1985), 921–32.

[4] W. P. Dove, C. S. Myers, and E. E. Milios, "An Object-Oriented Signal Processing Environment: The Knowledge-Based Signal Processing Package," RLE Technical Report 502, Cambridge, Mass.: MIT (1984).

[5] R. E. Crochiere and L. R. Rabiner, *Multirate Digital Signal Processing* (Englewood Cliffs, N.J.: Prentice Hall, 1983).

[6] J. S. Jaffe and J. M. Richardson, "A Code-Division Multiple Beam Imaging System," in *Proc. OCEANS '89* (1989):1015–20.

[7] A. V. Oppenheim and R. W. Schafer, *Discrete-Time Signal Processing* (Englewood Cliffs, N.J.: Prentice Hall, 1989).

[8] P. M. Peterson and J. A. Frisbie, "Interactive Environment for Signal Processing on a VAX Computer," in *Proc. ICASSP'87* (1987):1891–93.

[9] W. P. Dove, "Knowledge-based Pitch Detection," RLE Technical Report 518, Cambridge, Mass.: MIT (1986).

[10] E. Sacks, "Qualitative Mathematical Reasoning," in *Proc. of the Int. Joint Conference on Artificial Intelligence* (1985):137–39.

[11] The Mathlab Group, *MACSYMA Reference Manual*, Lab. for Computer Science, Cambridge, Mass.: MIT (1983).

[12] Symbolics, Inc., *Symbolics Common Lisp: Language Concepts*, Cambridge, Mass.: Symbolics, Inc. (1986).

[13] R. C. Singleton, "An Algorithm for Computing the Mixed Radix Fast Fourier Transform," *IEE Trans. Audio and Electroacoustics* 17 (June 1969):93–103.

[14] J. D. Markel, "FFT Pruning," *IEEE Trans Audio and Electroacoustics* 19 (December 1971):305–11.

[15] D. P. Skinner, "Pruning the Decimation in Time FFT Algorithm," *IEEE Trans. Acoustics, Speech and Signal Processing*, 34 (April 1976):305–11.

[16] C.-C. Hsiao, "Polyphase Filter Matrix for Rational Sampling Rate Conversions," in *Proc. ICASSP'87* (1987):1056–59.

[17] MathWorks, Inc., *MATLAB: User's Guide*, Natick, Mass.: The MathWorks, Inc. (1989).

[18] Signal Technology, Inc., *ILS: Interactive Laboratory System*, Goleta, Calif.

76 Computer-Aided Algorithm Design and Rearrangement   Chap. 2

APPENDICES

Appendix A.1 The sequence of transformations used in going from the 16-channel rectangular-window modulated filter bank to the pruned FFT structure shown in Figure 2.7.

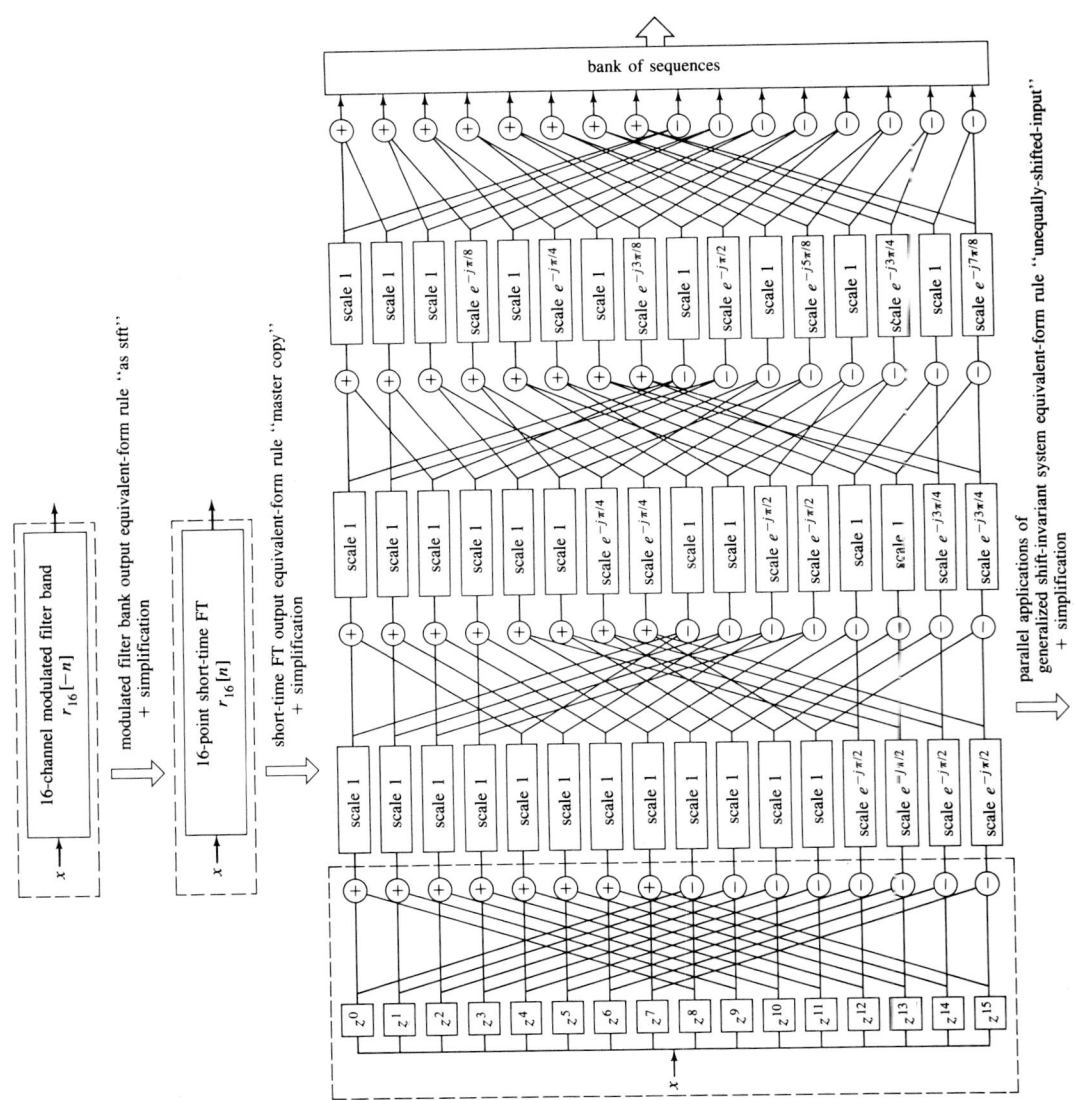

Chap. 2 Appendices

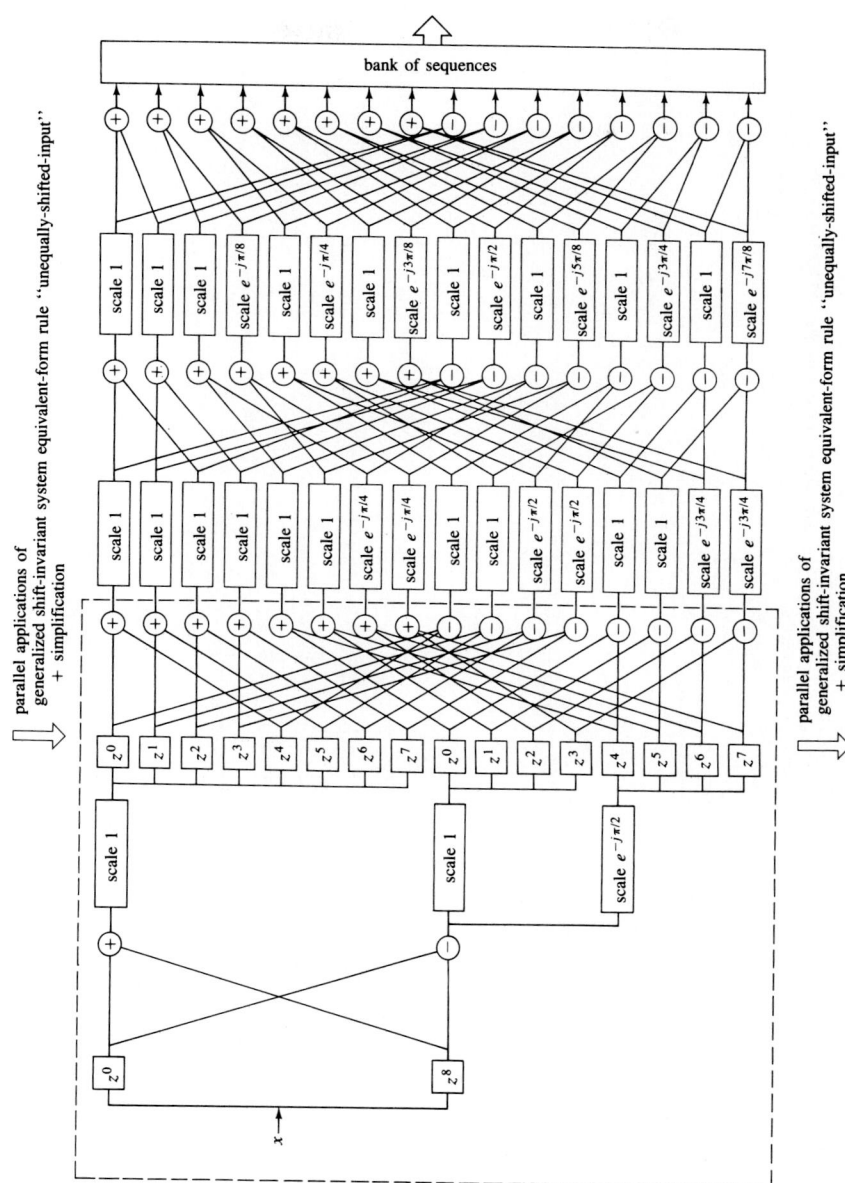

78 Computer-Aided Algorithm Design and Rearrangement Chap. 2

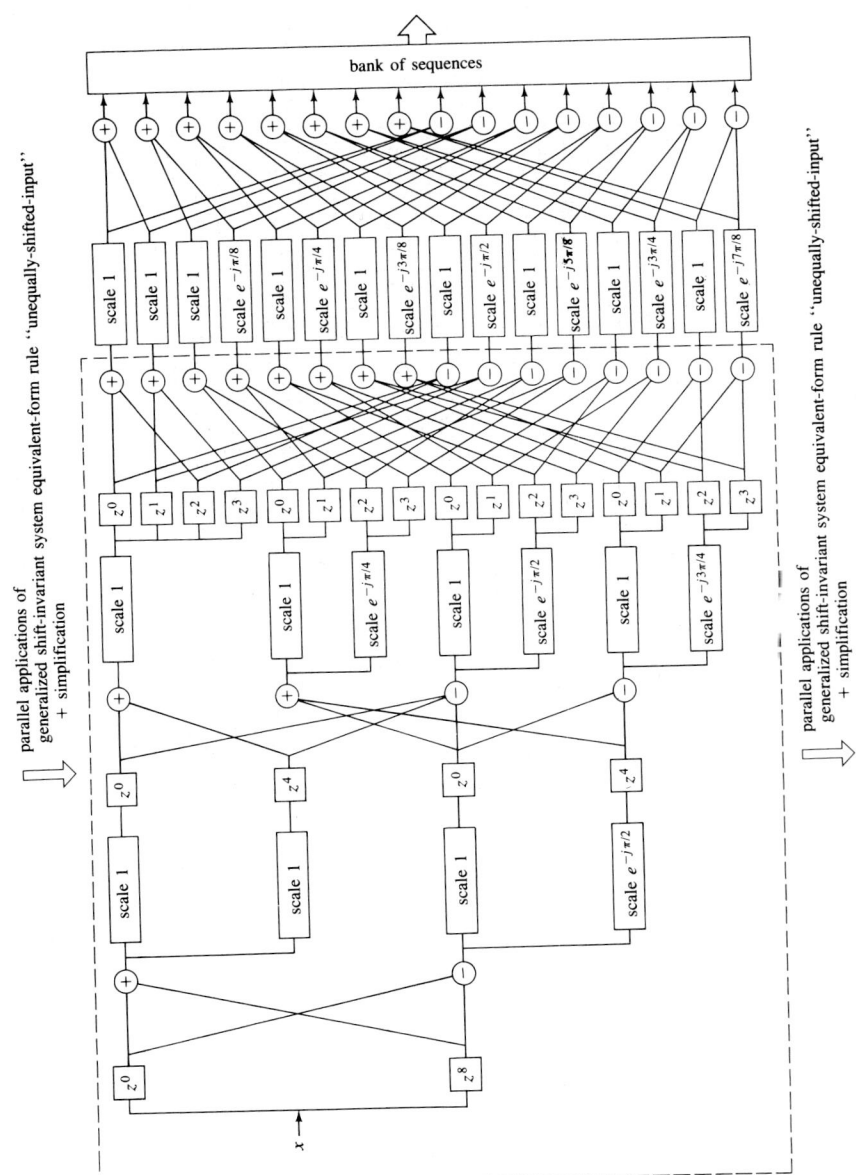

Chap. 2 Appendices

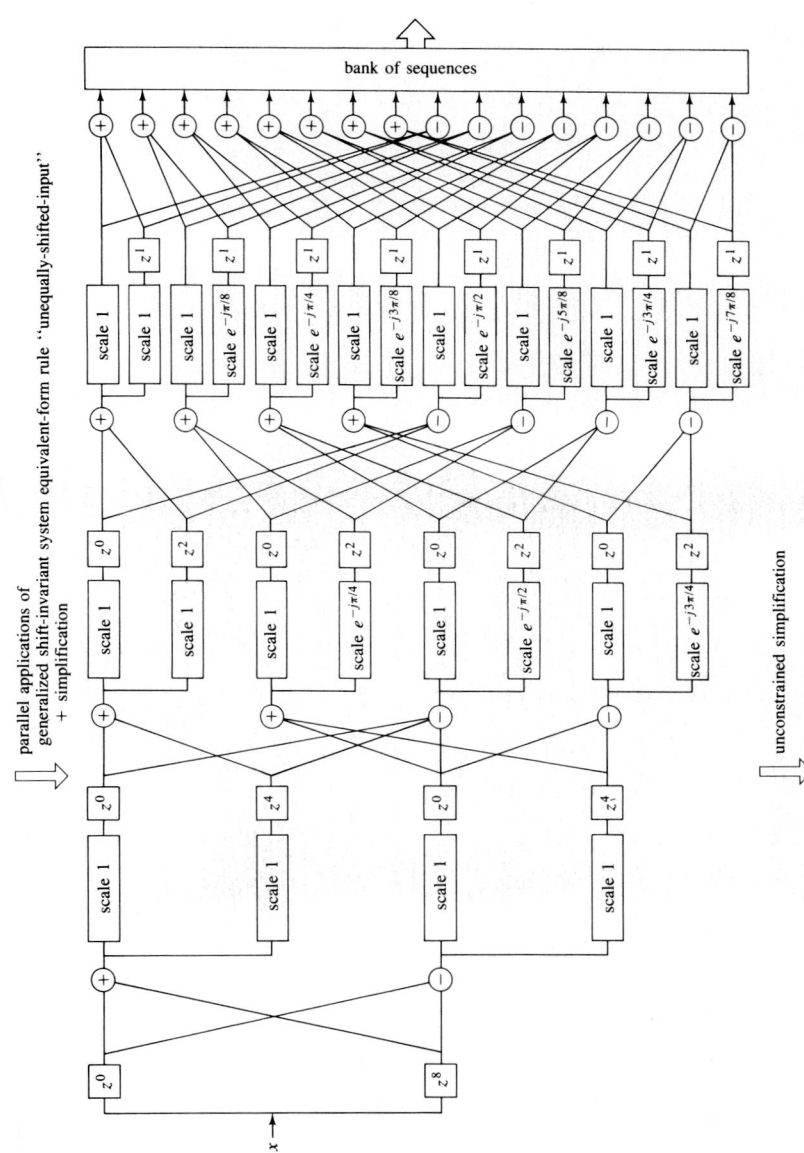

Pruned FFT structure shown in Figure 2.7

80   Computer-Aided Algorithm Design and Rearrangement   Chap. 2

Appendix A.2 The sequence of transformations used in going from the 16-point short-time Fourier transform with a 16-point Hanning window to the structure shown in Figure 2.8.

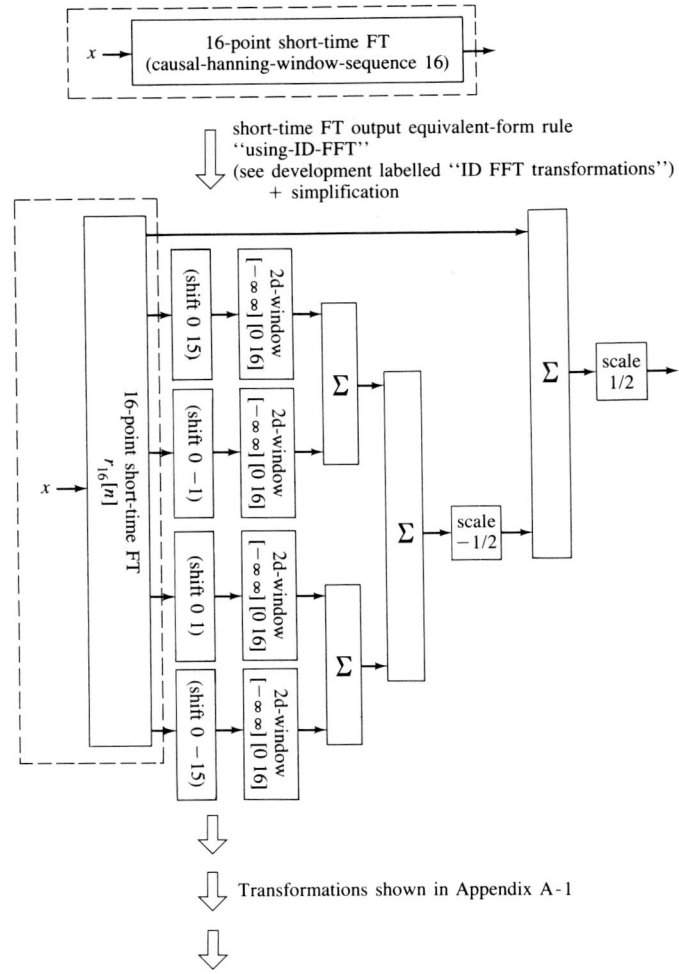

Structure shown in Figure 2.8

ID FFT transformations

Let "token" represent the abstract discrete-time sequence generated by the short-time Fourier transform output equivalent-form rule "using-1d-fft"

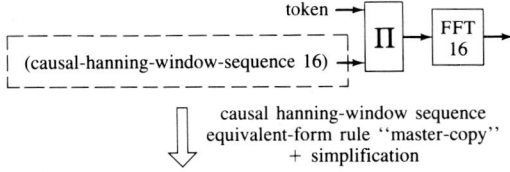

Chap. 2   Appendices

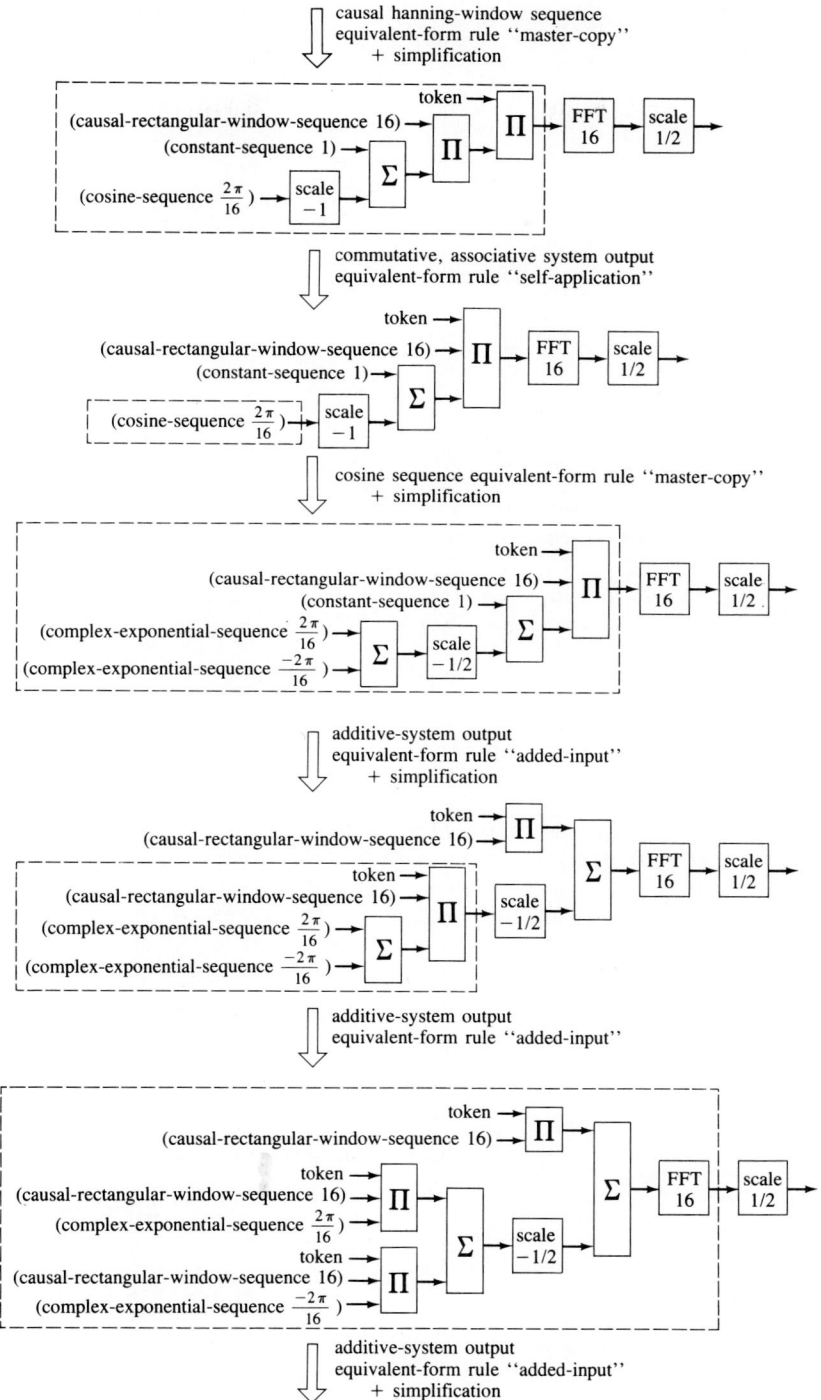

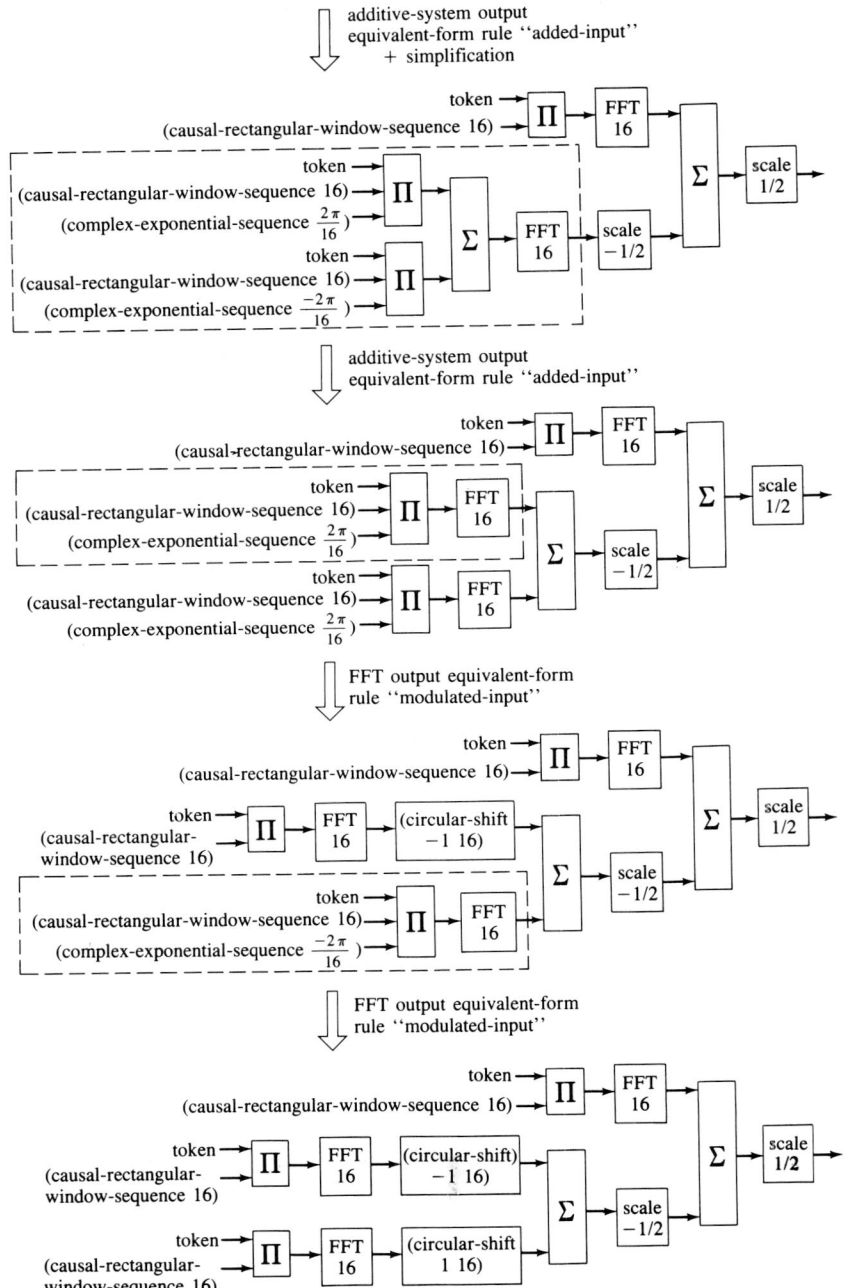

Chap. 2   Appendices

Appendix A.3 The sequence of transformations used in going from the 16-point short-time Fourier transform with a 32-point Hanning window to the structure shown in Figure 2.9.

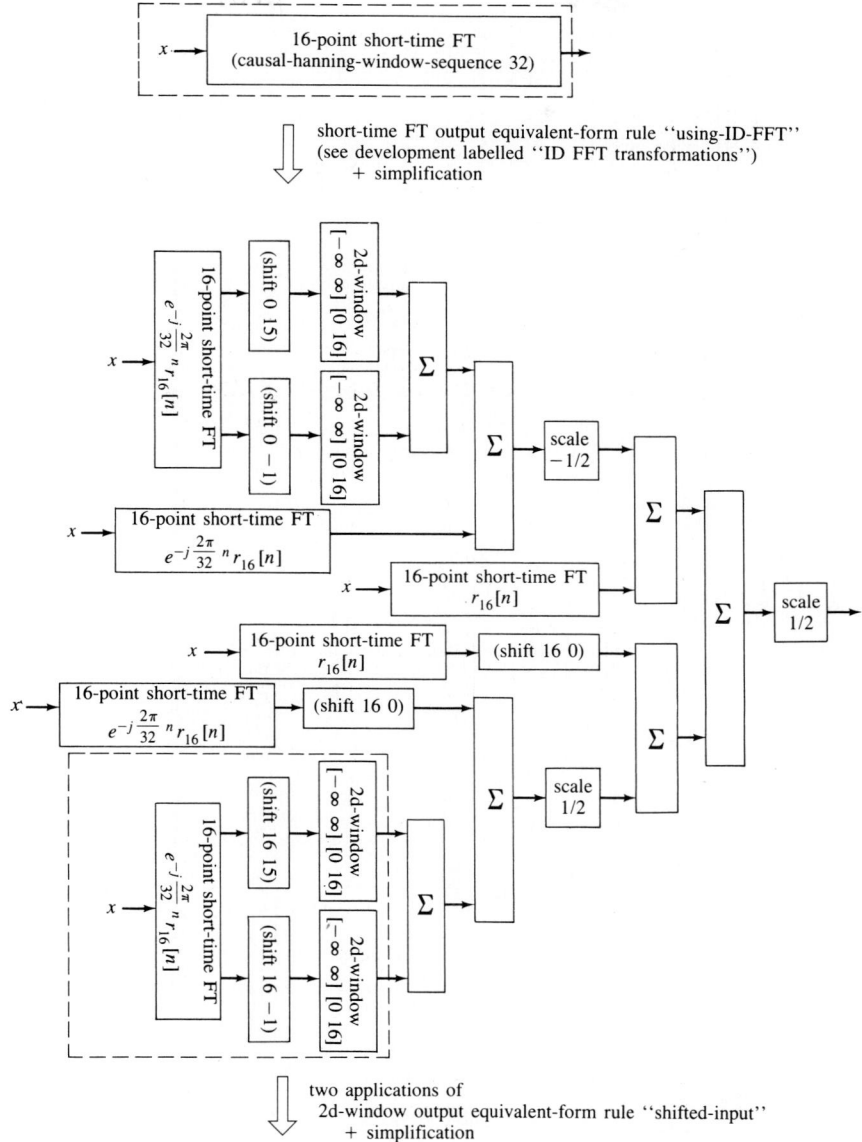

84   Computer-Aided Algorithm Design and Rearrangement   Chap. 2

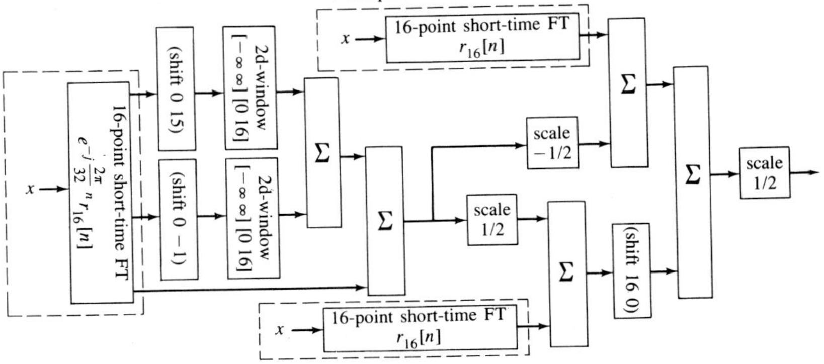

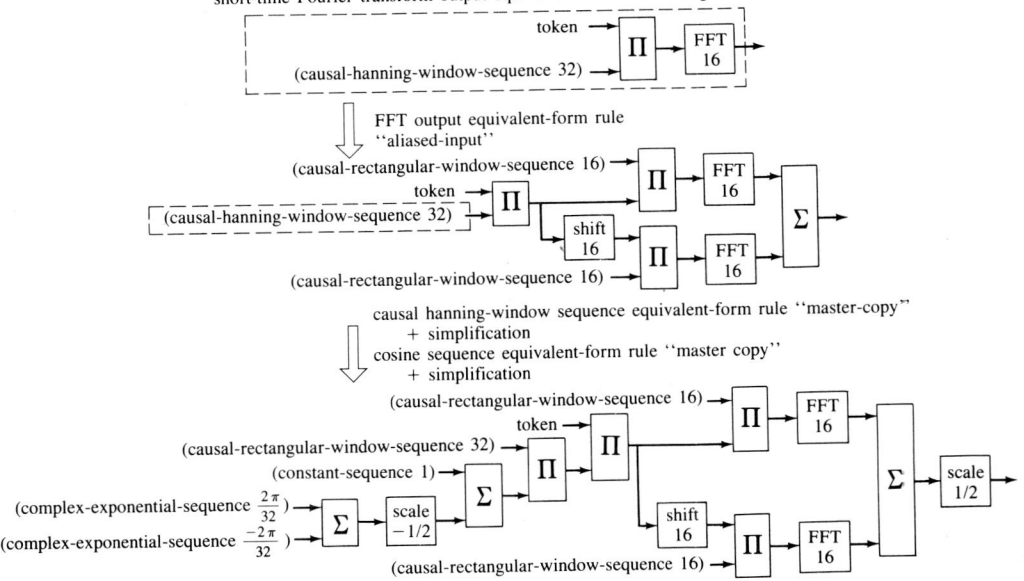

1D FFT transformations

ID FFT transformations continued: manipulation of structure below

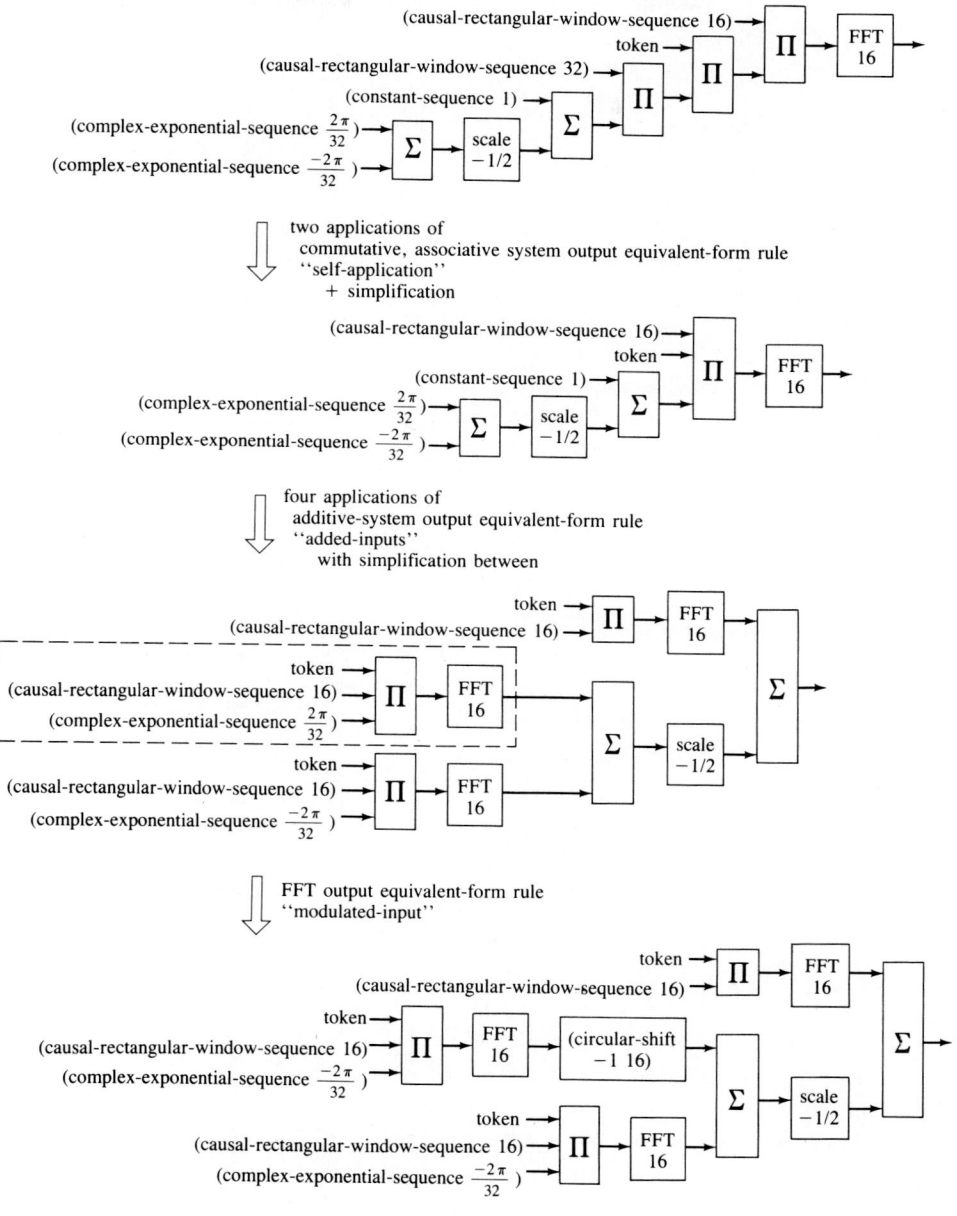

ID FFT transformations continued: manipulation of structure below

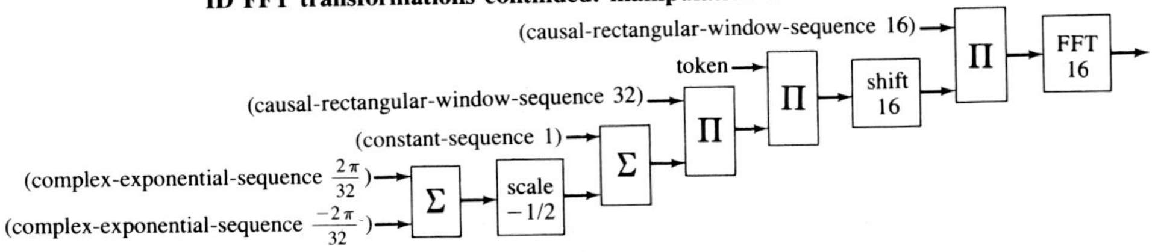

generalized shift-invariant system output equivalent-form rule
''single-shifted-input''

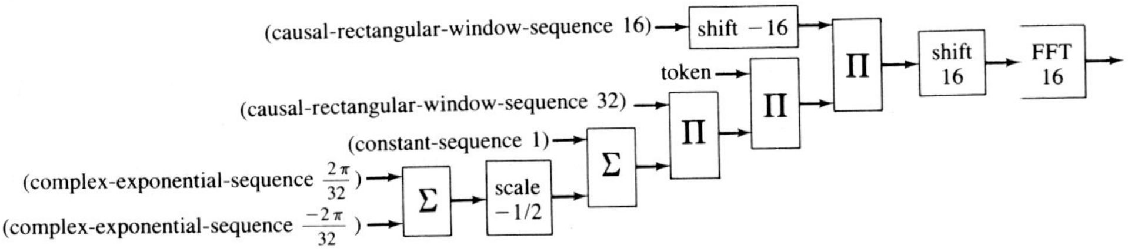

two applications of
commutative, associative system output equivalent-form rule ''self-application''
+ simplification

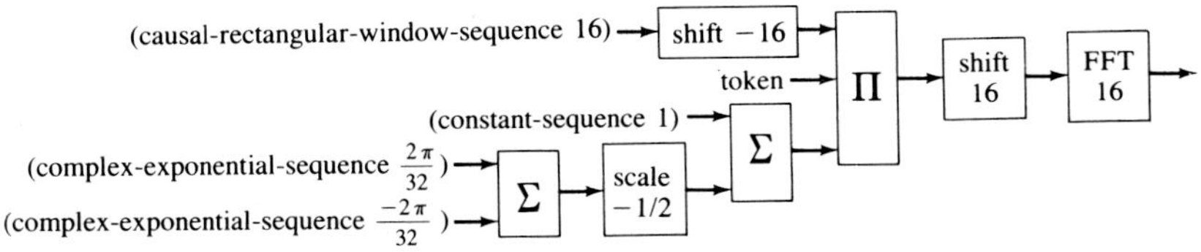

two applications of
additive-system output equivalent-form rule ''added-inputs''
+ two applications of
additive-system output equivalent-form rule ''shifted-added-inputs''
with simplification between

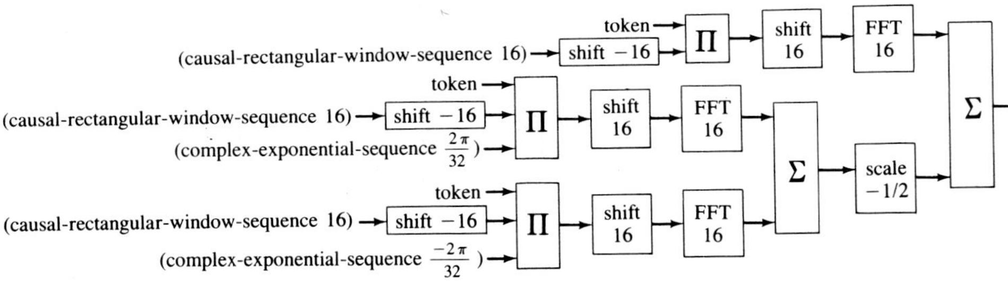

three applications of
generalized shift-invariant system output equivalent-form rule ''single-shifted-input''
+ simplification

Chap. 2 Appendices

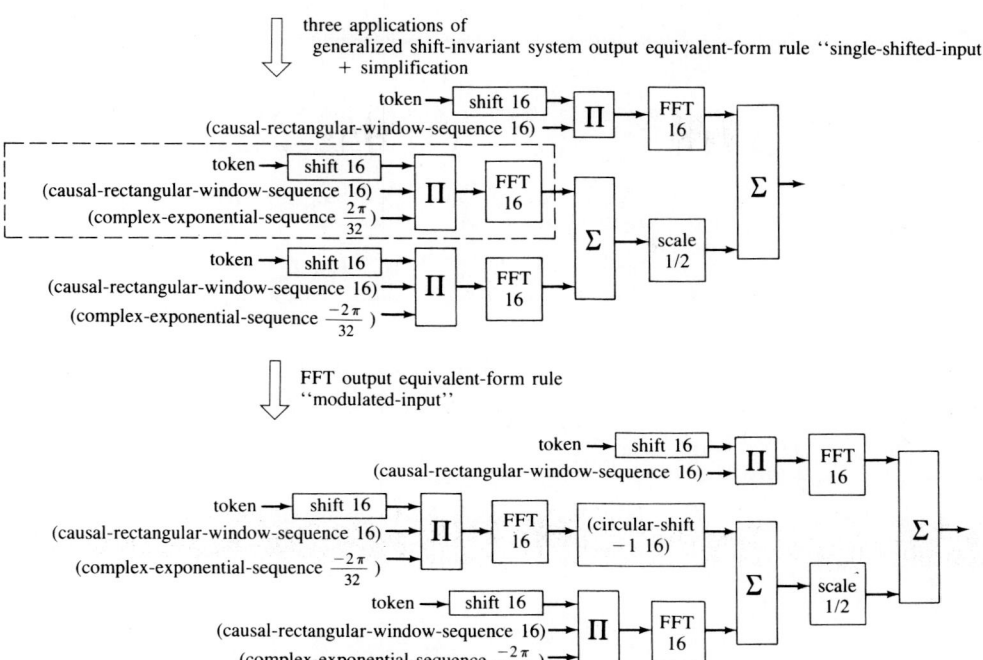

3

SYMBOLIC ANALYSIS OF SIGNALS AND SYSTEMS

• Brian L. Evans and James H. McClellan
Georgia Institute of Technology

The widespread availability of symbolic mathematics programs places high-powered mathematical functionality in the hands of students, educators, researchers, and developers. These environments are proficient at manipulating algebraic expressions and keeping track of intermediate calculations. They perform many symbolic operations well, especially integration, differentiation, and solving simultaneous equations. However, they do not always implement the functions and operations common in engineering. This chapter summarizes work that is aimed at making these environments very useful for engineers by using them to build powerful systems for symbolic and knowledge-based signal processing (KBSP).

The usual approach to building a KBSP system is to extend a high-level programming language like C or Lisp. In that case, a programmer would write procedures that operate on data structures. This approach, however, burdens the programmer (or researcher) whenever a new data type is added because that entails the creation of a new set of routines and possibly the modification of several existing ones. In some languages, a programmer could use object-oriented extensions to overcome this problem (e.g., Objective C or CLOS). Unfortunately, the programmer would still have no easy way to encode knowledge as conditional rules. By coding in an established language, though, a programmer could take advantage of existing source code and compiled libraries.

Several researchers have followed this approach by developing KBSP environments in Lisp, as discussed in the first, second, and fifth chapters of this book [8, 17, 20, 26]. The Algorithm Design Environment (ADE), for example, finds optimal implementations of signal processing expressions by intelligently applying its 600 simplification and rewrite rules [8]. ADE extends Common Lisp by adding a pattern matcher, a format for conditional rules, and an inference engine (for applying a collection of rules). In order to handle the propagation of constraints, ADE employs routines from two mathematical programs written in Lisp, *QM*™ and *MACSYMA*™.

Instead of starting with a high-level language, we have chosen to begin at an even higher level by using a symbolic mathematics environment. This approach is attractive because the starting point is already a very rich environment for the mathematical calculations needed in engineering. In the first section of this chapter, we highlight three candidate symbolic platforms—*Maple*™, *Mathematica*™, and *MACSYMA*™ (a forerunner of both). All three environments provide similar computational functionality. The primary attraction to using *Mathematica* is its *notebook* user interface. Notebooks integrate text, sound, graphics, animation, and "live" *Mathematica* code. Since we have chosen to work with *Mathematica*, we evaluate it as a platform for a KBSP system in section 3.2.

Our KBSP system helps students in solving elementary homework problems on transforms and convolutions and assists designers in developing algorithms by providing some of the capabilities of ADE. These extensions come in the form of *Mathematica* packages that define functions and operators common to signal processing but not available in the core of *Mathematica*. These include representations for signals and systems (section 3.3), tools for solving linear systems problems (sections 3.4 and 3.5), methods of transform-based analysis (section 3.6), and discrete and continuous convolution (section 3.7). Specifically, the signal processing packages can compute symbolic convolutions (continuous and discrete) and transforms (z, Laplace, Fourier, etc.). The symbolic transforms are general enough to work for multidimensional multilateral signals, which means that the z-transform and the Laplace transform routines track the region of convergence (ROC). The ROC is an example of a property that can be attached to a signal; it is important side information because it can be used to determine ranges of values on free parameters to guarantee stability. When taking transforms, solving difference equations, or performing analysis, the packages can justify every step of the process. The signal processing packages also provide the usual graphical representations for signals—sequence plots, pole-zero diagrams, frequency responses, and root loci.

As previously mentioned, the real advantage to using *Mathematica* is its notebook interface. A student can solve problems and document the solution in the same notebook. Professors can write interactive lessons and solution sets as notebooks.

Readers can tweak and evaluate the examples that contain "live" *Mathematica* code. We have developed notebooks for the topics of analog filter design, the z-transform, the discrete-time Fourier transform and piecewise convolution. In these tutorials, animations are used to illustrate difficult concepts such as the "flip-and-slide" approach to continuous-time convolution or the dependence of the frequency response on pole locations. Section 3.8 summarizes the contents of these notebooks and discusses their use in engineering education.

Mathematica provides extensive capabilities in solving small-scale problems (i.e., sets of equations), but it is a bit limited when combining these small-scale solutions into an overall solution of a design problem. However, this environment does permit the designer to derive a symbolic solution, from which numeric answers with arbitrary precision may be obtained. It can translate expressions into equivalent $\TeX$™, C, or Fortran code so that formulas can be translated into forms suitable for other environments. The signal processing packages extend these and other standard *Mathematica* abilities to formulas involving signals and systems. As a result, (algebraic) signal processing expressions can be simplified, rearranged, and expanded which are crucial in searching for optimal implementations of expressions. Section 3.9 shows how this system could be used to design and simulate complex systems.

A public domain version of the signal processing packages and notebooks is available via anonymous FTP to `gauss.eedsp.gatech.edu` (Internet Protocol #130.207.226.24). They reside in a compressed tar file ("Mathematica/SigProc2.0.tar.Z" at the time of publication). These extensions can also be ordered at a nominal fee from Wolfram Research Inc.

3.1 A TOUR OF SYMBOLIC MATHEMATICS PROGRAMS

Before we focus on the construction of a KBSP environment using a symbolic mathematics program, we describe some of the general features of three well-known symbolic mathematics programs: *Maple*, *Mathematica*, and *MACSYMA*. All three candidates are standardized, documented, portable environments that implement pattern matching, conditional rules, and a programming language. Because they embody extensive knowledge about mathematical structures and operations, their forte is the manipulation of formulas. For example, they compute $7x + 3x$ as $10x$. They are proficient at factoring polynomials and performing partial fractions expansions as well as differentiating and integrating functions. Many symbolic operations like integration and differentiation need to know which symbols to treat as variables. For example, the expression $C \tan(x)$ contains two symbols, C and x. Integrating with respect to x yields $-C \log(\cos(x))$, but integrating with respect to C yields $C^2 \tan(x)/2$. Symbolic mathematics programs can also represent and compute sequences from recursive formulas. At the very least, then, these platforms can be used as high-powered symbolic calculators.

All three environments maintain expressions in exact form until a user forces a numerical approximation to a specified precision. This means that they evaluate $\cos(\frac{\pi}{3})$ as $\frac{1}{2}$ but leave $\cos(1.0472)$ alone since 1.0472 only approximately (not exactly) equals $\frac{\pi}{3}$; that is, no exact rational number exists that represents the value of $\cos(1.0472)$ so it remains unchanged after evaluation. A user can force numerical evaluation to arbitrary precision so that $\cos(1.0472)$ would return some value close to $\frac{1}{2}$ like 0.499998 in the case of double precision floating point arithmetic.

Many signal processing operations can be expressed as formulas, so a symbolic mathematics environment could be extended to manipulate algebraic forms of signal processing operations. Formulas can compute signal values point by point or by array operations. In general, symbolic mathematics environments do not support streams or state-space computational models [16], although they can be made to do so. All three platforms implement some way of writing sets of rules that perform one-to-one transformations so that well-defined operations like linear transforms can be carried out. Unfortunately, the three candidate environments do not have a built-in mechanism to apply rules for transformations that are not one-to-one (e.g., searching for optimal implementations of an algorithm).

3.1.1 MACSYMA

The first large-scale symbolic mathematics program was *MACSYMA* (1970), developed at MIT. *MACSYMA* (Project MAC's SYmbolic MAnipulation system) is an interactive expert system and programming environment. After 100 man-years of development, *MACSYMA* now consists of over 300,000 lines of compiled Lisp code and supports more than 600 different mathematical operations [Pavelle]. *MACSYMA*'s preprocessor translates all input into a list-based representation called the *general form*. This general form can be converted into one of three other internal forms, although the user is generally unaware of the conversion. The general form computes values of formulas point by point and one of the other internal forms stores previously computed values in an array. *MACSYMA* provides simplification and manipulation rules for expressions based on these different internal representations.

MACSYMA provides many of the symbolic manipulation and reasoning abilities needed to implement the underlying mathematics of many different fields. Scientists, mathematicians, and engineers have applied this tool to problems in acoustics, algebraic geometry, antenna theory, general relativity, and number theory. Once a solution to a problem is obtained symbolically, the user can ask *MACSYMA* to generate the equivalent code in Fortran so that the formula can be transported to another environment. This means that *MACSYMA* can carry the solution process from the mathematical formulation of the problem to symbolic solutions to simulating those symbolic solutions in a compiled Fortran program.

One noticeable drawback to *MACSYMA* is its large size, although versions are available for PCs and Macintosh computers. Other versions exist for workstations like Sun's Sparc station, Digital's DECstation and DECsystem, and Hewlett-Packard's HP 9000. The focus of *MACSYMA*, however, is on formulas and not on abstract objects. Although properties can be attached to formulas, *MACSYMA* cannot deduce properties of an expression from the properties associated with its subexpressions. This means that *MACSYMA* cannot reason about signal properties such as bandwidth, extent, and symmetry. Even though *MACSYMA* is not a knowledge-based signal processing (KBSP) environment, the concepts behind it have strongly influenced such KBSP environments as E-SPLICE [20] and ADE [8].

3.1.2 Maple

The development of *Maple* began at the University of Waterloo in 1981, but *Maple* was not fully released until the fall of 1985 [4]. Its kernel is small in size (200 Kb to 500 Kb) but has a large library (7 Mb) of modules which are automatically loaded when referenced. Input syntax resembles Fortran, and the user can recall and edit previous commands. The output appears in text form. For example, the input expression

$$f := \text{int}(ln(1 + s*t)/(1 + t\char`\^2), t = 0 \ldots \text{infinity})$$

is displayed as

$$f := \int_{t=0}^{\infty} \frac{\log(1 + st)}{1 + t^2} dt$$

Maple can generate the equivalent Fortran and LATEX code for formulas. Like *MACSYMA*, *Maple* "knows" much about computational functions and formula manipulation. Its kernel performs arbitrary precision arithmetic, manipulates polynomials, and supports a programming language. The library functions provide additional abilities such as integration, linear programming, number theory, and statistics. *Maple* supports a total of 2,000 mathematical functions and runs on a wide variety of platforms including the IBM PC, Macintosh, Vax computers, Sun workstations, and the Cray 2 supercomputer [5]. With the release of version V in 1991, *Maple* now supports a sophisticated user interface for the X and Sunview windowing systems as well as versatile three-dimensional graphics capabilities. This new interface is not as powerful as *Mathematica*'s notebook facility (discussed below).

3.1.3 *Mathematica*

Mathematica, released in 1988, is a general mathematics platform like the other two, but it provides a better user interface [34]. The notebook front end provides pull-down menus, screen- and mouse-oriented editing of previous commands, a POSTSCRIPT™ driver for all graphics, and an interface to the machine's windowing system. Notebooks combine text, graphics, animation, sound, and code into one document. Because notebooks arrange information as a hierarchical grouping of cells, a user can quickly and randomly access information. A user can also choose to interact with these *living documents* by modifying the *Mathematica* code to try new examples. With notebooks, the off-line documentation is the same as the on-line documentation [18, 30, 34]. Unfortunately, the notebook facility is only available for the Macintosh, NeXT computers, and 386/486 machines running MicroSoft Windows 3.0 or higher. On other machines, the front end is nothing more than a teletype (tty) terminal interface. Chapter 4 explains the use of notebooks as interactive scientific documents, and section 3.8 examines their use as educational tools.

Although the user interface is machine-dependent, the back end *kernel* is the machine-independent computational engine. The kernel provides a powerful pattern matcher, the ability to express rules, arbitrary precision arithmetic, and a programming language. This programming language organizes code into packages (modules) much as Lisp does. Embedded in the kernel are packages that perform integration, differentiation, matrix manipulation, and linear programming. The kernel provides the more than 150 computational functions listed in Table 3.1 as well as powerful symbolic operations such as differentiation and integration. Users can customize *Mathematica* by loading additional packages as needed.

Without the notebook facility, it is difficult for a user to receive on-line help. The kernel offers one help function, `Help[]` abbreviated as `?`, which can display a list of objects that match a given pattern or usage information for one object. Even so, making the transition to *Mathematica* from a high-level language is not difficult. *Mathematica*'s preprocessor mimics most of the standard C operators as shown in Table 3.2. Many list-processing primitives are borrowed from Lisp, as well as the primitives `Print`, `Apply`, `Map`, and `MapAll` (see section 3.3.1). Users are also free to define their own preprocessor (including aliases and macros) so that they can use a more familiar syntax.

3.2 *Mathematica* PROGRAMMING FOR KBSP

Mathematica presents an attractive platform upon which an environment for KBSP can be built. As previously mentioned, the kernel provides ways to write programs and encode rules [18]. In fact, many different programming paradigms [31] are offered: procedural, functional, object-oriented, constraint-based, and rule-based. The `Block` construct is useful for writing procedures, and the `Function` primitive generates function pointers [18]. *Mathematica* does not implement data-directed programming per se but instead allows the programmer to attach data and code to an object. When using such an object-oriented approach, a programmer can write code that is incrementally extendible so that the addition of a new data type (tag) does not require modification of existing code. Constraint propagation, implemented by the `Solve` command, only applies for equality constraints. Rules in *Mathematica* are specified using an *if...then* form. When the conditional part of the rule is satisfied, its consequence is evaluated. The consequence expression can reference patterns matched during the evaluation of the conditional clause. This rule framework only supports reasoning with certainty.

TABLE 3.1 COMPUTATIONAL FUNCTIONS IN *MATHEMATICA* 1.2

These functions (objects) can take numerical or symbolic arguments. *Mathematica* will maintain these functions in exact form, but the user can ask for numerical values to arbitrary precision. For example, Cos[Pi] becomes −1 since cos(π) is exactly −1. Cos[3.1419] becomes 1. which is cos(3.1419) approximated to five digits (which is the precision of 3.1419). However, N[Cos[3.1419], 20] computes cos(3.1419) to 20 digits which is −0.99999995276909 [31].

Abs	Csc	HypergeometricOF1	Plus
AiryAi	Csch	Hypergeometric1F1	Pochhammer
ArcCos	Cyclotomic	Hypergeometric2F1	PolyLog
ArcCosh	Divide	HypergeometricU	Power
ArcCot	Divisors	Im	PowerMod
ArcCoth	DivisorSigma	InverseJacobiSN	Prime
ArcCsc	Dot	JacobiAmplitude	Quotient
ArcCsch	EllipticE	JacobiP	Random
ArcSec	EllipticExp	JacobiSN	Rationalize
ArcSech	EllipticF	JacobiSymbol	Re
ArcSin	EllipticK	LaguerreL	Round
ArcSinh	EllipticLog	LatticeReduce	Sec
ArcTan	EllipticPi	LCM	Sech
ArcTanh	EllipticTheta	LegendreP	SeedRandom
Arg	Erf	LegendreQ	Sign
Bernoulli	EulerE	LegendreType	Signature
BesselI	EulerPhi	LerchPhi	Sin
BesselJ	Exp	Log	Sinh
BesselK	ExpIntegralE	LogIntegral	SphericalHarmonicY
BesselY	ExpIntegralEi	Max	Sqrt
Beta	ExtendedGCD	Min	StirlingS1
Binomial	Factorial	Minus	StirlingS2
Ceiling	Factorial2	Mod	Subtract
ChebyshevT	FactorInteger	MoebiusMu	Tan
ChebyshevU	Floor	Multinomial	Tanh
Conjugate	Gamma	NBernoulli	Times
Cos	GCD	NonCommutativeMultiply	WeierstrassP
Cosh	GegenbaurC	PartitionsP	WeierstrassPPrime
Cot	HermiteH	PartitionsQ	Zeta
Coth			

Sec. 3.2 Mathematica Programming for KBSP

TABLE 3.2 MATHEMATICA OPERATORS THAT MIMIC C SYNTAX

mathematical operations	+	−	*	/	++	−−
	=	+=	−=	*=	/=	
relational operators	<	<=	>	>=	==	!=
bit operations	|	&				
logical operators	&&	||				

These programming styles offer two direct advantages over high-level languages for implementing signal processing operations. First, as a consequence of support of the functional programming paradigm, cascaded systems (operators) become nested calls to *Mathematica* objects. Second, *Mathematica* provides several ways to encode and apply rules to expressions. One way is to attach rules to object heads (function names). *Mathematica* will fire such rules whenever an expression contains that function. The relationship

$$\Re e\{x + y\} = \Re e\{x\} + \Re e\{y\}$$

can be rendered in *Mathematica* by attaching the rule to either the `Re` or `Plus` built-in primitives:

```
Re/:Re[x_ + y_]:= Re[x] + Re[y]
Plus/:Re[x_ + y_]:= Re[x] + Re[y]
```

Mathematica will check the first rule every time the real part of an expression is sought, even though the rule does not apply. This is an example of how attaching rules to primitives can slow computations down. This approach is useful for encoding transforms as recursive calls to the same function. Displaying the intermediate results, however, would only track the transformation of subexpressions as they are decomposed. Such a dialogue would not be beneficial to someone trying to understand the mechanics of the entire transformation process.

Another way to encode knowledge is to collect related rules in a list (called a rule base) and apply them recursively (e.g., using `ReplaceRepeated`). Since the rules are not attached to any function name (head), this approach does not slow down computations. Another benefit is that a programmer can control how the rules are applied. For one-to-one transforms, the programmer can take the expression to be transformed, fire the first applicable rule, and display the intermediate result. By repeating this method until the transform is computed, the resulting dialogue would mimic how a human would take the transform. From a user's point of view, this approach is desirable because these rules are fired only at the request of the user. The signal processing packages implement transforms using this rule-based approach.

Being a symbolic mathematics environment, *Mathematica* does more than apply rules. For example, two primitives—partial fraction decomposition (`Apart`) and power series expansion (`Series`)—are critical for taking inverse transforms. The ability to factor, expand, and rearrange polynomials plays another key role in computing symbolic transforms. When joined with a rule format, the result is a powerful and flexible way to encode the z-transform and other one-to-one transformations.

Besides implementing a wide variety of programming styles, this environment provides other advantages for KBSP:

1. dynamic data typing
2. diverse graphics capabilities
3. code generation

Dynamic data typing allows valueless symbols and permits variables to assume any data type. The kernel can graph one-dimensional functions and parametric relationships. It also supports scatter, density, and contour plotting as well as three-dimensional graphics. On general-purpose computers, *Mathematica*'s ability to generate the equivalent Fortran, C, and T_EX code for mathematical formulas (by default) and signal processing expressions (by extension) becomes useful, because the new code can be spliced into existing programs and documents.

Mathematica does come with some disadvantages for KBSP. First, the level of abstraction means that numerical computations are slower, although Version 2.0 introduces the notion of compiled functions to speed this up. Second, *Mathematica* can only chain rules in one direction; it is not an inference engine. Third, the rule rewriting mechanism applies a list of production rules in order to transform one expression into another. That is, *Mathematica* does not generate a tree of all possible new expressions, and therefore, heuristic search techniques through a solution space are not provided. Finally, common signal processing operators are, for the most part, not available in *Mathematica*. The signal processing packages overcome the last two deficiencies.

3.3 KNOWLEDGE REPRESENTATION

Computing environments often parse mathematical formulas into tree representations before evaluation. Nodes in such trees are the operators and the leaves are numbers. For example, the infix formula $1 + (5 \times 4)$ becomes

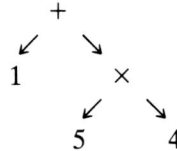

in tree form. The equivalent postfix, infix, and prefix representations result from a post-order, in-order, and pre-order (respectively) traversal of the expression tree. From the example, then, the prefix form is $\oplus(1, \otimes(5, 4))$ and the postfix form is $((4, 5)\otimes, 1)\oplus$ [28].

3.3 Knowledge Representation

Internally, *Mathematica* represents all expressions in prefix form (as does Lisp). The advantage of the prefix form is that operators can take a variable number of arguments. In *Mathematica*, one can attach rules to the prefix operator itself and to the head (tag) of any of the arguments. Luckily for the casual user, the kernel's preprocessor allows one to apply prefix, infix, and postfix operators in the same expression. For example, the kernel interprets $1 + (5 \times 4)$ as `Plus[1, Times[5, 4]]`, which becomes 21. The primitive `FullForm` is useful for seeing the internal prefix format.

The internal format, therefore, has the form

$$Head[slot_1, slot_2, \ldots]$$

where `Head` can be either a function name or a data tag and each slot is another expression. As expressions, then, data structures have the form

$$DataTag[datum_1, datum_2, \ldots]$$

and procedures have the form

$$FunctionName[argument_1, argument_2, \ldots]$$

Just as code and data are lists in Lisp, then, all code and data are expressions in *Mathematica*. This uniformity of knowledge representation is the key to *Mathematica*'s comprehensive pattern matching capability, which, in turn, is the key to *Mathematica*'s powerful and general rule format.

3.3.1 Common Data Structures

In *Mathematica*, a data structure is an expression that evaluates to itself. The most common data structure in *Mathematica* is the list, which is abbreviated as

$$\{element_1, element_2, \ldots\}$$

and internally represented as

$$\texttt{List}[element_1, element_2, \ldots]$$

Lists represent sets, intervals, sequences, vectors, and coordinates as well as trees and matrices. For plotting and other *Mathematica* routines, the interval $t \in [0, \pi]$ becomes `{t, 0, Pi}`. As another example, the tree

$$\begin{array}{c} a \to b \\ \searrow \\ c \to d \to f \\ \searrow \\ e \end{array} \Leftrightarrow \{a, b, \{c, \{d, f\}, e\}\}$$

Like a tree, a matrix is also a list of lists (vectors), but has a more regular structure [34]. For multidimensional functions, lists of variables like `{z1, z2, z3}` are sometimes used to specify dimensionality.

All of the usual operations on a single list (`First`, `Flatten`, `Rest`, `Reverse`, and `Sort`) as well as `Append` and `Prepend` are available. For multiple lists, *Mathematica* can perform the operations of `Intersection` and `Union` which treat the lists as sets. For vectors represented as lists and for matrices represented as a list of vectors, inner products (`Dot`), outer products (`Outer`), eigenvalue decomposition (`Eigensystem`), linear equation solving (`LinearSolve`), and linear programming (`LinearProgramming`) are also supported.

Two generalized functions, `Map` and `Apply`, are useful for manipulating general expressions. `Apply` takes its first argument (a function, function pointer, or data tag) and substitutes it for the head of the second argument. `Map` applies its first argument (a function, function pointer, or data tag) to the second argument to produce a new expression. The third argument specifies to which levels the function is applied (the default is the first level only). For example:

```
In[1] : = Apply[Plus, {a, b, c}]
Out[1] = a + b + c

In[2] : = Map[IntegerQ, {a, b, c}]
Out[2] = {False, False, False}

In[3] : = Map[IntegerQ, a + b + c]
Out[3] = 3 False
```

`IntegerQ` is a query (a.k.a. predicate or boolean) function that returns `True` if its argument is an integer and `False` otherwise. `Map` is very useful for applying rules to different levels of an expression tree.

3.3.2 Computational Functions for Signal Processing

Mathematica provides most of the common computational functions used in signal processing: sinusoids, exponentials, Bessel functions, and so on. In addition to the built-in computational functions shown in Table 3.1, the signal processing packages add the new *signal primitives* (i.e., functions) listed in Table 3.3, including step, impulse, sinc, and pulse functions. Since these packages provide an environment for both continuous-time and discrete-time signal processing, an impulse and a step function are defined in each domain. In the discrete domain, the Kronecker delta function `Impulse` and the discrete-step function `Step` are used, whereas in the continuous-time domain, the Dirac delta function `Delta` and the continuous-step function `CStep` are defined. Actually, `Delta[t]` and `CStep[t]` belong to a larger family of functions of a continuous-time variable known as unit functions, which are represented as `Unit[order][t]`. In the signal processing literature, unit functions are written as $u_n(t)$ so that $u_{-1}(t)$ denotes the continuous-time step function. Five of the rules (where n_ and t_ represent general patterns) encoding the behavior of unit functions are:

1. rewrite $u_{-1}(t)$: `Unit[-1][t_] := CStep[t]`

Sec. 3.3 Knowledge Representation

2. rewrite $u_0(t)$: `Unit[0][t_]:=Delta[t]`
3. derivative of $u_n(t)$ with respect to t:
   `Derivative[1][Unit[n_]][t_]:=Unit[n + 1][t]`
4. derivative of $u(t)$: `Derivative[1][CStep][t_]:=Delta[t]`
5. derivative of $\delta(t)$: `Derivative[1][Delta][t_]:=Unit[1][t]`

(`Derivative` is explained in section 3.3.3.) The signal processing packages also contain integration rules for `Delta` and `CStep` functions.

One function unique to the field of signal processing is `Dirichlet`. The Dirichlet kernel, or *aliased sinc* function, is internally represented as `Dirichlet[N, w]` and alternately represented as `AliasedSinc[N, w]` and `ASinc[N, w]`

TABLE 3.3 SIGNAL OBJECTS (NEW FUNCTIONS)

Object	Meaning
`CPulse`	continuous pulse: $$\sqcap_L(t) = \begin{cases} 0 & t < 0 \\ \frac{1}{2} & t = 0 \\ 1 & 0 < t < L \\ \frac{1}{2} & t = L \\ 0 & t > L \end{cases}$$
`CStep`	continuous step function: $$u_{-1}(t) = \begin{cases} 1 & t > 0 \\ \frac{1}{2} & t = 0 \\ 0 & t < 0 \end{cases}$$
`Delta`	Dirac delta function
`Dirichlet`	Dirichlet discrete kernel: $$d[N,\omega] = \frac{\sin(N\omega/2)}{N \sin(\omega/2)}$$ Also called the `ASinc` function
`Impulse`	Kronecker delta function
`LineImpulse`	m-D delta function, e.g., $\delta(t_1 - t_2)$
`Pulse`	discrete pulse: $$\sqcap_L[n] = \begin{cases} 0 & n < 0 \\ 1 & 0 \leq n \leq L - 1 \\ 0 & n \geq L \end{cases}$$
`Sinc`	$\text{sinc}(t) \equiv \sin(t)/t$ so $\text{sinc}(0) = 1$
`Step`	discrete step function: $$u[n] = \begin{cases} 1 & n \geq 0 \\ 0 & n < 0 \end{cases}$$
`Unit`	family of functions which includes `CStep` and `Delta`

(see section 3.4.3) [23]. It is the inverse discrete-time Fourier transform of an ideal low-pass filter:

$$\text{Dirichlet}[N, \omega] = \text{asinc}[N, \omega] = \frac{\sin(N\omega/2)}{N\sin(\omega/2)}$$

The Dirichlet kernel is one family of interpolation filters.

3.3.3 Operators for Signal Processing

Mathematica provides several operators that are useful in signal processing, e.g., Conjugate, Im, and Re. These operators do not require any side information, only a function (signal). Some signal processing operators require side information in the form of parameters. The shift operator, for example, needs to know the axis to shift and the amount of the shift. The signal processing packages add many of the common *parameterized* operators for signal processing, as shown in Table 3.4, having the form

$$OperatorName[parameter_1, parameter_2, \ldots][input_1, input_2, \ldots]$$

The primary benefit of this notation is that nested signal processing expressions become easier to write and read because the parameters are separated from the input signals.

Parameterized operators are different from computational functions and symbolic operations in that they typically do not modify their inputs. *Mathematica*'s lone built-in parameterized operator is Derivative, which supports the partial derivative object D. Derivative accepts a variable number of parameters and operates on one input—the function being differentiated. For example, D[Sin[x], x] becomes Derivative[1][Sin][x] which evaluates to Cos[x]. If the object unknown had no derivative definition in *Mathematica*, then the evaluation of D[unknown[x, y], x] would yield the expression Derivative [1, 0][unknown][x, y]. This expression represents the first derivative of unknown with respect to x and the zeroth derivative of unknown with respect to y. In the general case, the expression Derivative[n1, n2, . . .][f] *represents* a function obtained by differentiating f n1 times with respect to the first variable, n2 times with respect to the second variable, and so on. Like other operators, Derivative does not alter its arguments and instead *represents* an operation.

For all the new operators except Summation, the last parameter indicates which symbols are to be treated as variables. For example, the shift operator Shift[7, t] would have no effect on the function Cos[x] because Cos[x] does not depend on t, but

$$\text{Shift[7, t][Cos[t]]} \textit{ represents } \text{Cos[t − 7]}$$

By using a list of symbols in place of a variable, e.g., {t1, t2, t3} instead of t, the operator becomes multidimensional, and the length of the list specifies the operator's dimensionality. In multidimensions, all of the parameters in Table 3.4 become lists except for upsampling and downsampling factors, which become square integer matrices.

Sec. 3.3 Knowledge Representation

TABLE 3.4 NEW PARAMETERIZED OPERATORS FOR SIGNAL PROCESSING

These operators separate parameters (shown below) from input signal(s) (not shown), which makes cascaded systems easier to express.

`Aliasby[sc, w]`	aliases a continuous function of `w` giving it a period of $2\pi/$`sc` and divides by `sc`
`CConvolve[t]`	continuous convolution in `t`
`Convolve[n]`	discrete convolution in `n`
`DFT[L, n, k]`	the L-sample discrete Fourier transform of a function in `n` to a function in `k`
`DTFT[n, w]`	the discrete-time Fourier transform of a sequence in `n` to a continuous periodic function in `w`
`Difference[i, n]`	the *i*th backward difference in `n`
`Downsample[m, n]`	keep first sample in every block of `m` input samples; sampling rate decreases by a factor of `m`
`FIR[n, {h0, h1, ...}]`	all-zero digital filter with finite impulse response `h0, h1, ....`
`FT[t, w]`	continuous Fourier transform of a function of `t` into a function of `w`
`IIR[n, {a0, a1, ...}]`	all-pole digital filter with infinite impulse response (see text)
`Interleave[n]`	interleaves (combines) samples from each input function of `n` into one function
`InvDFT[L, k, n]`	inverse discrete Fourier transform (see `DFT`)
`InvDTFT[w, n]`	inverse discrete-time Fourier transform (see `DTFT`)
`InvFT[w, t]`	inverse continuous Fourier transform (see `FT`)
`InvL[s, t]`	inverse Laplace transform (see `L`)
`InvZ[z, n]`	inverse *z*-transform (see `Z`)
`L[t, s]`	Laplace transform of a function of `t`
`Periodic[p, v]`	argument is made periodic with period `p` with respect to variable `v`
`Rev[v]`	reverse the function of variable `v` (flip the `v` axis)
`ScaleAxis[sc, w]`	scale the `w` axis (variable) by `sc`
`Shift[v0, v]`	shift function of `v` by `v0` points
`Summation[i, ib, ie, inc]`	summation operator, `i` = `ib` to `ie` step `inc`
`Upsample[l, n]`	insert `l - 1` zeros after each input sample; sampling rate is increased by a factor of `l`
`Z[n, z]`	*z*-transform of a function of `n`

The signal processing packages represent downsampling by *m* with respect to the discrete variable *n* as

$$\text{Downsample}[m, n][\ \textit{input\_signal}\ ]$$

So, downsampling $\cos(\pi n/10)$ by two [9] is expressed as

$$\text{Downsample}[2, n][\ \text{Cos}[\text{Pi n / 10}]\ ]$$

and the two-dimensional downsampling of $\sin[\pi(n_1 + n_2)/10]$ by $m = \begin{bmatrix} 2 & 3 \\ 0 & 2 \end{bmatrix}$, which keeps one out every $det(m) = 4$ samples, is written as

$$\text{Downsample}[\{\{2, 3\}, \{0, 2\}\}, \{n1, n2\}][\text{Sin}[\text{Pi (n1 + n2) / 10}]]$$

This kind of downsampling is very important in multidimensional multirate structures, especially filter banks [2]. This expression only *represents* the two-dimensional downsampling operation and no samples are actually discarded.

Two parameterized operators IIR and FIR represent the impulse response of an all-pole filter (infinite in extent) and of an all-zero filter, respectively [19–Appendix D, 22]. When either IIR or FIR operate on an expression, that expression becomes the input to the filter (which is an impulse function by default). Therefore

$$\mathtt{IIR[n,\ \{1,\ 1/2,\ 1/4\}][a\char`\^ n\ Step[n]]}$$

represents a digital all-pole filter with a gain of 1/1 and feedback coefficients $-\frac{1}{2}$ and $-\frac{1}{4}$, whose input is $a^n u[n]$.

Neither the IIR nor the FIR operators actually compute signal values, but they can be converted into forms that do by applying TheFunction object. The FIR object is converted to a sum of delayed impulses. Converting IIR to a mathematical formula is more difficult because an infinite impulse response filter depends on previous values. So, we introduce a new object called IIRFunction which generates the corresponding recursive equation. The new recursive equation remembers previous values. For example

$$\mathtt{In[5] := TheFunction[\ IIR[n,\ \{1,\ 1/2,\ 1/4\}][a\char`\^ n\ Step[n]]\ ]}$$

$$\mathtt{Out[5] = IIRFunction[\ n,\ \{1,\ \tfrac{1}{2},\ \tfrac{1}{4}\},\ a\char`\^ n\ Step[n],}$$
$$\mathtt{TheFunction\ ->\ hiir1\ ]}$$

Now, signal values can be computed, plotted, and so on by substituting for the index n:

$$\mathtt{In[6] := Simplify[\ \%5\ /.\ n\ ->\ 5\ ]}$$

$$\mathtt{Out[6] = \frac{4 + 5\ a + 6\ a^2 + 8\ a^3 + 8\ a^4 + 16\ a^5}{16}}$$

TheFunction is also useful for rewriting other parameterized operators in terms of functions. That is, it provides "hooks" to the objects defined by other packages so that when new extensions are loaded, the signal processing packages can still take transforms of expressions involving new objects.

3.4 TRANSFORMS

The backbone of linear systems theory and signal processing is the linear transform. For continuous-time systems, the key transforms are of course the Laplace transform and the Fourier transform. The Laplace transform is a powerful tool for solving linear constant coefficient differential equations, so it is useful in analyzing passive networks and other linear time-invariant (LIT) systems [7, 24]. The Fourier transform is well suited for frequency-domain analysis [3]. For discrete-time systems, the important transforms are the z-transform, the discrete-time Fourier transform (DTFT), and the discrete Fourier transform (DFT) [23]. Since the z-transform is the generalized linear transform for sequences, it is useful for solving linear constant coefficient *difference* equations and therefore for analyzing linear time-invariant (LTI) digital systems. The z-transform, then, is the discrete-time analog of the Laplace transform. The continuous Fourier transform, however, has two discrete counterparts. The DTFT maps a sequence into the *continuous* frequency domain, whereas the DFT samples in both time and frequency. The DTFT is a tool for studying general frequency-domain behavior, and the DFT lends itself to hardware implementations since it is a purely numerical transform that has efficient implementations [10,23].

The signal processing packages implement all five of these linear transforms in their most general forms. That is, the multidimensional, multi-sided versions are supported. As a consequence, the region of convergence (ROC) is tracked for the forward z- and Laplace transforms to ensure that the inverse is uniquely specified. The transforms are implemented so that they can show each step of the transformation process, which can help in teaching the underlying mathematical operations and strategies to students. They also allow users to specify their own transform pairs.

3.4.1 Structure of the Transform Rule Bases

All of the transforms are implemented with the same structure: a preprocessor, a one-dimensional rule base, and a postprocessor. The preprocessor fills in missing arguments (if any) and calls the one-dimensional rule base once per dimension of the transform. The one-dimensional rule base applies six distinct sets of rules to transform the expressions with respect to one variable. This divide-and-conquer approach resembles the way a human might take a transform, as shown in Figure 3.1. The postprocessor simplifies the resulting expression and reports which parts (if any) could not be transformed.

Because the transform rule bases have the same structure, they have a similar arrangement of arguments and options. The transforms require two pieces of information—the function to be forward (inverse) transformed and the "time" (transform) variable(s). (The DFT also needs to know the length of the sequence.) If a third argument specifying the transform ("time") variable(s) is given, then all subsequent arguments are treated as options. Options are similar to Lisp keywords in that they override default settings. (The default settings for an *object* can be retrieved by the Options primitive, such as Options[ZTransform]). One important option is Dialogue which sets the level of justification: False for no justification, True for partial justification, and All for complete justification. Another important option is TransformLookup which allows users to specify their own transform pairs. For example, the option TransformLookup ->{ x[t] :> X[s] } would direct the forward Laplace transformer to transform x[t] as X[s]. A region of convergence say $-2 < \Re e(s) < -1$ could have been attached to X[s] by using the option

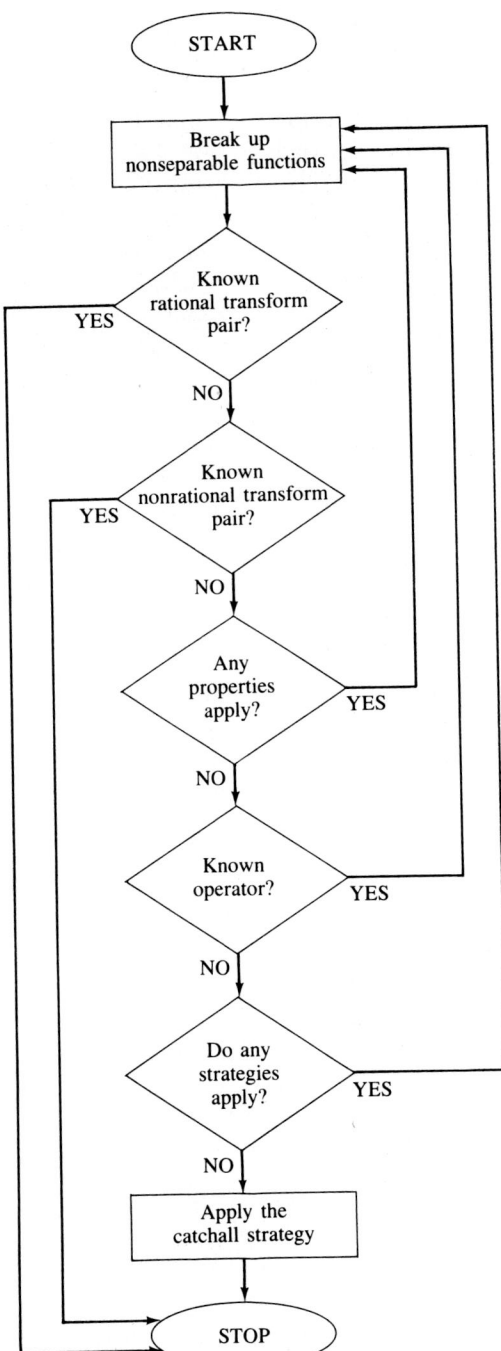

Figure 3.1 Structure of the One-Dimensional Transform Rule Bases. One-dimensional transform rule bases form the backbone of the more general multidimensional, multisided transforms. Each rule base calls a one-dimensional rule base once for each dimension of the transform.

Sec. 3.4 Transforms

```
                TransformLookup ->{ x[t]:>{ X[s], -2, -1}}.
```
For multidimensional transforms, the `TransformLookup` option becomes more complicated because a transform pair must be given for each dimension, e.g.,

```
In[7]:= LaPlace[ x[t1,t2], {t1,t2}, {s1,s2},
                TransformLookup ->{ x[t1,t2]:> X1[s1,t2],
                                    X1[s1,t2]:> X[s1,s2] }]
Out[7]= LTransData[ X[s1, s2], Rminus[{0, 0}],
                Rplus[{Infinity, Infinity}],
                LVariables[{s1, s2}] ].
```

Here, `X1[s1, t2]` denotes the Laplace transform of `x[t1, t2]` with respect to the first variable `t1`. `Dialogue`, `TransformLookup`, and the other options for the linear transforms are listed in Table 3.5 and explained by Table 3.6.

TABLE 3.5 OPTIONS FOR THE TRANSFORM RULE BASES

See Table 3.6 for the meaning of these options. The `TransformLookup` option (not shown below) defaults to an empty list.

Transform	Option	Default Value
CTFTransform	Dialogue	False
	Simplify	True
DFTransform	Dialogue	False
InvDFTransform	Definition	False
	Dialogue	False
	Terms	False
DTFTransform	Dialogue	False
LaPlace	Dialogue	True
	Simplify	True
InvCTFTransform	Apart	Rational
	Dialogue	False
	Simplify	True
	Terms	False
InvDTFTransform	Definition	False
	Dialogue	False
	Terms	False
InvLaPlace	Apart	Rational
	Dialogue	True
	Simplify	True
	Terms	10
InvZTransform	Dialogue	True
	Terms	10
ZTransform	Dialogue	True

TABLE 3.6 MEANING OF THE OPTIONS FOR THE TRANSFORM RULE BASES

The most important option is `Dialogue`, which sets the level of justification. Full justification is enabled by appending the code fragment `Dialogue -> All` as the last argument to a transform rule base. This justification option is also supported by the difference and differential equation solvers `ZSolve` and `LSolve` (see section 3.5) as well as the analysis routines `ASPAnalyze` and `DSPAnalyze` (see section 3.6).

Option	Possible Values	Meaning
`Apart`	`Rational, All`	Partial fraction decomposition only applies to polynomials with rational or real coefficients
`Definition`	`True, False`	Use the transform definition if all else fails to find the transform
`Dialogue`	`False, True, All`	Ascending levels of justification
`Simplify`	`True, False`	Apply `Simplify` to result
`Terms`	`False` or integer	Number of terms in series expansion (`False` means none)
`TransformLookup`	List of rules	Customized transform pairs

3.4.2 Symbolic z-Transforms

The forward z-transform produces an analytic function, $X(z)$, from a sequence, $x[n]$ [23]:

$$Z\{x[n]\} = \sum_{n=-\infty}^{\infty} x[n]z^{-n} = X(z) \tag{3.1}$$

This infinite summation is two-sided. The set of complex z values for which it converges is known as the region of convergence (ROC). The inverse z-transform requires the evaluation of a contour integral [23]:

$$Z^{-1}\{X(z)\} = \frac{1}{2\pi j} \oint_C X(z) z^{n-1} dz = x[n] \tag{3.2}$$

Here, C is a counterclockwise contour, within the ROC, encircling the origin. Equation (3.2) is rarely used to find the inverse z-transform because the Cauchy Residue Theorem reduces the contour integral to an algebraic evaluation of residues of $X(z)$. Note that the inverse z-transform is equivalent to the Laurent series expansion of $X(z)$.

Forward z-Transforms

The multidimensional z-transform can be obtained by simply applying the one-dimensional z-transform along each dimension. For example, the two-dimensional z-transform $X(z_1, z_2)$ of the function $x[n_1, n_2]$ is defined as follows [10]:

$$X(z_1, z_2) = \sum_{n_2=-\infty}^{\infty} \sum_{n_1=-\infty}^{\infty} x[n_1, n_2] z_1^{-n_1} z_2^{-n_2} \tag{3.3}$$

Sec. 3.4 Transforms

Letting Z_n be the one-dimensional z-transform with respect to the variable n:

$$X(z_1, z_2) = \sum_{n_2 = -\infty}^{\infty} \left(\sum_{n_1 = -\infty}^{\infty} x[n_1, n_2] z_1^{-n_1} \right) z_2^{-n_2}$$

$$= Z_{n_2}\{Z_{n_1}\{x[n_1, n_2]\}\} \qquad (3.4)$$

This last expression says that the two-dimensional transform is the same as the one-dimensional transform w.r.t. n_2 of the one-dimensional transform w.r.t. n_1. An example of a two-dimensional transform is shown in Figure 3.2.

The ZTransform rule base, then, consists of a call to a one-dimensional rule base for each variable to be transformed. The order in which the one-dimensional transforms are applied follows the order of the list of "time" variables passed to ZTransform. The one-dimensional rule base assumes that every term of the function to be transformed contains an impulse function $\delta[n - n_0]$, or a unit step function, $u[n - n_0]$ or $u[n_0 - n]$. Multiplication by the unit step $u[n - n_0]$ changes the forward z-transform into a one-sided infinite summation, which turns the bilateral z-transform into a sort of unilateral z-transform:

$$X(z) = \sum_{n = -\infty}^{\infty} x[n]u[n - n_0]z^{-n} = \sum_{n = n_0}^{\infty} x[n]z^{-n} \qquad (3.5)$$

Example: The two-dimensional z-transform of the separable sequence $x[n_1, n_2] = a^{n_1} b^{n_2} u[n_1, n_2]$ is

$$\frac{1}{(1 - az_1^{-1})(1 - bz_2^{-1})}$$

for $|a| < |z_1| < \infty$ and $|b| < |z_2| < \infty$. The equivalent of this transform is returned by ZTransform[a^n1 b^n2 Step[n1, n2], { n1, n2 }]:

```
ZTransData [         1
                --------------- ,
                ( 1 - a )( 1 - b )
                     z1        z2

            Rminus [{ Abs[a], Abs[b] }] ,

            Rplus [{ Infinity, Infinity }] ,

            ZVariables [{ z1, z2 }]]
```

The 2-D transform is taken as the 1-D transform with respect to n2 of the 1-D transform with respect to n1 of a^n1 b^n2 Step[n1, n2]. This is the general approach for 2-D transforms (see also Figures 3.9 and 3.11). Since the sequence is separable, the resulting transform is the product of two one-dimensional transforms [10].

Figure 3.2 Two-dimensional z-Transform Example.

Left-sided signals typically contain step functions in the form $u[n_0 - n]$, so that the signal is nonzero for $-\infty < n \leq n_0$.

The one-dimensional z-transform rules are divided into the six distinct sections shown in Table 3.7 (cf. Figure 3.1). The first section of the rule base (the "multidimensional hooks") tracks the multidimensional ROC of nonseparable functions, i.e., those containing line impulses and multivariable multinomials. Without these rules, the one-dimensional rule base would compute the correct z-transform function but the wrong ROC.

The second section encodes the rational transform pairs in Table 3.8 as rules, and the third section adapts the nonrational transform pairs in Table 3.9 into rules. These two sections are essentially lookup tables based on matching patterns of known transform pairs. Left-sided functions are transformed by the property rule

$$Z\{x[-n]\} = Z\{x[n]\}_{z \to z^{-1}} = X(z^{-1})$$

A strategy rule rewrites functions of $|n|$ into a sum of causal and anticausal components using step functions. Therefore, *only right-sided transform pairs need to be encoded*.

The fourth section of the rule base implements those z-transform properties listed in Table 3.10. For example, the mathematical definition of the shifting property

$$Z\{x[n + m]\} \leftrightarrow z^m Z\{x[n]\} \tag{3.6}$$

can be couched in terms of right-sided, left-sided, and impulsive functions:

$$Z\{f[n]u[n + m]\} = z^m Z\{f[n - m]u[n]\}$$
$$Z\{f[n]u[m - n]\} = z^{-m} Z\{f[n + m]u[n]\} \tag{3.7}$$
$$Z\{f[n]\delta[n + m]\} = z^m f[-m]$$

TABLE 3.7 ONE-DIMENSIONAL FORWARD Z-TRANSFORM RULES

These 64 rules are divided into 6 distinct sections. These sections are applied to an expression in the order listed. When a rule applies, the rule's consequence is evaluated and the rules are applied again to the resulting expression beginning with the first section. This is the general procedure of all of the transform rule bases, which is shown as a flow chart in Figure 3.1.

Rule Grouping	Rules	See Also
multidimensional hooks	5	
rational transform pairs	15	Table 3.8
nonrational transform pairs	6	Table 3.9
transform properties	13	Table 3.10
transforms of operators	20	Table 3.11
transform strategies	10	Figure 3.3

Sec. 3.4 Transforms

TABLE 3.8 RATIONAL Z-TRANSFORM PAIRS

Expression	z-Transform
0	0
$u[n]$	$\dfrac{1}{1-z^{-1}}$
$\sqcap_L[n+n_0]$	$z^{n_0}\dfrac{1-z^{-L}}{1-z^{-1}}$
$a\delta[n+n_0]$	$(a\|_{n\to -n_0})z^{n_0}$
$\sin[\omega n + b]\,u[n]$	$\dfrac{\sin(b)+\sin(\omega - b)z^{-1}}{1-2\cos(\omega)z^{-1}+z^{-2}}$
$\cos[\omega n + b]\,u[n]$	$\dfrac{\cos(b)+\cos(\omega - b)z^{-1}}{1-2\cos(\omega)z^{-1}+z^{-2}}$
$\sinh[\omega n + b]\,u[n]$	$\dfrac{\sinh(b)+\sinh(\omega - b)z^{-1}}{1-2\cosh(\omega)z^{-1}+z^{-2}}$
$\cosh[\omega n + b]\,u[n]$	$\dfrac{\cosh(b)+\cosh(\omega - b)z^{-1}}{1-2\cosh(\omega)z^{-1}+z^{-2}}$
$\binom{n}{k}u[n]$	$\dfrac{z^{-k}}{(1-z^{-1})^{k+1}}$, if $k>0$
$\binom{k}{n}a^n b^{k-n} u[n]$	$(b+az^{-1})^k$, if $k>0$
$\dfrac{(n+k-j)!}{n!}u[n]$	$\dfrac{z^{-j}k!}{(1-z^{-1})^{k+1}}$, if $k>0$
$\binom{n+k-j}{k}u[n]$	$\dfrac{z^{-j}}{(1-z^{-1})^{k+1}}$, if $k>0$

Adapted from [19,23].

The first two equations, which appear in Table 3.10, handle right-sided (causal) sequences and left-sided (anticausal) sequences, respectively. The third equation shows how to transform shifted impulses and is actually a rational transform pair (see Table 3.8). Note that the z^m term has been factored out in all three cases.

The z-transform rule base encodes the shifting property by encoding eq. (3.7) as rules:

```
ztransform1d[f_. Step[n_ + m_], n_, z_] :=
        z^m ztransform1d [ (f /. n -> n - m) Step[n], n, z ]
ztransform1d[f_. Step[m_ - n_], n_, z_] :=
        z^(-m) ztransform1d [ (f /. n -> n + m) Step[-n], n, z ]
ztransform1d[f_. Impulse[n_ + m_], n_, z_] :=
        transformpair[ ( f /. n -> -m ) z^m, 0, Infinity ]
```

In these rules, `ztransform1d` takes three arguments. The arguments are the function to transform, the "time" variable, and the transform variable. The left-hand

TABLE 3.9 NONRATIONAL Z-TRANSFORM PAIRS

Expression	z-Transform
$\dfrac{u[n]}{n!}$	$e^{z^{-1}}$
$\dfrac{u[n]}{2n+1}$	$\sqrt{z}\,\arctan(z^{-1})$
$\dfrac{u[n-1]}{n}$	$-\ln(1-z^{-1})$
$\dfrac{u[n]}{(2n)!}$	$\cosh\sqrt{z^{-1}}$
$\dfrac{(2n)!}{(2^n n!)^2}u[n]$	$\sqrt{\dfrac{1}{1-z^{-1}}}$
$\dfrac{\Gamma[r]}{\Gamma[n+1]\Gamma[r-n]}u[n]$	$(1+z^{-1})^{r-1}$ for real-valued $r > 0$

The last transform pair is from [6]; all others are from [19].

TABLE 3.10 Z-TRANSFORM PROPERTIES

Property	Expression	z-Transform
Homogeneity	$c\,x[n]$	$cX(z)$
Additivity	$x[n]+y[n]$	$X(z)+Y(z)$
	$\dfrac{x[n]+y[n]}{c[n]}$	$Z\left\{\dfrac{x[n]}{c[n]}\right\}+Z\left\{\dfrac{y[n]}{c[n]}\right\}$
Shifting	$f[n]u[n+m]$	$z^m Z\{f[n-m]u[n]\}$
	$f[n]u[m-n]$	$z^{-m}Z\{f[n+m]u[-n]\}$
Time reversal	$x[-n]$	$X\left(\dfrac{1}{z}\right)$
Multiplication-by-n	$n^m x[n]$	$-D^m X(z)$, where $D=-z\dfrac{d}{dz}$
Integration	$\dfrac{x[n]}{n}u[n-1]$	$\displaystyle\int_z^\infty \dfrac{X(u)-x[0]}{u^2}du$
Similarity	$c^{an+b}x[n]$	$c^b X\left(\dfrac{z}{c^a}\right)$
Trigonometric identities	$\cos[b+\theta n]x[n]$	$Z\{\cos[b]\cos[\theta n]x[n]-\sin[b]\sin[\theta n]x[n]\}$
	$\sin[b+\theta n]x[n]$	$Z\{\sin[b]\cos[\theta n]x[n]+\cos[b]\sin[\theta n]x[n]\}$
Modulation	$\cos[\theta n]x[n]$	$\dfrac{1}{2}[X(ze^{j\theta})+X(ze^{-j\theta})]$
	$\sin[\theta n]x[n]$	$-\dfrac{1}{2j}[X(ze^{j\theta})-X(ze^{-j\theta})]$

Adapted from [19,23].

Sec. 3.4   Transforms

side of each rule describes the patterns to be matched by using the underscore character (Lisp-based pattern matchers often use ?n instead of n_). Upon a successful match of all three arguments, the right-hand side is evaluated. The right-hand side can refer to symbols matched by the left-hand side of the rule. For example, all three right-hand sides perform algebraic substitutions for the variable n which is carried out by a combination of the infix operators /. ("replace all") and -> ("what to replace"). In the first rule, f /. n -> n - m means to take f and replace all occurrences of n with $n - m$.

The fifth section of the one-dimensional rule base takes the z-transform of signal processing operators, which enables the transform of impulse responses of FIR filters and of upsampled signals to be found (see Table 3.11). Because this rule base is recursive, z-transforms of cascaded systems are possible (e.g., the z-transform of an upsampled, time-reversed, impulse response of an FIR filter).

TABLE 3.11 Z-TRANSFORMS OF STRUCTURES (OPERATORS)

Structure (Operator)	z-Transform
$x^*[n]$	$X^*(z^*)$
$x[n] \star_n y[n]$	$X(z)Y(z)$
$\Delta_k x[n]$	$(1 - z^{-1})^k X(z)$, where Δ_k is the kth backward difference
$\downarrow_M x[n]$	$\dfrac{1}{M} \sum_{l=0}^{M-1} X(e^{\frac{-j2\pi l}{M}} z^{\frac{1}{M}})$
$FIR[n, \{h_0, h_1, h_2 \ldots\}][x[n]]$	$X(z)(h_0 + h_1 z^{-1} + h_2 z^{-2} + \ldots)$
$IIR[n, \{a_0, a_1, a_2, \ldots\}][x[n]]$	$\dfrac{a_0 X(z)}{1 - a_1 z^{-1} - a_2 z^{-2} - \ldots}$
$\Im m\{x[n]\}$	$\dfrac{1}{2j}[X(z) - X^*(z^*)]$
$\sum_{l=0}^{\infty} x[n + lk]$	$\dfrac{1}{1 - z^{-k}} X_k(z)$
	$x[n]$ periodic with period k
$\Re e\{x[n]\}$	$\dfrac{1}{2}[X(z) + X^*(z^*)]$
$Rev_n\{x[n]\}$	$X\left(\dfrac{1}{z}\right)$
$Shift_m\{x[n]\}$	$z^{-m} X(z)$
$f[n] \, \text{sign}[g[n]]$	$Z\{f[n]u[g[n]] - f[n]u[-g[n]]\}$
$\uparrow_M x[n]$	$X(z^M)$
$\sum_{j=0}^{n} x[j]$	$\dfrac{1}{1 - z^{-1}} X(z)$

Adapted from [9, 18, 22].

The last section of the one-dimensional rule base adds strategies for taking z-transforms (see Figure 3.3). These "tricks" are needed to make this collection of z-transform knowledge complete. Because the z-transform rules are fired sequentially in the order given by Table 3.7, strategy rules will be reached only if all the other rules have not completely given the transform. Unfortunately, the repeated application of any of the starred strategy rules in Figure 3.3 will cause an infinite loop. To prevent this, local state information is associated with every expression (and inherited by every subexpression) to prevent any of these rules from being applied more than once.

Inverse z-Transforms

The digital signal processing packages complement the forward z-transform rule base by providing a separate symbolic rule base, InvZTransform, which can inverse transform one-dimensional and *separable* multidimensional expressions. The restriction to separable is necessitated by the fact that the general m–D inverse requires multiple contour integrations that *Mathematica* cannot perform.

InvZTransform calls a one-dimensional rule base for each dimension to be inverse transformed. The 56 one-dimensional inverse z-transform rules (see Table 3.12) are partitioned in a manner similar to the one-dimensional forward z-transform

1. Rewrite affine step functions.
2. Break up any time-dependent absolute value function into a sum of a causal and an anticausal step function.
3. Collect exponential terms by changing an exponential divided by another into a ratio raised to a common power.
4. Expand $\frac{x[n]u[n]}{n}$ into $x[0]\delta[n] + \frac{x[n]}{n}u[n-1]$, if $\lim_{n \to 0} x[n] = 0$.
*5. Distribute products like $(n+1)(n+2)$ into $n^2 + 3n + 2$.
*6. Factor the denominator.
*7. Expand all numerator terms.
*8. Expand all denominator terms.
9. Try a two-sided transform.
*10. Substitute mathematical definitions for all terms.

---

*Once this rule is applied to an expression, that rule is disabled from being applied to any part of that expression.

Figure 3.3 Strategies for Forward z-Transforms
These strategy rules appear in the rule base in the order above. Since rules in rule bases are tested sequentially, exponential terms will be collected before a two-sided transform is attempted. These rules appear last in the rules base, so they are reached only after all other rules have failed to give the transform.

Sec. 3.4 Transforms

TABLE 3.12 ONE-DIMENSIONAL INVERSE Z-TRANSFORM RULES

The structure of the inverse z-transform rule base follows the general rule base structure (see Figure 3.1) so it follows the structure of the forward z-transform. The only difference is that `InvZTransform` does not invert nonseparable transforms so no "multidimensional hooks" are necessary (see Table 3.3).

Rule Grouping	Rules	See Also
rational transform pairs	27	Table 3.8
nonrational transform pairs	9	Table 3.9
transform properties	9	Table 3.10
transforms of operators	2	Table 3.11
transform strategies	10	Figure 3.4

rules. The only difference is that no multidimensional hooks exist for `InvZTransform` since the inverse of nonseparable transforms is not supported (compare Tables 3.7 and 3.12). Six of the inverse transform pairs concern `Impulse`, `Step`, and sinusoidal forms. The other pairs return the inverse transform of rational z-transforms (with first-order and second-order denominators) and nonrational z-transforms (containing logarithmic, factorial, and square root forms). Since the properties of the transform and the transforms of operators are the same for the inverse as for the forward transform, the rules in these sections of the `ZTransform` and `InvZTransform` rule bases are similar. Unlike the forward z-transform, however, z-transforms that correspond to exponentials in the "time" domain are inverse transformed by the multiplication-by-exponential property rule instead of by table lookup.

Some strategies for inverting z-transforms are similar to those applied in taking forward z-transforms (compare Figures 3.3 and 3.4), but some new ones are needed. Two such strategies are partial fractions and power series expansion. The goal of partial fractions is to break up rational polynomials into a sum of terms involving powers of first-order and second-order denominators. Then, each individual term can be easily transformed. Power series expansion can be used to approximate the first N polynomial terms of the inverse transform. Figure 3.5 demonstrates the use of a power series in computing the inverse z-transform of $\exp(\tan(z))$.

An example that demonstrates the inner workings of the inverse z-transform rule base is computing the complex cepstrum of a sequence—the inverse z-transform of the logarithm of the z-transform of the sequence [23]. If the sequence were a sum of two impulses, then the complex cepstrum would be the inverse z-transform of

$$\log(Z\{\delta(n) + a\delta(n-3)\}) = \log(1 + az^{-3})$$

which is `Upsample[3, n] [- (-1)^n (a)^n Step[n - 1] / n]`. In the first pass through the rule base, none of the transform pairs and none of the property rules apply. However, the upsampling structure rule does apply since the greatest common divisor of all exponents of z in `Log [ 1 + a z^-3 ]` is 3, an integer greater than one. This rule fires and rewrites the inverse transform as the upsampling by three of the

*1. Perform partial fraction expansion on rational polynomials (rule is actually located in the rational transform section).

*2. Perform partial fraction expansion (rule is actually located in the properties section).

3. Factor terms inside logarithmic functions.

4. Factor terms inside exponential functions.

5. Normalize the denominator of a rational polynomial: $1 + a_1 z^{-1} + a_2 z^{-2} + \cdots$ (2 rules).

6. Expand the definitions of all terms.

*7. Take a left-sided transform using right-sided rules by substituting z^{-1} for z, taking the inverse z-transform, and substituting $-n$ for n in the result.

8. Complex cepstrum: $Z^{-1}\{\log X(z)\} \to -\frac{1}{n} Z^{-1}\left\{\frac{z}{X(z)}\frac{d}{dz}X(z)\right\}$.

*9. Apply the inverse z-transform to the first N terms of a series expansion about $z = 0$ (when all else fails).

* Once this rule is applied to an expression, that rule is disabled from being applied to any part of that expression.

Figure 3.4 Strategies for Inverse z-Transforms.
The application and purpose of these rules is similar to the strategies for forward z-transforms (see Figure 3.3). Two of the key strategies are partial fractions decomposition and power series expansion.

```
In[6]: = InvZTransform[ Exp[Tan[z]], z, n,
                Terms -> 5,
                Dialogue -> True ]
```

Tan[z]
Breaking up the expression E
into its series representation
$1 + z + \frac{z^2}{2} + \frac{z^3}{2} + \frac{3 z^4}{8} + \frac{37 z^5}{120} + O[z]^6$
The inverse z-transform will now be
applied to the first 6 terms.

Out[6] = Impulse[n] + Impulse[1 + n] +
$\frac{\text{Impulse}[2 + n]}{2} + \frac{\text{Impulse}[3 + n]}{2} +$

$\frac{3 \text{ Impulse}[4 + n]}{8} + \frac{37 \text{ Impulse}[5 + n]}{120}$

Figure 3.5 Last Resort Rule for Inverse Transforms.
Since all other attempts at the inverse z-transform of $\exp(\tan z)$ failed, the series approximation was used as the last resort. This strategy can be disabled by using the option Terms -> False when calling InvZTransform (see Table 3.6).

inverse z-transform of Log [1 + a z^-1]. This last expression can be inverse transformed via table lookup [Muth]:

$$\log(1 + az^{-1}) \leftrightarrow \frac{(-1)^{n+1}a^n}{n}u[n-1]$$

In this case, the rule base determined the complex cepstrum without using any strategies, not even the complex cepstrum strategy rule.

3.4.3 Symbolic Discrete Fourier Transforms

There are two well-known versions of the Fourier transform used for discrete-time signals: the DTFT (discrete-time Fourier transform) and the DFT (discrete Fourier transform). The DTFT transforms an infinite or finite length discrete-time sequence into the normalized *continuous* frequency domain, where the frequency functions are periodic with a period of 2π radians. On the other hand, the DFT is a completely discrete transform, sampled in both time and frequency. For a finite length sequence, it corresponds to equally spaced samples of the DTFT at N points over a 2π interval. The *Mathematica* primitive Fourier implements the numerical DFT, whereas the new signal processing packages introduce the symbolic objects DFTransform for the DFT, and DTFTransform for the DTFT.

The Discrete-Time Fourier Transform (DTFT)

The one-dimensional DTFT $X(e^{j\omega})$ of the sequence $x[n]$ is defined as follows [23]:

$$X(e^{j\omega}) = \sum_{n=-\infty}^{\infty} x[n]e^{-j\omega n} \qquad x[n] = \frac{1}{2\pi}\int_{-\pi}^{\pi} X(e^{j\omega})e^{j\omega n}\,d\omega \qquad (3.8)$$

The forward DTFT equation is the same as the forward z-transform equation when $z = \exp(j\omega)$, so it is tempting to think that the DTFT can always be obtained from the z-transform. However, this is not always true for two reasons. First, some functions have a DTFT but not a z-transform (e.g., a pulse in the continuous frequency domain which corresponds to an *aliased sinc* function in the time domain). Second, some signals have a z-transform but not a DTFT; the substitution $z = \exp(j\omega)$ that converts a z-transform into a DTFT is only valid when the unit circle is in the region of convergence.

The DTFT rule base, DTFTransform, calls a one-dimensional DTFT rule base for each dimension to be transformed. The one-dimensional DTFT rule base contains transform pairs of forms that do not have z-transforms and rules that capture the frequency-domain behavior of downsampled and upsampled signals. The only redundancy with the z-transform rule base is the rewriting of three property rules (homogeneity, additivity, and multiplication-by-n) and one strategy rule (distribute products). If the first 24 one-dimensional DTFT rules cannot find the transform, then the last rule passes the expression to the z-transform rule base. From a valid z-transform expression, the DTFT is obtained by substituting $\exp(j\omega)$ for z. Forward DTFT examples are embedded in the digital signal analyzer examples of section 3.6 which are shown in Figures 3.10 and 3.11.

The inverse discrete-time Fourier transform rule base, `InvDTFTransform`, is implemented in a manner similar to the forward rule base. When the rule base cannot find the inverse, it consults the inverse z-transform rule base. In this case, the inverse DTFT rule base converts unresolved inverse transforms from the z-domain to the frequency domain and tries again to inverse transform them. If this fails, then all remaining unresolved frequency functions are plugged into the inverse DTFT definition. By specifying the option `Definition -> False` in the call to `InvDTFTransform`, a user can disable the use of the definition.

Like `InvZTransform`, `InvDTFTransform` cannot invert many nonseparable frequency responses. An example of this is the "diamond" tile shown in Figure 3.6, which arises in directional filtering [2]. Its representation as a function of the frequency variables $w1$ and $w2$ is

```
Pulse[2 Pi, w1 + w2 + Pi] Pulse[2 Pi, w1 - w2 + Pi]
```

which is rotation by 45 degrees of the separable frequency function

```
Pulse[2 Pi, w1 + Pi] Pulse[2 Pi, w2 + Pi].
```

```
DiamondTile[w1_, w2_] :=
        CPulse[2 Pi, w1 + w2 + Pi] CPulse[2 Pi, w1 - w2 + Pi]

Plot3D[ DiamondTile[w1, w2], {w1, -Pi, Pi}, {w2, -Pi, Pi}]
```

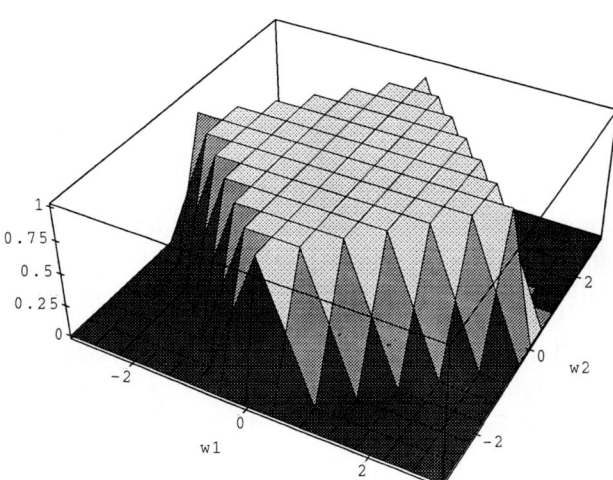

Figure 3.6 Diamond Tile in Two-Dimensional Frequency Domain. Its inverse DTFT cannot be found completely by the inverse DTFT rule base. Instead, the inverse DTFT is a two-dimensional discrete-time convolution of two line impulses.

The inverse DTFT of this last function is Sinc[Pi n1] Sinc[Pi n2], which is also separable (here, sinc[n] ≡ sin[n]/n). Although InvDTFTransform properly inverts this function, it cannot completely inverse transform the "diamond" tile. It can, however, invert each product term so that the actual inverse transform becomes a discrete-time convolution of two line impulses:

$$DTFT^{-1}$$
$$\text{Pulse[2 Pi, w1 + w2 + Pi]} \longleftrightarrow \text{Sinc[Pi n1] Impulse[n2 - n1]}$$
$$\text{Pulse[2 Pi, w1 - w2 + Pi]} \longleftrightarrow \text{Sinc[Pi n1] Impulse[n1 + n2]}$$

The first line impulse only has values along the line n1 = n2, the second along the line n1 = -n2. The final result is a nonseparable two-dimensional *sinc* function.

The Discrete Fourier Transform (DFT)

The N-point one-dimensional DFT $X[k]$ of the discrete function $x[n]$ is defined as follows [23]:

$$X[k] = \sum_{n=0}^{N-1} x[n] e^{-j\left(\frac{2\pi k n}{N}\right)} \qquad x[n] = \frac{1}{N} \sum_{k=0}^{N-1} X[k] e^{j\left(\frac{2\pi k n}{N}\right)} \qquad (3.9)$$

where N is the length of the sequence to be transformed, and both indices run from 0 to $N - 1$. From the definition of $x[\cdot]$ and $X[\cdot]$, then, only samples on the interval 0 to $N - 1$ are transformed. (Any N-point time series can be shifted so that it occupies $n \in [0, N - 1]$, which affects the phase but not the magnitude of the DFT.) Therefore, the symbolic DFT can be implemented as the DTFT of $x[n]$ over the range $n = 0, 1, \ldots, N - 1$ followed by a frequency sampling of the DTFT at N points: $\omega_k = 2\pi k/N$ for $k = 0, 1, \ldots, N - 1$.

The forward DFT rule base, DFTransform, calls the DTFT rule base and factors the result into a product of a real-valued amplitude and a phase function (if possible). For example, with I defined as $\sqrt{-1}$, the *Mathematica* code

```
DFTransform[Exp[I 2 n], 10, n, k]
```

computes the 10-point DFT of exp($j2n$) and returns

$$\text{DFTData}\left[E^{I\,(9\,-\,9/10\,\text{Pi}\,k)}\,\frac{\text{Sin}[10\,-\,\text{Pi}\,k]}{\text{Sin}[1\,-\,\frac{\text{Pi}\,k}{10}]},\right.$$
$$\left.\text{Start}[0],\,\text{Finish}[9],\,\text{FVariables}[k]\right].$$

In this case, the DFT is factored into a phase function (the E^(I (9 - 9/10 Pi k)) term) and a real-valued amplitude function (the ratio of two sinusoids). The amplitude function is in the form of an *aliased sinc* function, namely asinc [10, 2πk - 2]. In discrete-time, this function is the DTFT of an N-point rectangular window. Like the forward DFT, the inverse DFT rule base relies on a DTFT rule base. InvDF-

Transform operates on finite-extent sequences and requires three arguments—$X[k]$, N, and k. It rewrites the sampled frequency response into a continuous one (by the algebraic substitution $k = \frac{Nw}{2\pi}$) and then calls the inverse DTFT.

3.4.4 Laplace and Fourier Transforms

The analog signal processing packages provide the Laplace and Fourier transforms for continuous-time signals. Just as the z-transform is a generalization of the DTFT, the Laplace transform is a generalization of the continuous-time Fourier transform. For one-dimensional functions, the z-transform has an annular region of convergence (ROC) defined by $R_- < |z| < R_+, z \in C$, but the Laplace transform has a planar ROC given by $R_- < \Re e(s) < R_+, s \in C$ [7]. This means that the same data structure representing a z-transform object can also represent a Laplace transform object. The only difference is the forward transform data tag changes from ZTransData to LTransData and the data tag ZVariables becomes LVariables.

Once the rule bases were developed for multidimensional z-transforms and DTFTs, the same approach was applied to symbolic Laplace and Fourier transforms. The organization of the forward Laplace transform rule base is the same as that of the forward z-transform rule base (cf. Figure 3.1). That is, the Laplace transform rules are divided into the same six sections (see Table 3.7). Although this rule base contains more transform pairs and property rules than the z-transform rule base, the Laplace rule base currently has fewer structure (operator) rules. Its strategies are the same except that the z-transform strategy that rewrites a step function as a sum of an impulse and delayed step function does not apply in continuous time. And like the z-transform rule base, the Laplace transform rule base tracks the region of convergence because it is bilateral transform.

In the signal processing packages, the Laplace transform rule base LaPlace calls a one-dimensional Laplace transform rule base once per dimension and simplifies the resulting transform. For example, Figure 3.7 shows that the LaPlace rule base fires eight rules to find the Laplace transform of

$$A\,t\,e^{-\frac{1}{2}(a+b)t}\left[I_0\!\left(\frac{a-b}{2}t\right) - I_1\!\left(\frac{a-b}{2}t\right)\right]u_{-1}(t)$$

where I_k is the kth-order modified Bessel function and $u_{-1}(t)$ is the continuous-time step function, which is represented in *Mathematica* as CStep[t] (see Table 3.3). The first rule calls the one-dimensional rule base and the remaining seven rules find the transform. These seven one-dimensional rules are listed in Figure 3.7(b) in the order that they are applied.

The inverse Laplace transform, InvLaplace, has the same structure as InvZTransform but contains more transform pairs and fewer operator rules. This is because more Laplace transform pairs have been catalogued [7,22]. The continuous-time Fourier transform rule bases CTFTransform and InvCTFTransform are implemented in the same way as are DTFTransform and InvDTFTransform. This means that as a last resort, CTFTransform calls LaPlace and InvCTFTransform calls InvLaPlace.

Sec. 3.5   Solving Differential and Difference Equations

```
In[25]:= LaPlace[ A t Exp[- (a + b) t / 2]
                 (BesselI[0, (a - b) t / 2] -
                  BesselI[1, (a - b) t / 2] ) CStep[t], t, s]
Out[25]=
```

$$\text{LTransData}\,[\frac{1}{(a+s)^{3/2}\,\text{Sqrt}[b+s]},$$
$$\text{Rminus}\,[\,\text{Abs}[\frac{a-b}{2}]-\frac{\text{Re}[a+b]}{2}\,],$$
$$\text{Rplus}\,[\,\text{Infinity}\,],$$
$$\text{LVariables}\,[\,s\,]\,]$$

(a) *Mathematica* dialogue

$$A\,t\,e^{-\frac{a+b}{2}t}\left(I_0\!\left(\frac{a-b}{2}t\right) - I_1\!\left(\frac{a-b}{2}t\right)\right)u_{-1}(t)$$

$L\{A\,f(t)\}$	$\to\quad A\,L\{f(t)\}$	Property (homogeneity)
$L\{t\,f(t)\}$	$\to\quad -\dfrac{d}{ds}L\{f(t)\}$	Property (multiply-by-t)
$L\{e^{at}f(t)\}$	$\to\quad L\{f(t)\},\ s\to s-a$	Property (shifting)
$L\{(x(t)-y(t))z(t)\}$	$\to\quad L\{x(t)\,z(t)-y(t)\,z(t)\}$	Strategy
$L\{a(t)-b(t)\}$	$\to\quad L\{a(t)\}-L\{b(t)\}$	Property (additivity)
$L\{I_\nu(at)u_{-1}(t)\}$	$\to\quad \dfrac{(s-\sqrt{s^2+a^2})^\nu}{a^\nu\sqrt{s^2-a^2}}$	Transform pair
$L\{I_\nu(at)u_{-1}(t)\}$	$\to\quad \dfrac{(s-\sqrt{s^2+a^2})^\nu}{a^\nu\sqrt{s^2-a^2}}$	Transform pair

(b) Order of rules fired

Figure 3.7 Forward Laplace Transform Example.

Admittedly, the notation for the forward and inverse Laplace transforms is a bit awkward. The names were chosen so that they would not clash with the forward and inverse Laplace transforms bundled with *Mathematica*. *Mathematica* provides the objects `Laplace` and `InverseLaplace` which compute only one-dimensional one-sided transforms and cannot handle functions with one or more points of discontinuity (step, pulse, and delta functions) which are essential in signal processing.

3.5 SOLVING DIFFERENTIAL AND DIFFERENCE EQUATIONS

Because the signal processing packages implement the bilateral Laplace and z-transforms, the ability to solve linear constant coefficient differential and difference equations almost comes automatically. All that is missing is to adjust the driving function to account for the initial conditions. No linear systems of equations have to be solved to capture the homogeneous solution [23]. Other advantages of using a bilateral transform are that a left-sided solution can be obtained as easily as a right-sided one and the initial conditions do not have to take place at the origin.

The bilateral approach to solving difference equations is first to assume that all missing initial conditions equal zero. Then, augment the difference equation to account for the initial conditions. Finally, transform each side, solve for the transfer function $Y(z)$, and inverse transform to obtain the digital function $y[n]$. For example, the difference equation

$$y[n] - \frac{3}{2}y[n-1] + \frac{1}{2}y[n-2] = \left(\frac{1}{4}\right)^n u[n]$$

such that $y[-1] = 4$ and $y[-2] = 10$ \hfill (3.10)

provides all of the initial conditions so none have to be assumed zero. Substituting the initial conditions into (3.10) gives

for $n = -1$: $y[-1] - \frac{3}{2}y[-2] + \frac{1}{2}y[-3] = (4) - \frac{3}{2}(10) + \frac{1}{2}(0) = -11$

for $n = -2$: $y[-2] - \frac{3}{2}y[-3] + \frac{1}{2}y[-4] = (10) - \frac{3}{2}(0) + \frac{1}{2}(0) = 10$

The augmented difference equation, therefore, is

$$y[n] - \frac{3}{2}y[n-1] + \frac{1}{2}y[n-2] = \left(\frac{1}{4}\right)^n u[n] - 11\delta[n-1] + 10\delta[n-2]$$
\hfill (3.11)

which can be solved by taking the z-transform of both sides, solving for $Y(z)$, and inverse transforming the result to obtain $y[n]$.

The code that solves the example difference equation is

```
In[10]: = diffequ = y[n] - 3/2 y[n - 1] + 1/2 y[n - 2]
Out[10] = y[-2 + n]/2 - 3 y[-1 + n]/2 + y[n]

In[11]: = ZSolve[ diffequ == (1/4)^n Step [n],
              y[n], y[-1] ->4, y[-2] ->10]

Out[11] = (2 + (1/4)^n + 3(1/2)^n) Step[n] / 3
```

Once a solution has been obtained, it can be easily verified. One way to validate a solution is to (1) make y[n] into a function, (2) compute the initial conditions, and then (3) compare the values of both sides of the difference equation. The corresponding *Mathematica* dialogue follows (note that %11 represents the output associated with the eleventh input, which in this case is the solution to the difference equation):

```
In[12] := y[n_] := %11
In[13] := y[-1]
Out[13] = 4
In[14] := y[-2]
Out[14] = 10
In[15] := Table[diffequ == (1/4)^n Step[n], {n, 0, 5}]
Out[15] = {True, True, True, True, True, True}
In[16] := Clear[y]
```

In this case, the solution returned by ZSolve is perfectly legitimate. ZSolve, like the transform rule bases, supports the option Dialogue so that a user can view each step in the process of solving difference equations (see Table 3.6).

The object LSolve solves initial and boundary value linear constant coefficient differential equations. Like ZSolve, the difference equation solver, LSolve assumes that all missing initial conditions equal zero. This solver supports three levels of justification just as ZSolve does. Figure 3.8 shows how LSolve would solve a second-order initial value problem and how to verify the solution. In this case, the solver partially justifies its answer as a result of the Dialogue option being set to True. Complete justification (Dialogue -> all) would review each step of the problem solving process, including how to take the initial conditions into account.

3.6 ANALYSIS

The previous section discussed the use of linear transform rule bases to solve linear difference and differential equations. In this section, we examine transform-based techniques to analyze and characterize signals. Because the forward z-transform and Laplace transform rule bases track the region of convergence (ROC), information about the stability of a time-domain signal can be obtained. Analog signals are stable if the ROC returned by the Laplace transform lies entirely in the left-half of the s-plane. Digital signals are stable if the ROC returned by the z-transform contains the unit circle.

The object Stable, which performs this analysis, operates on a transform object. It returns True if the signal is stable and False if it is not. If the stability is dependent on values of free parameters, then Stable will return a set of stability conditions. For example, applying Stable to the z-transform of the two-dimensional separable function shown in Figure 3.2

$$a^{n_1} b^{n_2} u[n_1, n_2]$$

would return

$$\text{Abs}[a] < 1 \,\&\&\, \text{Abs}[b] < 1$$

```
In[17] :=
    LSolve [ y''[t] + 3/2 y'[t] + 1/2 y[t] =
             Exp[a t] CStep[t], y[t], y[0] -> 4,
             y'[0] -> 4, Dialogue -> All ]
```

Solving for y[t] in the differential equation

$$\frac{y[t]}{2} + \frac{3y'[t]}{2} + y''[t] = E^{a\,t} \text{CStep}[t]$$

subject to the initial conditions

(y[0] -> 4, y'[0] -> 4)

The Laplace transform of the left side is:

$$L(y[t]) \; (\frac{1}{2} + \frac{3\,s}{2} + s^2) - \frac{3\,y[0]}{2} - s\,y[0] - y'[0]$$

(In the general case, the bilateral Laplace transform of the nth derivative of y[t] is:

$$L\,(y^{(n)}[t]) = L\,(y[t])\;s^n + \frac{s^{-1+n}\,y[t0]}{E^{s\,t0}} + \ldots + \frac{s^{-2+n}\,y'[t0]}{E^{s\,t0}} + \ldots + \frac{y^{(-1+n)}[t0]}{E^{s\,t0}}$$

where t=t0 is the initial condition.)

The Laplace transform of the right side is:

$$\frac{1}{s - a}$$

Solving for the unknown transform yields

$$\frac{\frac{1}{s - a} + \frac{3\,y[0]}{2} + s\,y[0] + y'[0]}{\frac{1}{2} + \frac{3\,s}{2} + s^2}$$

Which becomes

$$\frac{10 + 4\,s + \frac{1}{s - a}}{\frac{1}{2} + \frac{3\,s}{2} + s^2}$$

Out[17] =
Inverse transforming this gives y[t]:

$$(2\,(-5 + 6\,E^{t/2} + E^{t + a\,t} - 16\,a + 22\,E^{t/2}\,a - 12\,a^2 + 16\,E^{t/2}\,a^2)\,\text{CStep}[t])/(E^t\,(1 + a)\,(1 + 2\,a))$$

```
In[18] :=
    yoft = %17 /. { CStep[t] -> 1 };
In[19] :=
    Simplify[ yoft /. t -> 0 ]
Out[19] =
    4
In[20] :=
    Simplify[ D[ yoft, t] /. t -> 0 ]
Out[20] =
    4
```

Figure 3.8 Interaction with the Differential Equation Solver. LSolve solves an initial value differential equation using the bilateral Laplace transform. Because the Dialogue option was All, LSolve completely justified its answer. The commands numbered 18 through 20 verify the solution of *y(t)* found by LSolve. Command 18 defines yoft as *y(t)* for $t > 0$; since $u_{-1}(t) = 1$ for $t > 0$, the continuous-time step function is replaced 1. Command 19 verifies that $y(0+)$ is indeed 4 and the last command checks the value $y'(0+)$.

Furthermore, applying Stable to the Laplace transform of

$$A\,t\,e^{-\frac{a+b}{2}t}\left(I_0\!\left(\frac{a-b}{2}t\right) - I_1\!\left(\frac{a-b}{2}t\right)\right)u_{-1}(t)$$

(which is shown in Figure 3.7) would return

Abs[(a - b)/2] - Re[(a + b)/2] < 0.

Stable works well for one-dimensional and separable multidimensional functions. It often works well for nonseparable multidimensional functions, as shown in Figure 3.9.

Sec. 3.6 Analysis

In[36]: =

```
ZTransform[(n1 + n2)! a^n1 b^n2 Step[n1,n2] / (n1! n2!),
    {n1, n2}, {z1, z2}]
```

Out[36] =

$$\text{ZTransData}\left[\frac{1}{1 - \frac{a}{z1} - \frac{b}{z1}},\ \text{Rminus}[\{\text{Abs}[a],\ \text{Abs}[b]\ \text{Abs}[\frac{1}{1 - \frac{a}{z1}}]\}],\right.$$

$$\left.\text{Rplus}[\{\text{Infinity, Infinity}\}],\ \text{ZVariables}[\{z1,\ z2\}]\right]$$

In[37]: =

```
Stable[%]
```

Out[37] =

$$\text{Abs}[a] < 1\ \&\&\ \text{Abs}[\frac{1}{1 - a}]\ \text{Abs}[b] < 1$$

Figure 3.9 Stability Analysis of a Non-Separable Two-Dimensional Signal. This is the stability analysis of a two-dimensional non-separable function of variables n1 and n2 having the form

$$\frac{(n_1 + n_2)!}{n_1! n_2!} a^{n_1} b^{n_2} u[n_1, n_2]$$

The function is not separable, so neither is the ROC. Stable is able to eliminate the transform variable z1 in the ROC by replacing it with 1 since the unit circle $|z_1| = 1$ must be in the ROC for the function to be stable. According to the stability checker, the function is stable if $|a| < 1$ and $|b|/|1 - a| < 1$. Since $|a| < 1$, the second condition can be safely rewritten as $|b| < |1 - a|$. The second condition $|b| < |1 - a| < 1 - |a|$ becomes $|a| + |b| < 1$ which is identical to that reported by [10].

By combining the abilities of the transform rule bases with stability checker, and by extending graphics capabilities, signal analyzers for one- and two-dimensional signals were developed—DSPAnalyze for digital signals and ASPAnalyze for analog signals. DSPAnalyze and ASPAnalyze require two arguments: the expression to analyze and the variable(s) in the expression. The third and fourth arguments, which are optional, specify the interval over which to plot the expression in the "time" domain. DSPAnalyze relies on SequencePlot, and ASPAnalyze calls SignalPlot, to display the signal in the time-domain. DSPAnalyze calls the z-transform and ASPAnalyze calls the Laplace transform and then checks stability by passing the result to Stable. For one-dimensional and separable two-dimensional expressions, the signal analyzers plot the pole-zero diagram(s), and for non-separable two-dimensional expressions, they plot the pole and zero root locus in each dimension. If the generalized transform exists, then the signal analyzers compute the Fourier transform by a substitution of transform variables—z = Exp[I w] for DSPAnalyze and s = I w for ASPAnalyze. If the generalized transform does not exist, then the frequency response is obtained via `DTFTransform` or `CTFTransform`, respectively. `MagnitudePhasePlot` is called to plot the magnitude and phase response of the signal. The rest of this section will present four examples (two in discrete time and two in continuous time—see Figures 3.10–3.13) of output from these signal analyzers.

In[41]:=

    DSPAnalyze[Upsample[3,n][Sinc[n]], n, -20, 20]

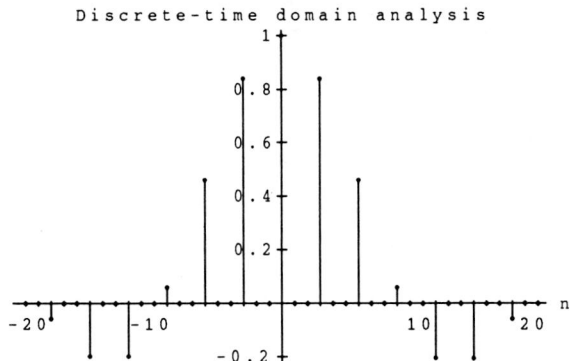

ZTransform::notvalid:
   The forward z-transform could not be found.

```
Upsample    [Sinc[n]]
       3,n
has the following frequency response:
ScaleAxis    [Pi (-CStep[-1 + w] + CStep[1 + w])]
           3,w
```

Out[41]=

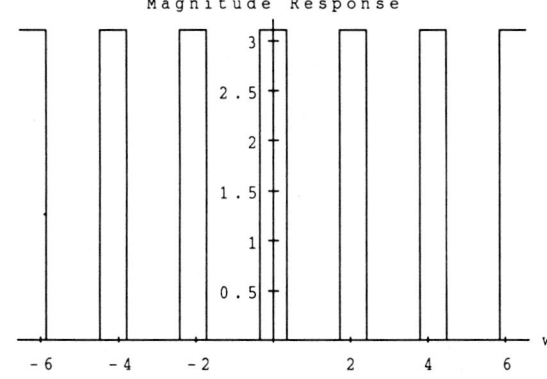

-Incomplete z-Transform-

Figure 3.10 One-Dimensional Digital Signal Analysis. This is a 1-D function in n to be plotted from n = −20 to n = 20. First, `DSPAnalyze` will plot the expression at integer values of n for n ∈ [−20, 20] (recall that $\text{sinc}(t) \equiv \frac{\sin t}{t}$ in the signal processing packages). Since this expression does not have a z-transform, the signal analyzer does not display a pole-zero diagram nor does it print a z-transform. The expression does have a Fourier transform, so the magnitude and phase responses are still plotted. The frequency response is normally periodic with period 2π. Upsampled signals, however, scale the frequency axis by the upsampling factor. In this case, the magnitude response is pulse function with a width of $\frac{2}{3}$ rad replicated every $\frac{2}{3} \pi$ rad. This figure omits the phase plot generated by the analyzer because the phase is zero for all frequencies.

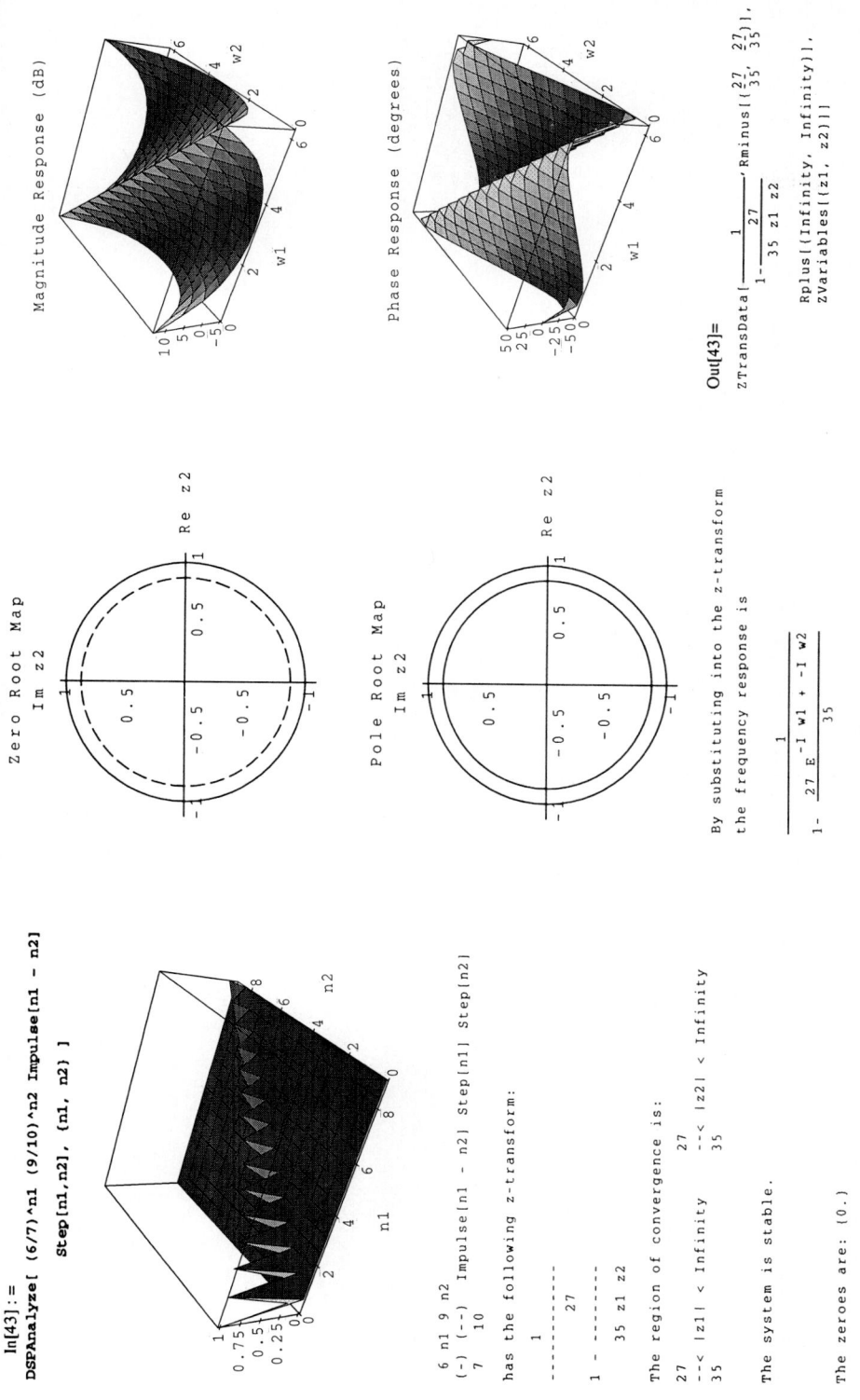

Figure 3.11 Analysis of a Two-Dimensional Line Impulse. This is a two-dimensional nonseparable function with nonzero values only when $n_1 = n_2$. In this example, the impulse function is actually a line impulse which is 1 for $n_1 = n_2$ and 0 elsewhere. The discrete time-domain plot captures this behavior. Because this expression only has values along one dimension, the z-transform is the one-dimensional transform with respect to n, where $n = n_1 = n_2$, adjusted by the substitution $z = z_1 z_2$. A pole-zero root map was drawn instead of a pole-zero diagram because the z-transform is not separable. Since the z-transform $F(z_1, z_2) = F(z_2, z_1)$, one pole root map is sufficient to show the behavior of the poles, which are always inside the unit circle.

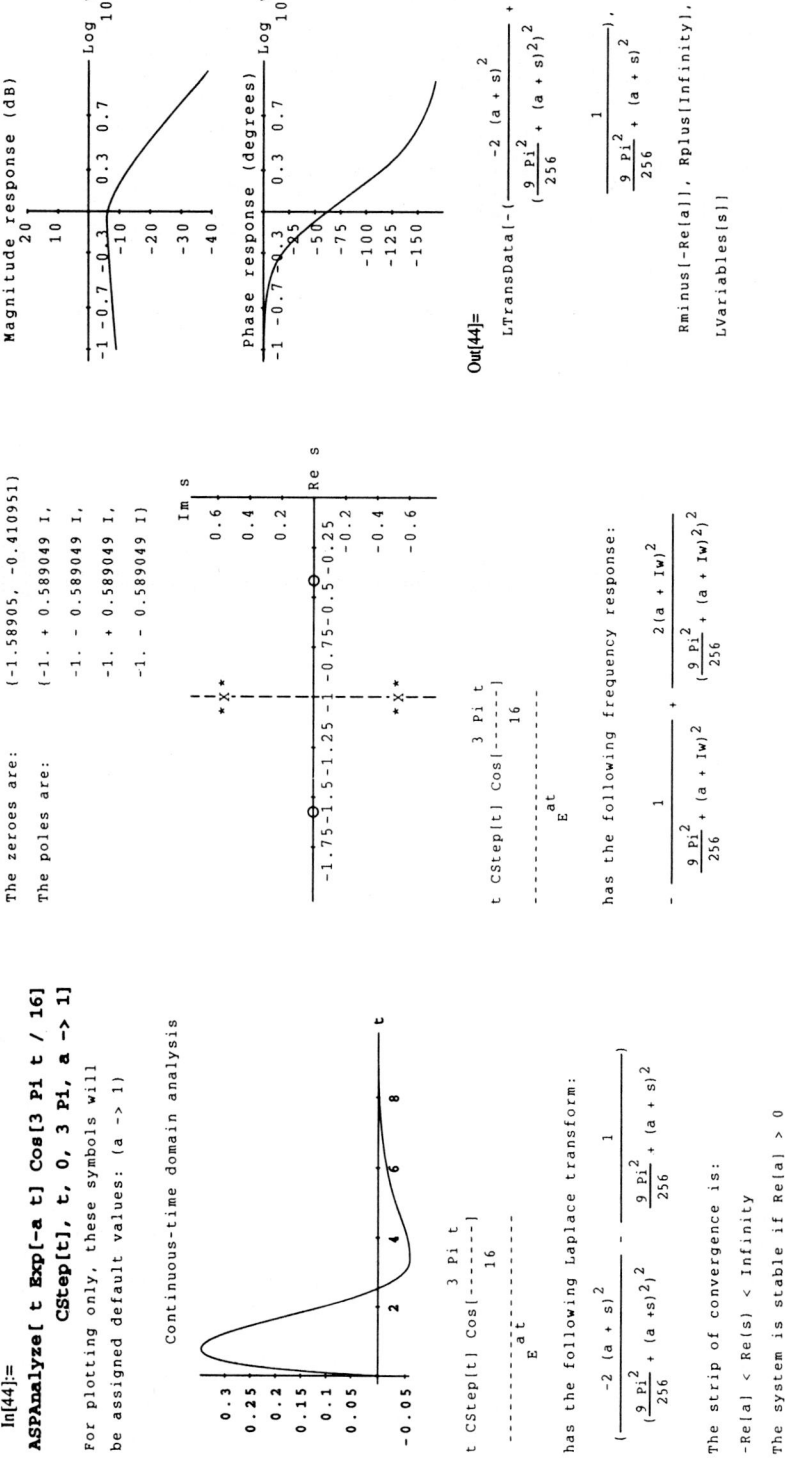

Figure 3.12 One-Dimensional Analog Signal Analysis. The fourth example is a function of the continuous variable t to be plotted over the interval [0, 3π]. This damped sinusoid ramp function cannot be plotted as is, because the parameter a is unknown. Before graphing the expression as a function of a continuous variable, ASPAnalyze will substitute a default value of 1 for a (because of a – >1). The Laplace transform gives rise to a rational polynomial in s with two poles, each with multiplicity 2. In a pole-zero diagram, a single zero is represented as O and a single pole as X. A pole or zero with multiplicity greater than 1 is plotted as *X* or *O*, respectively.

Sec. 3.6   Analysis

In[45]:=

    ASPAnalyze[Delta[t - 1/2] +
                Delta[t - 3/2],t, 0, 2]

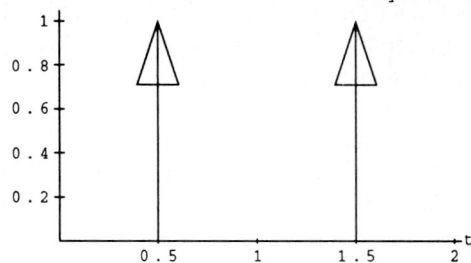

$$\text{Delta}[-(\tfrac{3}{2}) + t] + \text{Delta}[-(\tfrac{1}{2}) + t]$$

has the following Laplace transform:

$$\frac{1 + E^{s}}{E^{(3s)/2}}$$

The strip of convergence is:

-Infinity < Re(s) < Infinity

The system is stable.

PoleZeroPlot::notrational:
    Transform is not a rational polynomial.

PoleZeroPlot::noplot:
    A pole-zero plot cannot be generated.

$$\text{Delta}[-(\tfrac{3}{2}) + t] + \text{Delta}[-(\tfrac{1}{2}) + t]$$

has the following frequency response:

$$E^{(-3 I)/2\, w} (1 + E^{Iw})$$

ASPAnalyze::notinteresting:
    *Could not determine the important
    section of the frequency response.*

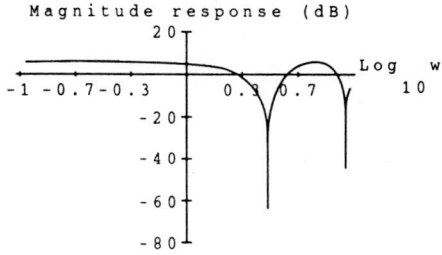

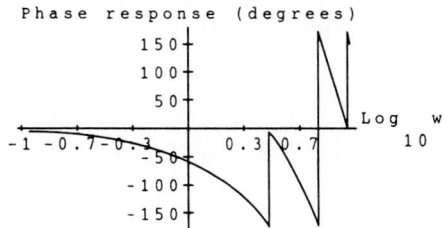

Out[45]=

  LTransData[$\frac{1 + E^{s}}{E^{(3s)/2}}$, Rminus[-Infinity],
             Rplus[Infinity],
             LVariables[s]]

Figure 3.13 Signal Analysis of Dirac Delta Functions. This last example shows the flexibility of the analog signal analyzer. In this example, the time-domain response over $t \in [0, 2]$ is a pair of infinite height, area 1 delta functions located at $t = \frac{1}{2}$ and $t = \frac{3}{2}$. This time-domain plot represents delta functions with arrowheads, which is a common engineering convention. This function has a Laplace transform but no poles or zeros so a pole-zero diagram is not displayed. The magnitude and phase responses are plotted as per usual.

3.7 CONTINUOUS AND DISCRETE CONVOLUTION

The convolution operation, which for linear time-invariant systems is nothing more than filtering, is a basic building block in signal processing. Understanding convolution is crucial for understanding signal processing theory, yet convolution is a very difficult concept for students to grasp. Students first work convolution problems for continuous functions that are defined piecewise where each interval supports a simple function. The solution for such a piecewise convolution is accomplished by considering the graphical interpretation of convolution (i.e., a "flip-and-slide" approach) and then performing the integration over those intervals where the functions overlap. The answer would be expressed in a piecewise manner as well.

Continuous convolution of two functions, $f(t)$ and $g(t)$, is defined as [7]:

$$f(t) * g(t) = \int_{-\infty}^{\infty} f(\tau) g(t - \tau) d\tau$$

In discrete time, the integral is replaced by summation [23]:

$$f[n] * g[n] = \sum_{m=-\infty}^{\infty} f[m] g[n - m]$$

In both cases, one-dimensional convolution can be viewed as flipping g in time and sliding it across f.

The signal processing packages introduce `PiecewiseConvolution`, which performs discrete convolution, as well as convolution for piecewise functions. `PiecewiseConvolution` accepts f and g as expressions or as lists of *F-intervals*. Each F-interval is represented as a list of three elements:

{*function*, *left-endpoint*, *right-endpoint*}

F-intervals can be finite or infinite in extent, but the right endpoint should always be greater than the left endpoint. As a list of F-intervals, then, `t CPulse[10, t] + 5 Delta[t - 20]` becomes `{ {t, 0, 10}, {Area[5], 20, 20} }`.

`PiecewiseConvolution` implements a general strategy for computing convolutions by

1. converting f and g to lists of F-intervals,
2. convolving each F-interval in f with each F-interval in g, and
3. simplifying the result by
   a. reordering (sorting) the F-intervals, and
   b. merging overlapping F-intervals (when possible).

Convolving one F-interval with another generates three F-intervals, as shown in Table 3.13. This is a consequence of the "Square Matrix Rule" for convolution [35]. Using a *Mathematica* notebook, we have illustrated the mechanics of this "flip-and-slide" approach to convolution, as shown in Figure 3.14. Note that the default

Sec. 3.8 *Mathematica* as an Educational Tool

TABLE 3.13 ENDPOINTS FOR CONVOLUTION INTERVALS

Convolving one finite F-interval $\{f, l_f, u_f\}$ with another finite F-interval $\{g, l_g, u_g\}$ generates three intervals with nonzero functions, as shown for (a) continuous-time and (b) discrete-time convolution. Care must be taken in applying these formulas when one or more of the endpoints is infinite.

Interval	Value
$t \in (-\infty, l_f + l_g)$	0
$t \in [l_f + l_g, l_f + u_g)$	$\int_{l_f}^{t-l_g} f(t)g(t-\tau)d\tau$
$t \in [l_f + u_g, u_f + l_g)$	$\int_{t-u_g}^{t-l_g} f(t)g(t-\tau)d\tau$
$t \in [u_f + l_g, u_f + u_g)$	$\int_{t-u_g}^{u_f} f(t)g(t-\tau)d\tau$
$t \in [u_f + u_g, \infty)$	0

(a)

Interval	Value
$n \in (-\infty, l_f + l_g]$	0
$n \in [l_f + l_g + 1, l_f + u_g]$	$\sum_{m=l_f}^{n-l_g} f[m]g[n-m]$
$n \in [l_f + u_g + 1, u_f + l_g]$	$\sum_{m=n-u_g}^{n-l_g} f[m]g[n-m]$
$n \in [u_f + l_g + 1, u_f + u_g]$	$\sum_{m=n-u_g}^{u_f} f[m]g[n-m]$
$n \in [u_f + u_g, \infty)$	0

(b)

domain for `PiecewiseConvolution` is `Continuous` which can be overridden by specifying the `Domain -> Discrete` or the `Domain -> Continuous` option.

3.8 *Mathematica* AS AN EDUCATIONAL TOOL

The last section demonstrated how animation could be used to teach the "flip-and-slide" method of performing convolution. In addition to animation, the *Mathematica* notebook facility supports other interactive capabilities including playing sound and interpreting code. Notebooks can be effective teaching tools because they combine these interactive features with tables, graphs, and formatted text. On-line tutorials can be more than mere adaptations of textbooks because they can now take full

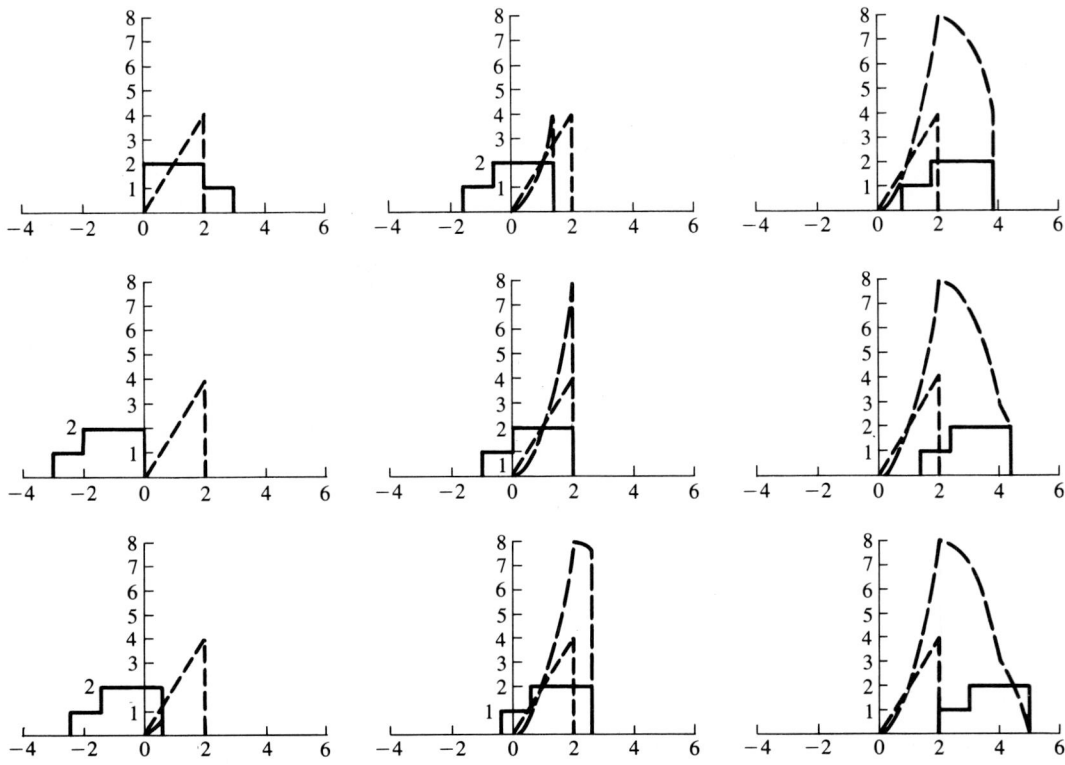

Figure 3.14 Animation of the "Flip-And-Slide" View of Convolution. Above are nine out of thirty-five frames in an animation sequence illustrating the "flip-and-slide" approach to convolution. The first frame superimposes the triangular function $x(t)$ and the piecewise rectangular function $h(t)$. The second frame shows $h(t - \tau)$ for $t = 0$, which is $h(\tau)$ "flipped" about the τ axis. A side benefit of the second frame is that it separates the two functions being convolved. The rest of the frames illustrate the sliding of $h(t - \tau)$ across $x(\tau)$ for different values of t. The final frame shows the function resulting from the convolution.

advantage of the multimedia capabilities and express formulas and models in terms of *Mathematica* code for evaluation and investigation. Besides learning from notebooks, students can use the notebook facility to solve homework problems while documenting the solution process in the same environment.

· Some colleges and universities are introducing *Mathematica* into freshmen curricula primarily to aid in the teaching of calculus. One concern with this approach is that freshmen may lose some of their symbolic manipulation abilities, as there was concern that a dependence on a calculator would detract from one's ability to manipulate numbers [32]. Another concern is that *Mathematica* be integrated into the curriculum so as to motivate students while not overwhelming their computer skills. Section 3.8.1 highlights attempts at the Rose-Hulman Institute of Technology and the University of Illinois at Urbana-Champaign to use *Mathematica* as the primary learning tool in first-year classes.

For a symbolic mathematics platform to be useful in courses beyond the first-year curriculum, it will probably need extension. The Department of Music at Stanford University, for example, has defined several common windowing functions [14] so that students can investigate their temporal and spectral characteristics. Besides analyzing functions, students also learn mathematical procedures. Sometimes a mathematical procedure can be implemented so that it is robust or intuitive, but not both. As an example, the Risch algorithm for symbolic integration is robust but its intermediate calculations would reveal little about how to perform integration manually. For teaching purposes, then, mathematical procedures should be implemented in ways that mimic how humans might do it. This is how the transform and other high-level routines in the signal processing packages are written. These packages and the accompanying notebooks can help students learn linear systems theory, as discussed in section 3.8.2.

3.8.1 *Mathematica* in the Freshman Curriculum

One reason to use a symbolic mathematics environment in the freshman curriculum is to shift the emphasis away from *rote numerical and symbolic manipulation* (e.g., find the derivative of this function and solve these four equations in four unknowns) and toward *problem formulation and solution interpretation* (e.g., derive the function whose minima must be found and formulate the equations that must be solved to determine when objects collide given that one or more objects is in motion). This approach promotes problem formulation as opposed to searching for the "plug" that fits the problem. If students can translate problem statements into functions and equations, then computer solutions are obtained rapidly and accurately. Students must be taught to verify solutions by analyzing them to see if they make sense. When inconsistencies arise, students cannot blame algebraic mistakes (e.g., a dropped minus sign); instead, their formulation of the problem is called into question. This forces them to rethink their understanding of the problem and reformulate the underlying mathematics.

The rest of this section briefly discusses two attempts at integrating *Mathematica* into the freshman curriculum. The University of Illinois approach fits *Mathematica* into a traditional calculus course structure, whereas the Rose-Hulman approach uses *Mathematica* as part of a workstation for a new interdisciplinary curriculum. In both cases, the educators hoped that *Mathematica* would encourage students to explore, observe, and perceive concepts on their own. In exploring calculus, for example, students can plot the secant line between two points on a curve and note that as the horizontal interval between the two points approaches zero, the secant line approaches the tangent line to the curve. They can perform this experiment either in single frames or as an animated sequence. For vectors, they can plot two vectors' sum, their difference, the projection of one onto the other, their projection onto a plane, and their cross product.

Dr. Jerry Uhl and others at the University of Illinois have been developing calculus courseware in *Mathematica* since its release in June 1988. They approach first-year calculus as a laboratory course with 30 minutes of lecture per week. The students spend the rest of their time interacting with lessons in the form of notebooks (60 in all). Each notebook follows the same format: introduction, basic problems, tutorial problems, partially worked problems, homework problems, and review. Students can peruse notebooks at their own pace. Before they can move on to the next one in the series, however, they must first complete the partially worked problems and solve the homework problems in the current notebook. Although professors assume a supporting role in this approach, they still administer tests (including a final exam). The professor can write these tests to ensure that the students are learning the rudimentary manipulations of calculus as well as problem-solving techniques. The biggest difficulty was writing notebooks so that they are more than mere adaptations of a written text [32].

Instead of integrating *Mathematica* into a standard set of classes, Rose-Hulman has developed an interdisciplinary freshman curriculum [13] in which *Mathematica* plays a key role. This new curriculum integrates the calculus, chemistry, computer science, engineering design, and physics courses into one 12-credit course, which is taken for three quarters. The primary goal of this curriculum was to make students aware of the links between related topics in the first-year engineering courses, which many students do not perceive because they are exposed to them in radically different contexts.

Each student has access to a NeXT workstation. The students take notes, solve homework problems, conduct laboratory experiments, and even take tests on the workstation. During a lecture on electrostatics, for example, students can use *Mathematica* to plot the electrostatic potential as a function of two variables and observe the surface (or the equipotential contours) instead of waiting for the teacher to sketch an approximation on the blackboard. *Mathematica* gives them the ability to import lab data, plot the data, and fit functions to the data. In the case of studying pendulum position as a function of time, the students would fit the data to a damped sinusoidal function. Since the fit is excellent, this procedure reinforces the application of sinusoidal and exponential functions. The teachers have developed special front ends to *Mathematica* that quiz students. One front end tests the student's ability to identify a function, its first derivative, and its second derivative from three plots. Using the mouse, the student matches each graph with what it represents. During the 1990–1991 academic year, 39 of the 60 volunteer students completed the first integrated curriculum. Their progress in future classes will be monitored.

3.8.2 *Mathematica* in the Engineering Curriculum

Symbolic mathematics platforms typically do not define many of the functions and operations common in engineering, which limits their use in an engineering curricu-

Sec. 3.8 Mathematica as an Educational Tool

lum. Two sets of packages extend *Mathematica* so that it can be integrated more easily into the engineering curriculum: *nodal*™ [27] and the signal processing packages [12]. *Nodal* gives *Mathematica* extensive circuit analysis abilities similar to those of SPICE™, except that the analysis can be carried out symbolically. *Nodal* provides circuit models, Smith charts, and other tools commonly used in electronics classes. For more complex problems, students can define and evaluate their own circuit models.

The signal processing packages are well suited for courses in linear systems theory, including controls, signal processing, and communications classes. As discussed in sections 3.4 and 3.7, the new packages contain rule bases that *symbolically* compute convolutions and all the transforms common to signal processing: Fourier, Laplace, z, and so on. Based on the transforms, these packages can solve linear constant coefficient difference and differential (LCCD) equations (see section 3.5) as well as perform simple stability tests (see section 3.6). Students can learn from the output of the transform rule bases and the LCCD equation solvers because these routines can fully justify their answers.

Accompanying these extensions are several notebooks that are intended to further motivate students to learn linear systems theory. One notebook provides an introduction to *Mathematica* and signal processing, as well as an overview of the signal processing packages. Another summarizes all of the new definitions, and a third discusses a small set of examples. A fourth notebook is an adaptation of the paper "*Mathematica* as an Educational Tool for Signal Processing" [12] that allows the reader to tweak the examples. A fifth serves as a tester for the forward and inverse bilateral Laplace transform. The remaining notebooks are tutorials discussing piecewise convolution, analog filter design, the DTFT, and the z-transform.

The piecewise convolution notebook shows the student how to solve one-dimensional convolution problems using the piecewise convolution package (discussed in section 3.7). After reviewing the data structures used by the convolution packages, the notebook introduces the new routines and then focuses on the `Piecewise-Convolution` object, which performs discrete-time and continuous-time convolution. The notebook helps the student identify overlapping intervals in piecewise convolution by proceeding step-by-step through a simple problem. This is reinforced by an animation sequence which further illustrates the "flip-and-slide" approach to convolution (nine frames in the sequence are shown in Figure 3.14). Each frame in the sequence shows an important interval over which to integrate, and the reader can control the frame rate.

Linear filters are certainly the workhorse of signal processing. The analog filter design notebook introduces the topic of filter design and guides the reader through several design examples. These examples include a low-pass design example for each classical filter type (Butterworth, Chebyshev type I, Chebyshev type II, elliptic, and Bessel) as well as band-pass, band-stop, and high-pass design examples. Using animation, the student gets a chance to visualize the changes in the filter's magnitude response as one design parameter varies (two animations are provided—one for the ripple control parameter ϵ, the other for the filter order N). The last section of the notebook examines the effects of precision in implementation—the student can truncate the coefficients of the filter and observe the frequency response [12].

The DTFT notebook introduces the topic of Fourier analysis for sequences. The notebook begins by showing a Fourier series approximation of a square wave and relates it to the discrete Fourier transform (DFT). It then defines the DTFT and discusses its relationship to the continuous-time Fourier transform. It concludes with a library of DTFT examples.

The z-transform notebook comes in three parts. Part I defines the bilateral z-transform and discusses the importance of the ROC in uniquely defining the transform. Discussing the ROC leads naturally to introducing stability and causality of a signal based on the location of its poles relative to the ROC. Part II links the z-transform with the DTFT and includes several animations relating pole-zero diagrams to magnitude frequency responses. Part III shows how to choose poles and zeroes to construct filters and animates the effect on the magnitude response when filter pole locations are perturbed. Scattered throughout the notebook are several examples and exercises with solutions.

Self-paced teaching via notebooks can supplement classroom lectures and laboratory experience. Students can use the signal processing packages to tackle homework problems in linear systems while documenting the solution as they go. From the sample notebooks provided with the new packages, students can learn concepts in signal processing, as well as the way they are implemented in *Mathematica*. Among the schools that are currently using these extensions are the Georgia Institute of Technology, the Pennsylvania State University, the Rose-Hulman Institute of Technology, Stanford University, and the University of Pennsylvania. This approach to the linear systems and signal processing curriculum is just starting, but the trend toward using such computer-based mathematical environments will certainly grow in the years to come.

3.9 *Mathematica* AS A DESIGN TOOL

An environment such as *Mathematica* has enormous potential as a tool for designing DSP systems. The availability of the signal processing packages, which extend the basic *Mathematica* platform into the realm of DSP, is a first step in making this design capability a reality. With these packages, a designer can express certain algorithms in the language of symbolic transforms and symbolic signals. At this level, various mathematical manipulations and transformations can be carried out on the symbolic expressions. In general, these transformations will generate different realizations for the algorithm. This type of approach has already been investigated in the development of E-SPLICE [20] and ADE [8]. However, a true design environment will require much more work, especially in the area of knowledge representation, because most design problems require an enormous amount of domain-specific exper-

Sec. 3.9 Mathematica as a Design Tool

tise. Before continuing this discussion, we first pause to mention some of the general characteristics of the design process.

The design process for DSP is basically the activity of taking algorithms expressed in a high-level (mathematical) form and turning them into working systems, either hardware or software. At present, the typical target for such implementations is a DSP microprocessor chip, but another common target is a C language program that might run on a DSP machine. The responsibility of the designer is to produce an efficient implementation by trading off the characteristics of equivalent forms within a class of possible solutions. In our discussion of design, we will touch on the following points:

1. "Design" turns algorithms into implementations.
2. Symbolic expressions are convenient intermediate representations of an algorithm.
3. Design trade-offs need to be expressed in symbolic form during the knowledge acquisition phase.
4. Graphical representations are a variant of "symbolic" representations that are useful to the designer (e.g., signal flow graphs, block diagrams, etc.).
5. Expressions can be optimized in the symbolic domain by applying rearrangement rules.

In the context of this chapter, the crucial question is "what role can symbolic mathematics play in the design process?"

Since the design process amounts to producing implementations, we remind the reader that programs such as *Maple*, *MACSYMA*, and *Mathematica* can generate computer source code from an algebraic expression. So, after some extension, these environments could generate complete programs in Fortran (as well as C in the case of *Mathematica*) which can then be compiled onto a target architecture. This capability could be extended to automatic code generation in assembly language for DSP chips, if necessary. Of course, one issue with this approach is how to manipulate the symbolic expressions into a form that is suitable for automatic code generation. For example, in the implementation of a second-order section for an IIR filter, the rational transfer function $B(z)/A(z)$ is sufficient to describe the input/output characteristics of the section, but it does not have enough information to describe the internal workings of a filter implementation. This is where "knowledge" about different filter forms would be needed, so a KBSP approach linked to symbolic representations of second-order transfer functions would be needed.

If we recognize the strengths of *Mathematica* for manipulating complex mathematical expressions, then a topic area such as number theoretic transforms (and related convolution algorithms) is a prime candidate for exploration. One roadblock for design in a difficult area such as number-theoretic transforms (NTTs) is the manipulation of residue polynomials that are needed to break down the convolution algorithms. However, programs like *Mathematica* can easily perform residue polynomial operations, e.g., $P(z) \bmod Q(z)$. Thus, the theory underlying NTTs and polynomial transforms can be expressed in a convenient symbolic form. Indeed, the early development of short convolution algorithms by Agarwal and Cooley at IBM was done with the aid of a symbolic math program called SCRATCHPAD [1].

The graphical representation of a system is essential for supporting a designer's view of the implementation, which often comes in the form of block diagrams, filter structures, and the like. This sort of design visualization is already being offered in two symbolic circuit analysis packages: the *nodal*™ package [27] for *Mathematica* and the symbolic SPICE program *Sspice*. For DSP design, this sort of graphical presentation would be useful if it is extended to frequency responses at the inputs and outputs of different blocks within a complex system.

The last, and perhaps the most powerful, feature is that of rearrangement, which has been implemented by several Lisp-based systems [8, 20, 26]. This requires a representation for system properties that is best accomplished by object-oriented programming.

3.9.1 Representing System Properties

Two systems that can find optimal implementations in the symbolic domain are E-SPLICE [20] and ADE [8]. These systems search for optimal forms by rearranging subexpressions in an algorithm. If rules were written to describe every possible interaction between operators, including how to simplify and rewrite combinations of them, then a combinatorial explosion would result. Therefore, these systems encode simplification and rewrite rules in terms of system properties so that adding a new operator can be as simple as defining the properties of that operator.

The system properties used in both ADE and E-SPLICE are additivity, associativity, commutativity, homogeneity, linearity, memorylessness, and shift-invariance. In the *Mathematica* implementation, we have recognized the need for two additional properties, CONTINUOUS and DISCRETE, in order to display warning messages when discrete signals are passed into continuous systems and vice-versa. Table 3.14 summarizes the definition for each of these system properties. The signal processing packages, however, offer additional flexibility for users to define their own properties.

In general, *Mathematica* does not provide a mechanism to attach properties to operators, although it does support commutativity for the purposes of pattern matching. The signal processing packages provide an object called DefineSystem that associates system properties with operators:

DefineSystem[< operator > , < $property_1$ > , < $property_2$ > , ...]

Using DefineSystem, the signal processing packages assign properties to the signal processing operators in Table 3.4 as well as the *Mathematica* operators Times and Plus. For example, the shift operator is linear and shift-invariant and works

TABLE 3.14 IMPORTANT SYSTEM PROPERTIES

These are the systems properties about which E-SPLICE reasons [19]. The signal processing packages add CONTINUOUS and DISCRETE properties.

System Property	Meaning
ASSOCIATIVE	can change grouping of inputs
ADDITIVE	distributes over addition
COMMUTATIVE	can change order of inputs
CONTINUOUS	inputs are continuous signals
DISCRETE	inputs are discrete signals
HOMOGENEOUS	scaled input gives scaled output
LINEAR	additive and homogeneous
MEMORYLESS	output does not depend on previous inputs or outputs; if a single-input system, then SHIFTINVARIANT
SHIFTINVARIANT	shifted input gives shifted output

in both the discrete and continuous domains. The appropriate specification appears in the signal processing packages as

```
DefineSystem[Shift[n0,n], CONTINUOUS, DISCRETE,
        LINEAR, SHIFTINVARIANT]
```

Although the additive and homogeneous properties were not explicitly listed among the properties of the shift operator, the linearity property implies both additivity and homogeneity. Therefore, the above code attaches the following properties to the Shift operator: ADDITIVE, CONTINUOUS, DISCRETE, HOMOGENEOUS, LINEAR, and SHIFTINVARIANT. This means that the evaluation of DISCRETE[Shift[n0,n]] is True. DefineSystem, however, does not assume that all unspecified properties are False. Continuing with the example of the shift operator, the expression COMMUTATIVE[Shift[n0,n]] would evaluate to itself, which is neither True nor False. This allows a user to add properties to a previously defined system primitive at any time. For the sake of example, suppose that the shift operator exhibits a property named DSPNESS. A user could add this property to the existing properties of the shift operator by evaluating

```
DefineSystem[Shift[n0, n], DSPNESS]
```

3.9.2 Signals as Objects

Signal properties can also be captured via object-oriented programming [17]. The signal primitives (see section 3.3.2) add the necessary computational functions for signal processing to *Mathematica*, but they do not capture information about signal properties like bandwidth, data type, and symmetry. In order to simulate the behavior of systems and analyze their properties, a higher level of abstraction is required. In the signal processing packages, objects composed of data *slots* and programmable *methods* are used [15, 20]. This approach extends the concept of signal from a mathematical function to that of an object whose data, properties, and characteristics are self-contained. This provides a way to cache signal values, for example.

For the most part, every signal has a data type (e.g., real, complex, phoneme, etc.), dimensions, domain (e.g., frequency or time), and a history (of generation). Some signals have bandwidth, and some exhibit symmetry (real-valued signals can be odd or even with respect to a center point and the real and imaginary parts of complex-valued signals can be conjugate symmetric or conjugate antisymmetric). The use of such signal objects is primarily for simulation and property resolution.

3.9.3 New Applications

An application area where symbolic representations are likely to have an impact on the design problem is that of multirate systems (i.e., filter banks). The explosion of published information in this area has led us to the point where there are a myriad possibilities for implementing filter bank systems. The novice designer would be hopelessly lost trying to determine an optimal implementation within this class of systems (or one of its subclasses, such as perfectly reconstructing filter banks). When extensions to the two-dimensional case are included, the theory leads to filter structures that are quite difficult to visualize and generate.

Therefore, this design process must rely on the more general ability of finding optimal implementations of multirate structures by working in the symbolic domain. To do this efficiently requires heuristic algorithms that apply simplification and rewrite rules in an intelligent manner. A rich set of simplification and rewrite rules has already been identified for (multirate) signal processing in the one-dimensional case [8, 9, 20, 23] but few have been generalized for the multidimensional case [2, 10, 33]. For one-dimensional filter banks, a very general, and flexible, algorithm [20] holds the promise of a general solution via traditional optimization methods. However, for the case of multidimensional filter banks, many new developments are needed to produce better methods of design, analysis, and simulation. The use of symbolic representations of these more complex structures will be a key part of this research.

3.10 CONCLUSIONS

Over the last twelve years, several researchers have extended Lisp so that it has become a knowledge-based signal processing environment [8, 17, 20, 26]. Such environments contain much knowledge about signal processing, but they are missing detailed mathematical knowledge. This chapter describes a different approach. We have extended a powerful symbolic mathematics platform, *Mathematica*, so that it becomes a KBSP environment that is a useful design and educational tool. Part of this development is a new set of packages that enable *Mathematica* to perform a variety of analyses on signals and systems based primarily on transform theory. Several graphical analysis tools, like pole-zero plots and frequency response plots, display signals in different domains. By combining these different abilities, high-level tools for analyzing one- and two-dimensional signals and systems have been developed.

Accompanying the signal processing packages are a collection of notebooks that explain aspects of signal processing theory and how they are implemented in *Mathematica*. For example, one of the notebooks teaches the mechanics of convolution in the time domain primarily by animation. Other notebooks teach the z-transform, the OTFT, and the classical methods of analog filter design. These notebooks represent a first attempt at supplementing classes in linear systems theory. We hope that these and other interactive multimedia tutorials will stimulate students to learn.

The signal processing packages represent first steps in making a symbolic mathematics environment into a viable KBSP research and design tool. KBSP systems based on symbolic mathematics environments enable users to manipulate signals and systems as algebraic expressions. Thus, the signal processing packages provide mechanisms to rewrite and simplify algebraic forms of signal processing expressions. This ability to rearrange subexpressions is crucial in finding optimal implementations of algorithms [8, 20] which the signal processing packages will eventually be able to perform. Once an algorithm has been developed, the packages can generate the equivalent C, Fortran, and TEX code. A future goal is to have *Mathematica* generate complete programs so that an algorithm can ultimately be compiled for a specific architecture. These extended abilities of the signal processing packages will be focused on the design and analysis of multidimensional multirate signal processing structures. After a complete set of simplification and rewrite rules for these structures are identified, the next step would be to capture them in a KBSP environment and encode heuristic techniques to apply them in an intelligent manner.

ACKNOWLEDGMENTS

The authors would like to thank Dr. David Schwartz for suggesting the use of *Mathematica* as a symbolic signal processing tool, as well as Dr. Malcolm Slaney for his useful feedback on the code and the paper.

The authors would like to thank the *Mathematica Journal* and Wolfram Research Inc., for widely distributing the signal processing packages and notebooks. The authors appreciate the time and effort that Dr. Silvio Levy, Dr. Paul Abbot, Dr. Bruce Sawhill, and others at Wolfram Research Inc., have spent in evaluating our work.

The authors would also like to thank NeXT Inc., and the *NeXT On Campus* magazine for giving our work good exposure, especially for including a notebook overview of the signal processing packages in the Higher Education Mailbox.

The authors would also like to thank Wallace McClure for his programming and debugging efforts, Jim Proctor for help in writing the DTFT notebook, and Lina

Karam for her work on the z-transform notebook. The authors especially want to thank Kevin West for writing the convolution package and the accompanying notebook.

The authors would like to thank Sam Liu, Craig Richardson, and Ayhan Sakarya, for their proofreading.

The development of the signal processing packages and notebooks for *Mathematica* at the Georgia Institute of Technology was supported in part by the Joint Services Electronics Program contract #DAAL-03-90-C-0004.

Support for the integrated curriculum at the Rose-Hulman Institute of Technology has been provided by The Lilly Endowment Inc., the National Science Foundation (grants #USE-895 3553, #USE-895 1290, and #USE-895 0669), the General Electric Foundation, the Westinghouse Educational Foundation, and the Arvin Foundation.

REFERENCES

[1] R. C. Agarwal and J. W. Cooley, "New Algorithms for Digital Convolution," *IEEE Trans. on Acoustics, Speech, and Signal Processing*, ASSP-25, no. 5 (October 1977), 392–410.

[2] R. Bamberger, "The Directional Filter Bank: A Multirate Filter Bank for the Directional Decomposition of Images (Ph.D. thesis, Georgia Tech, November 1990).

[3] R. N. Bracewell, *The Fourier Transform and Its Application* (New York: McGraw-Hill, 1978).

[4] B. Char et al., *Maple Reference Manual*, 5th ed. (Waterloo, Canada: WATCOM Publications, 1988).

[5] B. Char et al., *First Leaves: A Tutorial Introduction to Maple*, 5th ed., (Waterloo, Canada: WATCOM Publications, 1988).

[6] M. A. Clements and J. W. Pease, "On Causal Linear Phase IIR Digital Filters," *IEEE Trans. on Acoustics, Speech, and Signal Processing*, ASSP-37, no. 4 (April 1989), 479–84.

[7] R. V. Churchill, *Operational Mathematics* (New York: McGraw-Hill, 1958).

[8] M. M. Covell, "An Algorithm Design Environment for Signal Processing, RLE Technical Report 549, Cambridge, Mass.: MIT (December 1989).

[9] R. E. Crochiere and L. R. Rabiner, *Multirate Digital Signal Processing* (Englewood Cliffs, N.J.: Prentice Hall, 1983).

[10] D. E. Dudgeon and R. M. Mersereau, *Multidimensional Digital Signal Processing* (Englewood Cliffs, N.J.: Prentice Hall, 1984).

[11] B. L. Evans, J. H. McClellan, and W. B. McClure, "Symbolic Transforms with Applications to Signal Processing," *The Mathematica Journal*, 1, no. 2 (Fall 1990), 70–80.

[12] B. L. Evans, J. H. McClellan, and K. A. West, "*Mathematica* as an Educational Tool for Signal Processing," *IEEE Southeastern Conference* (April 1991), 1162–66.

[13] J. E. Froyd and B. J. Winkel, "A New, Integrated, First-Year Core Curriculum in Engineering, Mathematics, and Science: A Proposal," *Proceedings of the 18th Annual Frontiers in Education Conference* (October 1988), 92–97.

[14] F. J. Harris, "On the Use of Windows for Harmonic Analysis with the DFT," *IEEE Proceedings* (January 1978), 51–83.

[15] M. Karjalainen, "DSP Software Integration by Object-Oriented Programming: A Case Study of QuickSig," *IEEE ASSP Magazine*, 7, no. 2 (April 1990), 21–31.

[16] G. Kopec, "The Representation of Discrete-Time Signals and Systems in Programs" (Ph.D. thesis, MIT, 1980).

[17] G. Kopec, "The Signal Representation Language SRL," *IEEE Trans. on Acoustics, Speech, and Signal Processing,* ASSP-33, no. 4 (August 1985), 921–32.

[18] R. E. Maeder, *Programming in Mathematica* (Redwood City, Calif.: Addison-Wesley, 1989).

[19] E. J. Muth, *Transform Methods* (Englewood Cliffs, N.J.: Prentice Hall, 1978).

[20] C. S. Myers, "Signal Representation for Symbolic and Numeric Processing, RLE Technical Report 521, Cambridge, Mass.: MIT (August 1986).

[21] K. Nayebi, "General Time Domain Analysis and Design of Exactly Reconstructing FIR Analysis/Synthesis Filter Banks," *International Symposium on Circuits and Systems* (May 1990), 2022–25.

[22] F. Oberhettinger and L. Badii, *Tables of Laplace Transforms* (New York: Springer-Verlag, 1973).

[23] A. V. Oppenheim and R. W. Schafer, *Discrete-Time Signal Processing* (Englewood Cliffs, N.J.: Prentice Hall, 1989).

[24] A. V. Oppenheim and A. Willsky, *Signals and Systems* (Englewood Cliffs, N.J.: Prentice Hall, 1983).

[25] R. Pavelle, *Applications of Computer Algebra* (Hingham, Mass.: Kluwer Academic Publishers, 1985).

[26] C. H. Richardson and R.W. Schafer, "Symbolic Manipulation and Analysis of Morphological Expressions," *IEEE International Conference on Acoustics, Speech, and Signal Processing* (April 1990), 2173–76.

[27] A. Riddle, "A Nodal Circuit Analysis Program," *The Mathematica Journal*, 1, no. 1 (Summer 1990), 62–68.

[28] R. Sedgewick, *Algorithms* (Redwood City, Calif.: Addison-Wesley, 1988).

[29] S. S. Skiena, *Implementing Discrete Mathematics: Combinatorics and Graph Theory with Mathematica* (Redwood City, Calif.: Addison-Wesley, 1990).

[30] M. Slaney, "Interactive Signal Processing Documents," *IEEE ASSP Magazine*, 7, no. 2 (April 1990), 8–20.

[31] H. Abelson and G. Sussman, *Structure and Interpretation of Computer Programs* (Cambridge, Mass.: MIT Press, 1985).

[32] D. Brown, H. Porta, and J. Uhl, "Calculus & *Mathematica*: Courseware for the Nineties," *The Mathematica Journal*, 1, no. 1 (Summer 1990), 43–50.

[33] P. P. Vaidyanathan, "The Role of Smith-Form Decomposition of Integer Matrices in Multidimensional Multirate Systems," *IEEE International Conference on Acoustics, Speech, and Signal Processing* (May 1991), 1777–80.

[34] S. Wolfram, *Mathematica* (Redwood City, Calif.: Addison-Wesley, 1988).

[35] Z. Zuhao, "The Square Matrix Rule of the Convolution Integral," *IEEE Transactions on Education* (November 1990), 369–72.

4

THE SYMBOLIC MANIPULATION AND ANALYSIS OF MORPHOLOGICAL ALGORITHMS

• Craig H. Richardson and Ronald W. Schafer
Georgia Institute of Technology

4.1 INTRODUCTION

Mathematical morphology is a set theoretic approach to the representation of images and image processing systems [1–7]. The attractive feature of this approach is its inherent ability to quantitatively describe and analyze geometric structure in images. The types of information that can be analyzed include size or shape descriptions, spatial relationships between objects, and topological properties such as the number of connected components. Because of its suitability for analyzing geometric features, mathematical morphology has become increasingly important in image analysis applications and low-level machine vision. The range of morphological applications includes particle extraction, defect analysis, noise suppression, size/pattern analysis, and structure determination. Many applications of morphological systems are described in [2, 6, 7].

Mathematical morphology differs from traditional signal processing in that the basic operations do not satisfy an additive superposition property and consequently are not linear. Because linear filters are not designed from a structural perspective, they tend to obscure or blur image features. Mathematical morphology addresses

Sec. 4.1   Introduction

this issue through nonlinear superpositions of signals using the operations of union (or maxima) and intersection (or minima). The signals in morphological systems are viewed as sets rather than functions in order to emphasize the union/intersection perspective instead of the addition/subtraction perspective. In fact, the mathematical origins of mathematical morphology include set theory, stereology, topology, integral geometry, and geometric probability [1, 2].

Despite the successful use of morphological systems in many applications, an effective design methodology has not yet emerged. Morphological systems are often generated using trial and error techniques to combine a series of transformations according to their local or individual characteristics. Frequently however, no direct consideration is given for the properties of the complete transformation resulting from the sequence of operations. Consequently, there may be some inherent structure in the expression that can be exploited to reduce the number of calculations in the overall morphological algorithm.

One solution for easing the development and exploration of morphological algorithms is to build a knowledge-based environment that is capable of manipulating and analyzing these algorithms. The foundation of this environment and the essence of the knowledge consists of rules and information about signals and morphological systems, as well as heuristics to help perform difficult manipulation and analysis tasks. The combination of these types of knowledge permits the computer to perform different types of manipulations that traditionally have been done by a human analyst. The types of manipulations that may be performed include the rearrangement of compound expressions and the analysis of signal and system properties. Expression rearrangement involves the generation of input/output equivalent expressions that are simpler or have some other desired property. An example of signal and system property analysis would be the determination of the computational cost incurred in computing a signal, or in finding the region where the signal is known exactly without any edge effects.

Although rewrite systems have long been used in computer science related fields such as symbolic mathematics, theorem proving systems, and expert systems [8–12], it is only recently that these ideas have been applied to signal processing [13–22]. This chapter describes how the idea of expression rewriting and analysis may be applied and extended to morphological signal processing. This discussion includes the requirements for these types of manipulations and presents examples from the MetaMorph environment, which is a Common Lisp implementation of these ideas.

The contents of this chapter include a brief review of the fundamental operations of morphology and the associated notation. Following the introduction to morphological systems, the requirements for representing morphological systems, the types of signal and system properties that are desirable to analyze, and a definition of computational cost will be given. The rearrangement of compound expressions is shown and extended to the idea of system property analysis. A discussion and summary of the capabilities and limitations of the ideas offered here concludes the chapter.

4.2 FUNDAMENTALS OF MORPHOLOGICAL SYSTEM THEORY

This section describes some of the primitive operations from the theory of mathematical morphology. A more detailed treatment of this material may be found in [2]. Morphological transformations are based on the operations of set union and intersection, which, as will be seen, form the basis for dilation and erosion. Typically, morphological transformations are composed of an operation and structuring element pair where the structuring element defines the neighborhood of image pixels that are used in the transformation and the operation determines the interpretation of these pixels. In essence, the structuring element acts as a probe to uncover or analyze structure in an image.

4.2.1 Binary Signals

The definitions of erosion and dilation are easily interpreted from a set theoretic perspective when the signal and structuring element are binary. In this case, a signal having values zero and one can be represented by simply specifying the set of points, X, where the value is one. The set of points where the signal is zero is therefore just the set complement, X^c. Following the convention of Serra [2], the eroded set is defined as the centers of all the shifted structuring elements, B, that fit completely inside the input set, X. Denoting the *erosion* operation with the symbol $\ominus$, this definition is:

$$Y = X \ominus B = \{x \mid B_x \subset X\} = \{x \mid \forall b \in B, x + b \in X\} \quad (4.1)$$

where B_x means "the set B translated by the vector x." This is equivalent to

$$X \ominus B = \bigcap_{b \in B} X_{-b} = \bigcap_{-b \in B} X_b \quad (4.2)$$

and may be informally interpreted as a shrinking operation on the input set X, as shown in Figure 4.1 for a circular structuring element.

Dilation is the dual operation to erosion. The dual, $\Psi^*(\cdot)$, of an operation $\Psi(\cdot)$ is the operation applied to the complement of the input image and then the result is complemented. This is defined as:

$$\Psi^*(X) = [\Psi(X^c)]^c \quad (4.3)$$

where $[\cdot]^c$ denotes set complement. Duality refers to the relationship that exists between the foreground and the background of an image. Intuitively, for the erosion operation, shrinking the foreground of an image is the same as expanding the background.

Using duality, the dilated set may be defined as the centers of all the shifted structuring elements that just intersect the input set, X. Denoting the *dilation* with the symbol $\oplus$, this definition becomes:

$$X \oplus B = \left[\bigcap_{b \in B} (X)^c_{-b} \right]^c = \bigcup_{b \in B} X_{-b} \quad (4.4)$$

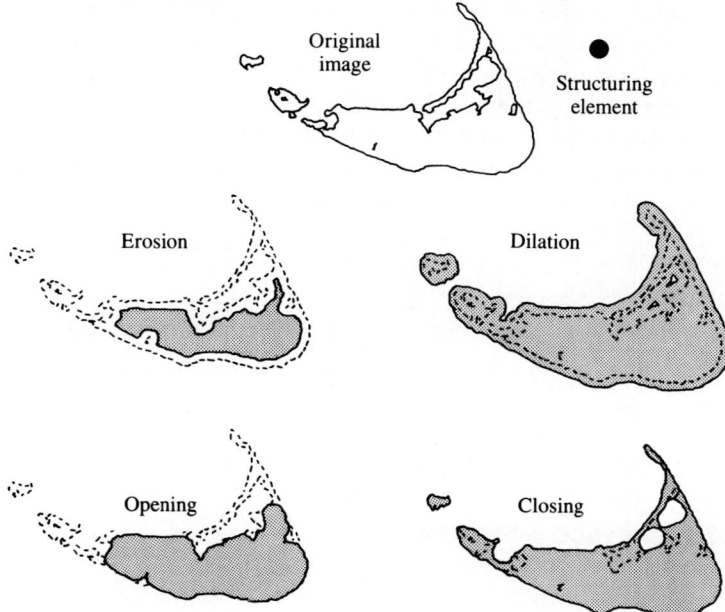

Figure 4.1 Binary erosion, dilation, opening, and closing of X with a circular structuring element B. The shaded areas correspond to the interior of the transformed images. The dark solid lines are the boundaries of the transformed images. The dotted boundary of the original image has been superimposed to show the effects of the transformations.

Figure 4.1 gives an example of the dilation operation using a circular structuring element.

Two closely related operations to erosion and dilation are Minkowski subtraction ($\ominus$) and Minkowski addition ($\oplus$). These two operations are the same as the erosion and dilation, respectively, except for a transposition of the structuring element (or a rotation of 180 degrees in two dimensions). When the structuring element is symmetric, there is no distinction between the Minkowski subtraction (resp. addition) and erosion (resp. dilation). The definitions for these operations are:

$$X \ominus B = \bigcap_{b \in B} X_b \qquad (4.5)$$

$$X \oplus B = \bigcup_{b \in B} X_b \qquad (4.6)$$

It is easily shown that Minkowski subtraction and Minkowski addition are dual operations just as erosion and dilation are. By defining the set transposition operator as $\check{B} = \{-b \mid b \in B\}$, the Minkowski subtraction (resp. addition) and erosion (resp. dilation) are related as:

$$X \ominus B = X \ominus \check{B} \qquad (4.7)$$

$$X \oplus B = X \oplus \check{B} \qquad (4.8)$$

Although the erosion and dilation operations are duals, this does not imply that erosion and dilation are inverses of each other. Once a set X is eroded by B it is not possible, in general, to recover X by dilating the eroded set. This dilated set returns a simpler subset of the original set X that is called an opening. The opening performs a smoothing type of operation on the foreground. To be more precise, the *opening* ($\circ$) is defined as an erosion followed by Minkowski addition, and has the form

$$X \circ B = (X \ominus B) \oplus B = (X \ominus \check{B}) \oplus B \qquad (4.9)$$

Exploiting duality again, the *closing* ($\bullet$) may be defined as a dilation followed by a Minkowski subtraction and has the form:

$$X \bullet B = (X \oplus B) \ominus B = (X \oplus \check{B}) \ominus B \qquad (4.10)$$

Figure 4.1 illustrates the difference between opening and closing operations. The opening removes objects smaller than the structuring element and the closing fills in holes smaller than the structuring element. It is readily seen that the order of application of the erosion and dilation operators is significant.[1]

4.2.2 Greyscale Signals

The above operations have been extended to greyscale signals by many researchers [2, 4, 5]. The most common formulation consists of a greyscale signal, f, and a binary structuring element, B. Using a threshold decomposition approach, eqs. (4.2) and (4.4) lead to the expressions:

$$(f \ominus B)(x) = \min_{b \in B} f(x + b) = \min_{y \in B_x} f(y) \qquad (4.11)$$

$$(f \oplus B)(x) = \max_{b \in B} f(x + b) = \max_{y \in B_x} f(y) \qquad (4.12)$$

These are respectively the pointwise minima and maxima of the signal over the region of support of the translated structuring element. Figure 4.2 shows an example of the erosion, dilation, opening, and closing of a greyscale function using a 5-point symmetric (binary) line.

Although the most general formulation of erosion and dilation occurs when both the signal and structuring elements are greyscale functions, this case is not too common because the resulting transformation is more sensitive to the illumination characteristics of the image and is more computationally expensive. The general greyscale operations are defined as:

$$(f \ominus g)(x) = \min_{y \in B_x} \{f(y) - g(y - x)\} \qquad (4.13)$$

$$(f \oplus g)(x) = \max_{y \in B_x} \{f(y) + g(y - x)\} \qquad (4.14)$$

where g is a greyscale structuring element with region of support given by B.

---

[1] Recently there have been some differences in notation and definitions among researchers. Although we have aligned ourselves with the notation given by Serra [2, 6], others [4, 5, 7] have defined their dilation as our Minkowski addition, and consequently their opening as an erosion followed by a dilation and closing as a dilation followed by an erosion. This difference in definition is only significant when the structuring element is not symmetric.

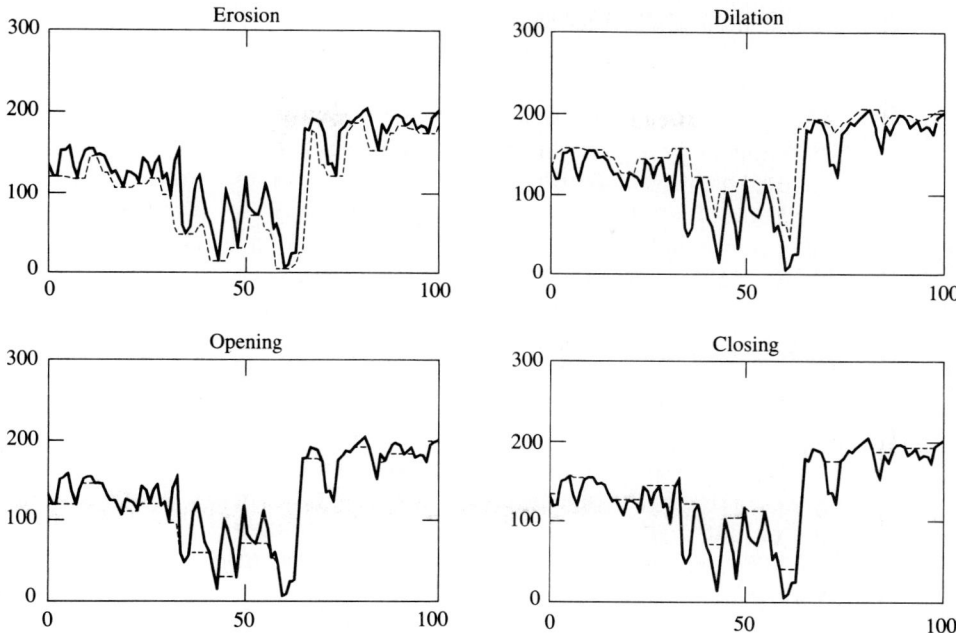

Figure 4.2 Greyscale erosion, dilation, opening, and closing using a 5-point symmetric, centered line structuring element. The solid line corresponds to the input function while the dashed line is the resulting output function.

It is interesting to note that erosion and dilation are composed of unions and intersections in the binary case, and maxima and minima in the greyscale case. Except for the most general form in eqs. (4.13) and (4.14), there are no additions or subtractions and there are no multiplications in any of the expressions.

Although only the fundamental operations have been presented, more complex operations are defined in terms of these basic operations. Examples include a morphological peak/valley extractor which is the difference of an image and its opening [2, 23, 24], morphological edge detection which is a sequence of differences between an image and its erosions and dilations [25], and a morphological skeleton which is a sequence of differences between erosions and their openings [2, 26, 27]. In addition to these examples, Maragos has discovered a rich representation theory that relates morphological systems to linear system theory and rank order filtering. He has shown how many familiar operations (such as median filtering) can be implemented as unions of erosions or intersections of dilations [3, 28–30].

To remain consistent with Serra [2] the more common notation of $\oplus$ and $\ominus$ [with a transpose of B where necessary as indicated by eqs. (4.7) and (4.8)] will be used instead of the symbols ⓓ and ⓔ. Throughout the rest of this chapter a Lisp-based notation will often be used to represent morphological algorithms. An example of the types of operators and their Lisp notation is given below for the intersection operation:

$$(X \cap Y) \leftrightarrow (\text{SIGNAL-INTERSECT } X Y)$$

4.3 SYMBOLIC MANIPULATIONS

As an aid in the design of morphological systems, our work has focused on the specification and creation of an environment capable of manipulating morphological expressions. The underlying motivation is to generate expressions that have a lower computational cost than the original expressions. Combining the capabilities that are described hereafter into a morphological analysis tool increases productivity and helps optimize image processing solutions. This environment allows one to take advantage of the computer to extend the designer's capacity for doing signal processing and free him/her for more important tasks. The following paragraphs describe the features of the environment.

Reducing Complex Expressions Into Less Complex Forms

Reducing the complexity of an algorithm simplifies the expression by exploiting the interaction of signals and systems. These simplifications do not modify the input/output characteristics of the system, but rather remove redundant computations. They are reductions in the sense of lowering the computational cost of the algorithm.

Generating New Forms from Old Expressions

The generation of new forms from old forms is performed through truth preserving equivalent form transformations. Generating new forms allows the discovery of potentially better ways of organizing the given computation that may lead to more efficient implementations. In addition, by expressing algorithms in different forms, more insight into the algorithm may be found.

Analyzing Signal Properties

Signal property analysis allows one to determine the signal properties of the output of a compound expression by analyzing the properties of the components. As signals propagate through systems, their properties are modified as determined by the characteristics of the system. This type of analysis is very useful in finding the computational cost of an expression.

Finding System Properties of Compound Expressions

System property analysis allows one to precisely describe systems that are built from simpler systems. System property analysis attempts to determine the system properties of compound systems from the properties of their component systems. Once compound systems are created, it is important to understand the properties associated with them since this will make subsequent manipulations easier. As part of the analysis of system properties, it may be necessary to find bounding relationships between signals. Inclusion analysis approaches this problem by attempting to answer whether $\Psi_1(A) \subseteq \Psi_2(B)$ for arbitrary inputs A and B and operations Ψ_1 and Ψ_2.

The following subsections describe some of the requirements for these types of manipulation and analysis tasks.

4.3.1 Manipulation Requirements

In order to perform the manipulation and analysis tasks previously described, many representational issues must be addressed. The fundamental issues include [21, 22]:

- a representation of signals, structuring elements, and their properties,
- a representation of morphological system information,
- a description of the interaction between signals and systems,
- a mechanism to support symbolic inferencing.

The next four subsections will discuss each of these issues in turn.

4.3.2 Signals, Structuring Elements, and Their Properties

Signals and structuring elements are the basic building blocks within the MetaMorph environment. Signals are used as inputs to systems and are produced as outputs. Structuring elements are used to fine tune, or parameterize, morphological transformations. For the purposes of manipulation and analysis, we have chosen to represent signals from an abstract view as a collection of property values. Included are such properties as the type of symmetry or the number of points in the domain. A complete list of signal properties is given in Table 4.1. This abstract view allows the representation of entire classes of numeric signals that have a set of properties—for instance all real signals, or all signals with a certain region of support. By not storing the underlying sample values, these signals do not represent any particular signal but rather represent all signals that have those properties. Signals that are represented this way are called abstract signals [15, 17].

Figure 4.3 shows how an abstract signal, X, would be defined in our implementation of the MetaMorph environment. X is defined to be an abstract signal that is greyscale and has a square region of support of size 256×256. The object `# < Signal: DISCRETE-TIME-SEQUENC X >` is returned and may be used in subsequent manipulations.[2]

---

[2] Within the examples, user input is distinguished from the system's response through the following convention: user input will be prefaced by a prompt of the form `MORPH:` while the output from the system will be prefaced by `---->`.

TABLE 4.1 EXAMPLES OF SIGNAL PROPERTIES THAT ARE USED IN MATHEMATICAL MORPHOLOGY

Property Name	Description
closed-wrt	Signals that the defined signal is morphologically closed with respect to.
compact	This indicates whether the signal is compact.
convex	This indicates whether the signal is convex.
dimension	The dimension of the defined signal.
end	The end of the support of the signal.
greyscale	This indicates whether the signal is a grey level signal or a binary set.
include-origin	This indicates whether the origin is included within the support of the signal.
local-knowledge	The region where the transformation is known exactly without edge effects.
number-points	The number of points in the signal.
open-wrt	Signals that the defined signal is morphologically open with respect to.
range	The allowable signal amplitude values.
signal-type	The value of the numerical type of the signal (real, complex, ...).
start	The start of the support of the signal.
support	The region of support of the signal.
symmetry	The type of symmetry that the signal possesses.

Structuring elements may be represented in the same way as signals. The properties are encapsulated within an abstract structuring element that permits entire classes of structuring elements to be represented. Figure 4.4 defines B, an abstract structuring element that is symmetric, has six points, is morphologically open with respect to LIN000, and includes the origin.

Structuring elements have the same properties as signals with the additional property of decomposability. This added property is analogous to the separability of linear filter kernels. A structuring element, S, is decomposable if it can be represented as the Minkowski addition of several sets: $S = J_1 \oplus J_2 \oplus \cdots \oplus J_M$ for some collection of sets $\{J_i\}$ [2, 31]. Signals do not require the decomposability property since, because of their generality, real signals cannot normally be decomposed.

As will be shown in section 4.5, the decomposition of structuring elements allows for a more efficient implementation of morphological operations since there are usually fewer points in the collection of $\{J_i\}$ than in the original set S [2, 31, 32]. Within the manipulation environment, a structuring element is classified as decomposable if a decomposition has been found from previous manipulations, or if the structuring element has been specifically given the decomposable property.

```
MORPH: (define x (a-member-of 'discrete-time-sequence
                               :signal-type :real
                               :greyscale   t
                               :start       (0 0)
                               :end         (255 255)))
----># < Signal: DISCRETE-TIME-SEQUENCE X >
```

Figure 4.3 Defining an abstract signal, X.

Sec. 4.3   Symbolic Manipulations

```
MORPH: (define b (a-member-of 'structel-class
                               :symmetry          :symmetric
                               :number-points     6
                               :open-wrt          lin000
                               :include-origin    t))
---->#⟨STRUCTEL: B⟩
```

Figure 4.4   Defining an abstract structuring element, B.

Because of the significance of the geometric characteristics of structuring elements, another level of representation is permitted. In addition to specifying structuring elements through a collection of properties, the user may want to specify the elements of the structuring element exactly. Examples of some commonly used structuring elements are shown in Figure 4.5, while Figure 4.6 shows the decomposition of a 21-point CIRCLE structuring element Figure 4.5 in terms of four three-element components such that CIRCLE = $B_1 \oplus B_2 \oplus B_3 \oplus B_4$. Large structuring elements can be formed from small structuring elements by successively growing the structuring element, i.e., a structuring element B scaled by n is defined as:

$$n \cdot B = \underbrace{B \oplus B \oplus \cdots \oplus B}_{n-1 \text{ Minkowski additions}}$$

where n is the size of the scaled structuring element.

In addition to the structuring elements in Figure 4.5, the morphological algorithm designer may want to use arbitrary-shaped structuring elements for particular applications. The mixed use of specific structuring elements that are defined by their points and abstract structuring elements that are defined by their properties

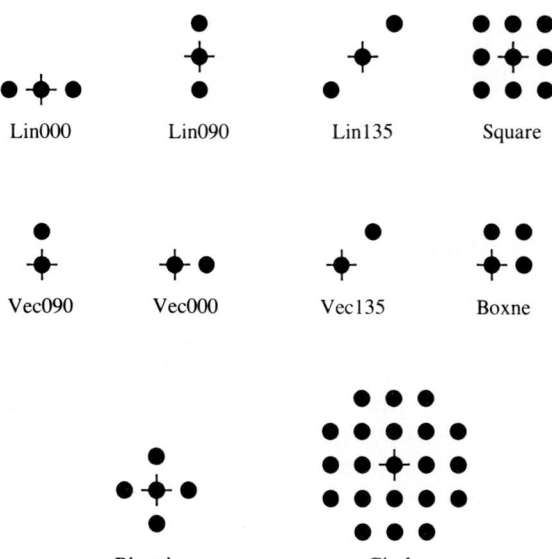

Figure 4.5   Common structuring elements that are used in morphology.

```
MORPH: (decompose circle :terms 3)
The decomposition of CIRCLE is :
    ((0  0) (-1  1) (-1  0))
    ((0  0) ( 0 -1) (-1 -1))
    ((0  0) (-1  1) ( 0  1))
    ((0  0) (-1 -1) (-1  0))
```

Figure 4.6 The decomposition of the CIRCLE into four three-element components.

allows the designer to specify as much detail about the structuring elements as desired (or available) to guide the manipulation process. Although abstract structuring elements have been used in the MetaMorph environment for initial prototyping, a uniform inquiry mechanism ensures that the two types of structuring elements are externally viewed in the same way and can be used together.

4.3.3 Morphological System Information

The system information that is required for symbolic manipulations consists of two primary components: system properties and system specification. System properties (such as those in Table 4.2) are a collection of facts that describe individual systems. As an example of the utility of system properties, if the MetaMorph environment knows that an operator is idempotent, then unnecessary applications of this operator may be removed. A further reason for explicitly representing system properties is to permit information that applies to a variety of systems (such as all anti-extensive systems) to be defined.

Although system properties represent facts, some descriptive information is also stored with them in the MetaMorph environment. This information includes the functional definition of the property and the interaction of the property with other system properties. Figure 4.7 shows how this information is specified with the definitions of the idempotent and anti-extensive system properties. The functional definition gives the mathematical form that defines the system property, such as $\Psi(X) \subset X$ for the anti-extensive property. The interaction of between system properties may be described through a hierarchy such as the one in Figure 4.8. This figure shows the relationships associated with the idempotent property and its interaction with the suppotent and subpotent properties. From the figure, idempotency is implied by being both suppotent *and* subpotent, and prohibits the properties of strict extensivity and strict anti-extensivity.

The specification of systems allows a description of how systems are defined in terms of other systems, the interaction of signal and system properties, and rules for manipulating the expressions that the system generates. Since it is not feasible

Sec. 4.3 Symbolic Manipulations

TABLE 4.2 MORPHOLOGICAL SYSTEM PROPERTIES

X, X_1, X_2 are input signals and B, B_1, and B_2 are structuring elements. Within the property name the phrase *wrt* means "with respect to," *se* means "structuring element," and x refers to the "input signal." Using this convention, the property *increasing-wrt-x* indicates that the system is increasing with respect to the input signal.

Property Name	Description
additive	$\Psi(X + Y) = \Psi(X) + \Psi(Y)$
anti-extensive	$\Psi(X) \subseteq X$
associative	$\Psi(X, \Psi(Y, Z)) = \Psi(\Psi(X, Y), Z)$
commutative	$\Psi(X, Y) = \Psi(Y, X)$
commute-with-thresh	$[\psi(X)]\|_a = \Psi(X\|_a)$
decreasing-wrt-se	$B_1 \subseteq B_2 \Rightarrow \Psi_{B_2}(X) \subseteq \Psi_{B_1}(X) \,\forall X$
decreasing-wrt-x	$X_1 \subseteq X_2 \Rightarrow \Psi_B(X_2) \subseteq \Psi_b(X_1) \,\forall B$
extensive	$X \subseteq \Psi(X)$
linear	$\Psi(a_1 X + a_2 Y) = a_1 \Psi(X) + a_2 \Psi(Y)$
local knowledge	$\forall Z_1, \exists \text{ bounded } Z : [\Psi(X \cap Z)] \cap Z_1 = \Psi(X) \cap Z_1$
lower semicontinuous	$\{X_i\} \rightarrow X \Rightarrow \Psi(X) \subset \underline{\lim} \Psi(X_i)$
idempotent	$\Psi(\Psi(X)) = \Psi(X)$
increasing-wrt-se	$B_1 \subseteq B_2 \Rightarrow \Psi_{B_1}(X) \subseteq \Psi_{B_2}(X) \,\forall X$
increasing-wrt-x	$X_1 \subseteq X_2 \Rightarrow \Psi_B(X_1) \subseteq \Psi_B(X_2) \,\forall B$
shift-invariant	$\Psi(X_t) = (\Psi(X))_t$
upper semicontinuous	$\{X_i\} \rightarrow X \Rightarrow \Psi(X) \overline{\lim} \Psi(X_i)$

to predict all potentially useful transformations, it must be possible to define composite systems from existing systems.

Figure 4.9 shows the definition of the intersection system which accepts an arbitrary number of input signals. The intersection has the anti-extensive, increasing-wrt-x, translation-invariant, usc, commutative, and associative system properties. The interaction of the system with signals is given in the `:signal-property-rules` portion of the specification. In addition, a fetch method is provided in order to compute the value of output signals that are defined in terms of numerically valued input signals. The `:simplification-rules` part of the specification defines simplification rules that can be used to symbolically simplify the expression. Other types of rules such as equivalent form and inclusion rules can be defined in the same way. The `(:>system)` notation is used in the pattern matching process to allow the matching variables to be bound to the actual symbols in the input expressions.

There are presently about 200 rules used to describe more than 50 systems within the MetaMorph environment. More information about the use of these rules is given in section 4.4.

4.3.4 Symbolic Inferencing Mechanism

The core of any manipulation environment must consist of relationships that enable symbolic manipulations. These relationships can be expressed in several different ways including rules, semantic networks, frames, and procedural forms [33–36]. For

```
(define-system-property idempotent
   ()
   "Successive applications of the transformation (after the first)
   do not further modify the output."
   (equal (system x) (system (system x)) )
   ( :example-system     'opening
     :exclude-property   (strict-extensive strict-anti-extensive)
     :implied-by-property (and subpotent suppotent)
     :implies-property        (extensive anti-extensive)
     :simplification-rules
     (
       (:name       Idempotent-system
        :form       ((:>system) ((:<system) (:@ signals)))
        :test       t
        :result     ((:<system) (:@ signals))
       )
     )
   )
)

(define-system-property anti-extensive
 (generalized-anti-extensive subpotent)
 "The output has less (or equal) non-zero extent than the input."
 (subset (system x) x)
 (:example-system 'opening
  :exclude-property (strict-extensive)
  :inclusion-rules
   (
     (:name       Anti-extensive-inclusion
      :form       ((:>system) (:>signal))
      :test       t
      :result     (:<signal)
     )
   )
  )
 )
)
```

Figure 4.7 The introduction of the *idempotent* and *anti-extensive* system properties into the MetaMorph system. The definition of the property allows it to be used as a descriptive attribute of systems.

the manipulation and analysis of morphological expressions, a rule representation has been chosen. This choice was motivated by the uniformity and modularity of the rule structure and the separation of the control structure from the knowledge. For both organizational purposes and execution efficiency, the collection of rules has been partitioned into rule bases. The rule bases contain relationships related to specific types of manipulations. There are rule bases for simplifications, equivalent

Sec. 4.3   Symbolic Manipulations

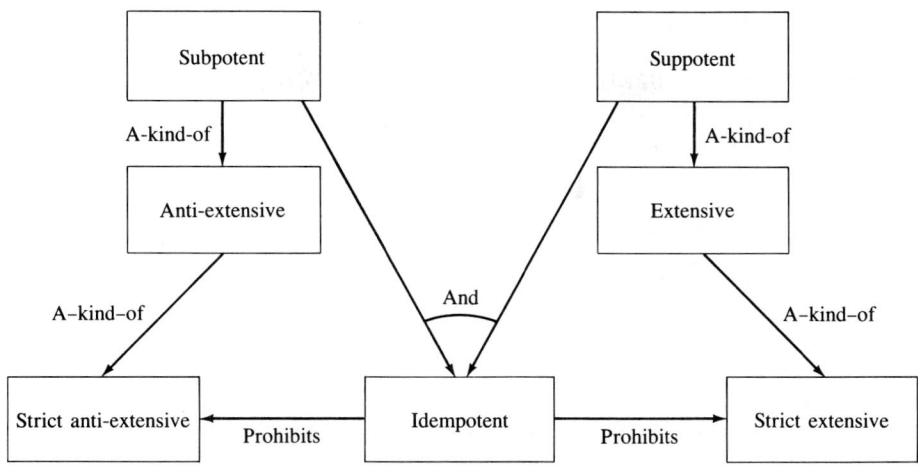

Figure 4.8 The relationships that exist between system properties.

forms, inclusion forms, and for the analysis of system/signal property interaction. Although the rule bases are physically separate, they are used together. For instance, signal property analysis often occurs transparently during expression rewriting to provide information to the inference process.

The use of a rule representation requires a pattern matching scheme and inference and control mechanisms. The inference system, or inference engine, is what guides the problem solving approach. Problems are usually solved by working forward from the known facts or information (forward chaining), working backward from known goals (backward chaining), or some combination of both forward and backward chaining (means–end analysis). The control structure is responsible for conflict resolution when multiple rules are applicable at a particular stage of the inferencing process. Because of the exactness of the signal processing relationships in the expression analysis and rewriting, it has not been necessary to use inexact reasoning methods.

4.3.5 The Interaction Between Signals and Systems

The analysis of signal properties is concerned with finding the effect that systems have on the input signal properties (such as those properties in Table 4.1). As signals propagate through systems, their properties are modified. In order to analyze these modifications it is necessary to describe how signals and systems interact. One way of expressing this information is to associate a signal property rule with each type of system, either as part of the system object or as a separate entity.

As an example of describing the signal and system interaction, the signal property propagation characteristics of the intersection system are shown in Figure

```
(define-system signal-intersect (&rest signals)
 (anti-extensive increasing-wrt-x translation-invariant usc
  commutative associative)
 ''The intersection of signals''
 ((type-of (first signals)))
 ()
 (
 :signal-property-rules
 (
  :atomic-type     (signal-type (first signals))
  :convex          (some #'signal-convex signals)
  :dimension       (apply#'$min (mapcar #'signal-dimension signals))
  :end             (signal-end (signal-support self))
  :greyscale       (some #'signal-greyscale-p signals)
  :number-points   (apply #'$max (mapcar #'signal-number-points signals))
  :start           (signal-end (signal-support self))
  :support         (apply #'region-intersect (mapcar #'signal-support signals))
  :structel        (every #'signal-structel-p signals)
  )
  :fetch          ((&rest indices)
                   (apply #'$min (mapcar #'(lambda (a-signal)
                                             (apply #'signal-fetch
                                                    (cons a-signal indices)))
                                          signals)))

:simplification-rules
(
(:name Intersect-with-complement
       :form    (signal-inersect (:>sig1) (signal-complement (:<sig 1)))
       :test    t
       :result  (null-space (signal-dimension (:< sig1)))
)
(:name Intersect-with-self
       :form (signal-intersect (:> sig1) (:<sig1))
       :test t
       :result (:<sig1)
    )
              ...more rules...
 )
 )
 )
```

Figure 4.9 The definition of the intersection system.

Sec. 4.4   Expression Rewriting   157

4.9. Within the system definition of Figure 4.9, the variable *signals* is bound to the list of signals that are passed as arguments to the intersection system. Given an expression to analyze, such as (SIGNAL-INTERSECT SIGNAL-1 SIGNAL-2), the variable SIGNALS would be associated with the list (SIGNAL-1 SIGNAL-2). As an example, determining whether the output is a greyscale signal would cause an inquiry into whether either SIGNAL-1 or SIGNAL-2 were greyscale. When SIGNAL-1 and SIGNAL-2 are signal objects, the desired property will be readily available by sending an inquiry to the signal object. If SIGNAL-1 is a compound expression consisting of other morphological signals, the inquiries would continue requesting the property from these subexpressions recursively until the information was found either from previous manipulations or directly from signal objects found at the end of the recursive inquiry path.

4.4 EXPRESSION REWRITING

Expression rewriting provides a means for modifying the structure of an expression without modifying the overall effective transformation. Examples of this include the generation of equivalent forms, expression simplifications, and dual forms.

4.4.1 Generation of Equivalent Forms

Equivalent forms imply other structures for representing a given expression, some of which may be more computationally efficient than the original structure. For instance, $X \oplus (B \oplus B) = (X \oplus B) \oplus B$, is an example of an equivalent form. This is the familiar definition of the Minkowski addition with a scaled structuring element, B, of size 2. Table 4.3 lists some equivalent form rules.

Equivalent forms may be generated by recursively going through an expression looking for all equivalent forms of the subexpressions. As new forms are generated,

TABLE 4.3 EXAMPLES OF MORPHOLOGICAL EQUIVALENT FORM RULES THAT ARE ALWAYS APPLICABLE

Rule Name	Rule Equation
dilation-of-se-union	$A \oplus (B_1 \cup B_2) = (A \oplus B_1) \cup (A \oplus B_2)$
dilation-of-union-input	$(A_1 \cup A_2) \oplus B = (A_1 \oplus B) \cup (A_2 \oplus B)$
dilation-chain-rule	$A \oplus (B_1 \oplus B_2) = (A \oplus B_1) \oplus B_2$
erosion-of-se-union	$A \ominus (B_1 \cup B_2) = (A \ominus B_1) \cap (A \ominus B_2)$
erosion-chain-rule	$A \ominus (B_1 \oplus B_2) = (A \ominus B_1) \ominus B_2$
erosion-within-mask	$(A \cap Z) \ominus B = (A \ominus B) \cap (Z \ominus B)$
expand-open	$A \circ B = (A \ominus \breve{B}) \oplus B$

they are used as seeds for new equivalent forms. This approach will generate all equivalent forms for an expression that are possible using the available rules.[3]

Unfortunately, the number of forms that can be potentially generated is described by a combinatoric function of the number of subexpressions. Thus, even simple expressions have a large number of potential forms, and because of the large number of permutations, most of these forms are not interesting. Covell introduced a heuristic for preventing this explosion that not only substantially reduced the potential number of generated forms, but did it in a way that corresponds to the way a human would search for equivalent forms [17, 18].

Figure 4.10 shows the definition of the CLOS-OPEN system which has been used in edge detection algorithms. By defining the CLOS-OPEN in terms of the familiar OPENING and CLOSING operations, MetaMorph generated more than 60 equivalent forms. Using the three-point decomposition that was described in Figure 4.6 and labeling the components as DECOMP-CIRCLE-3-1 through DECOMP-CIRCLE-3-4 results in the third form.

4.4.2 Generation of Simplifications

Simplifications are the reduction of complicated expressions into simpler equivalent expressions. As an example, the expression $(B \oplus C) \circ B \Rightarrow B \oplus C$, states that the Minkowski addition of the structuring elements B and C is open with respect to B. Simplifications are equivalent forms that are always simpler (i.e., have lower computational cost) than the original expression [15]. Table 4.4 lists some of the simpli-

```
    MORPH: (define-composite-system clos-open
                          :args (x b)
                          :form (opening (closing x b) b))
        ----> CLOS-OPEN
    MORPH: (all-equivalent-forms (clos-open x circle))
      1. (OPENING (CLOSING X CIRCLE) CIRCLE)
      2. (DILATE (ERODE (CLOSING X CIRCLE) CIRCLE) CIRCLE)
      3. (OPENING (CLOSING X DECOMP-CIRCLE-3-1 DECOMP-CIRCLE-3-2
                             DECOMP-CIRCLE-3-3 DECOMP-CIRCLE-3-4)
                  DECOMP-CIRCLE-3-1 DECOMP-CIRCLE-3-2
                  DECOMP-CIRCLE-3-3 DECOMP-CIRCLE-3-4)
                                  :
     25. (MINKADD (ERODE (MINKADD X CIRCLE)
                         (HOMOTHETIC CIRCLE 2)) CIRCLE)
                                  :
     60. (OPENING (ERODE (MINKADD CIRCLE X) CIRCLE) CIRCLE)
```

Figure 4.10 Equivalent form generation for the CLOS-OPEN system.

---

[3] One must be careful to avoid creating infinite loops of manipulations.

Sec. 4.4 Expression Rewriting

TABLE 4.4 EXAMPLES OF MORPHOLOGICAL SIMPLIFICATION RULES AND THE CONDITIONS FOR THEIR APPLICABILITY (T MEANS ALWAYS TRUE)

Rule Name	Rule Equation	Conditional
comp-of-comp	$(X^c)^c \Rightarrow X$	T
dilate-with-null	$A \oplus \emptyset \Rightarrow \emptyset$	T
hit-or-miss	$(X \ominus B) \cap (X \oplus \check{C})^c \Rightarrow \emptyset$	$B \cap C \neq \emptyset$
intersect-with-self	$A \cap A \Rightarrow A$	T
intersect-with-comp	$A \cap A^c \Rightarrow \emptyset$	T
open-idempotence	$(A \circ B_1) \circ B_2 \Rightarrow A \circ B_1$	$B_1 \circ B_2 = B_1$
open-idempotence	$(A \circ B_1) \circ B_2 \Rightarrow A \circ B_2$	$B_2 \circ B_1 = B_2$

fication rules that are built into the MetaMorph environment. Many of the simplifications are used to prevent infinite expansions of equivalent forms such as $(\cdots ((X^c)^c) \cdots)^c$.

Simplifications are easier to generate than equivalent forms because the inference process is less complex. By their definition, simplifications are always useful to apply, which means that one need only apply any rule that is applicable and then continue to simplify all the components of the expression.

Having defined signal objects in Figures 4.3 and 4.4, it is possible to simplify an expression defined in terms of these signals. The simplification example in Figure 4.11 shows that the MetaMorph system finds that since B is open with respect to Lin000, it is also open with respect to the predefined structuring element Vec000 (see Figure 4.5). The simplification process recursively goes through the expression looking for rules to apply at all levels of the expression. If a rule is applied at any level within the expression, the process will start again at the top level with the reduced expression. The process ends when there are no more applicable rules.

```
MORPH: (simplify (signal-intersect (opening (Dilate x b) vec000) x))
  Trying to simplify (SIGNAL-INTERSECT X (OPENING (DILATE X B) VEC000))

    Trying to simplify (OPENING (DILATE X B) VEC000)
    Applying SIMPLIFICATION rule OPEN-IDEMPOTENCE to
    (OPENING (DILATE X B) VEC000)

      Trying to simplify (DILATE X B)

    Trying to simplify (SIGNAL-INTERSECT (DILATE X B) X)

    Applying SIMPLIFICATION rule INTERSECTION-WITH-EXTENSIVE-OP to
    (SIGNAL-INTERSECT (DILATE X B) X)
----> X
```

Figure 4.11 An example of expression simplification.

4.4.3 The Generation of Dual Forms

Dual form generation is another type of expression rewriting that is concerned with determining the effect of a transformation on the background of a signal [2]. This type of manipulation is useful for understanding the dual nature of morphological algorithms and for gaining insight into how signals and systems interact. The generation of dual forms can be easily constructed by using the definition introduced in eq. (4.3) and simplifying the result.

4.5 AN EXAMPLE OF SIGNAL PROPERTY ANALYSIS: COMPUTATIONAL COSTS

In addition to the signal properties in Table 4.1, the computational cost is perhaps the most important signal property when it comes to comparing different forms of an expression. Therefore, the main subject of this section is the specification of *what* types of costs should be represented and *how* these costs should be represented and combined. It should be noted that the cost definitions described here are not meant as absolute complexity measures, but rather as relative measures to allow a computational complexity comparison among different algorithms. An absolute measure would be tied closely to the underlying architecture and the ingenuity of the algorithm implementor. As shown by the Abingdon Cross test for comparing image processing architectures [37], an absolute measure of performance is difficult to determine.

The selection of the types of costs to represent has been influenced by commonly used cost measures such as the number of additions and multiplies. Generalizing this to the study of morphological algorithms, we include the number of maxima, minima, unions, and intersections. In addition to these costs, an elementary estimate of the number of signal accesses has been useful. To accommodate the varied types of cost measures, a cost vector has been chosen (similar to Myers [15] and Covell [17]) to represent the total algorithmic cost.

Because of the data-dependent iterative nature of some algorithms, it is not always possible to generate an absolute measure of the computational cost. This has motivated the representation of the cost in terms of two components of cost measures, one for the fixed (noniterative) portion and another for the iterative portion of the algorithm. By parameterizing the iteration cost with a variable representing the number of iterations, these two costs can be combined into a single vector. Figure 4.12 shows a simple example of combining the cost components for the cascade of two iterative systems.

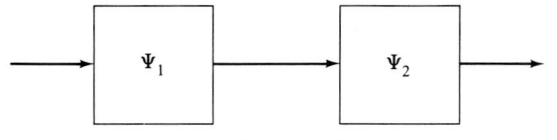

$$\text{Cost}(N_1, N_2) = (N_1 \Psi_1^{iterative} + \Psi_1^{fixed}) + (N_2 \Psi_2^{iterative} + \Psi_2^{fixed})$$

Figure 4.12 The cascade of two iterative systems, Ψ_1 and Ψ_2.

4.5.1 Computational Cost Definitions

Although the methods for evaluating the computational cost depend on the particular architecture upon which the algorithm will be implemented, the current research effort has concentrated on representing the costs for a general digital computer with several variants, as described below. Future work may emphasize specialized architectures such as Connection machines and the Cytocomputer [38].

Representing Alternate Architectures

By modularizing the representation and aggregation of the computational costs associated with an algorithm, the possibility exists for representing some of the features of different types of architectures. A generalization of the costs for a general-purpose architecture allows the way that costs are determined and accumulated to vary. In particular, the maximum number of inputs to the arithmetic, multiplier, and logic units are allowed to vary from a minimum of two to a user specified maximum. The computational cost of an expression will generally be a decreasing function of the maximum number of inputs allowed to the processing units. In addition to varying the number of inputs, a figure of merit may be calculated by mapping the cost vector with a user defined emphasis, or weighting, function to give a greater weight to a given architectural feature. For instance, in optical implementations, a large weighting factor may be associated with maxima but not with binary union. In general, the most efficient form of an expression will depend on the underlying architecture.

Addition and Multiplication Cost

The addition cost of an algorithm represents an estimate of the number of additions and subtractions associated with the computation of the algorithm. As with the cost descriptions to follow, the addition cost depends on the number of inputs that are compatible with the arithmetic unit. For the simplest case of two input signals, f and g, defined as the functions $f: D_1 \rightarrow R_1$ and $g: D_2 \rightarrow R_2$, with possibly different numbers of points in their region of supports, $|D_1|$ and $|D_2|$, the addition cost may be defined as:[4]

$$\text{ADDITION-COST } (f + g) = \max\{|D_1|, |D_2|\} \qquad (4.15)$$

The characteristics of the machine architecture become important when there are more than two signals being accumulated. For example, consider the calculation of the sum, $f_1 + f_2 + f_3$. If we assume that these signals have been sorted according to the number of points in their region of support, then $|D_1|$, $|D_2|$, and $|D_3|$ form a nondecreasing sequence. Since the addition cost has been defined as a function of the maximum number of points in the region of support, the sorting allows one to

---

[4] For an $N \times N$ image, $|X|$ would be N^2.

TABLE 4.5 COST DEFINITIONS

Expression Form	2-Input System
MULTIPLY–COST $(f_1 \times f_2)$	$\max\{|D_1|, |D_2|\}$
MAXIMA–COST $(f_1 \vee f_2)$	$\max\{|D_1|, |D_2|\}$
MINIMA–COST $(f_1 \wedge f_2)$	$\max\{|D_1|, |D_2|\}$
UNION–COST $(X_1 \cup X_2)$	$\max\{|X_1|, |X_2|\}$
INTERSECT–COST $(X_1 \cap X_2)$	$\max\{|X_1|, |X_2|\}$

add the signals most efficiently. If the arithmetic unit accepts three inputs, the cost would be given by:

$$\text{ADDITION–COST } (f_1 + f_2 + f_3) = \max\{|D_1|, |D_2|, |D_3|\} = |D_3| \qquad (4.16)$$

If only two inputs to the arithmetic unit were allowed (as is usually true of a general-purpose architecture), the addition would be done in two stages, i.e.

$$\text{ADDITION–COST } (f_1 + f_2 + f_3) = |D_2| + |D_3| \qquad (4.17)$$

The other costs can be computed similarly and are presented in Table 4.5 for the two-input case.

Union and Maxima Cost

As described earlier, the cost of combining some number of signals (or structuring elements) with the union operator depends on the number of inputs that are allowed to the logic processor. For the case of a logic processor that allows only two inputs, the cost of performing the union of two signals is given by:

$$\text{UNION–COST } (X_1 \cup X_2) = \max\{|X_1|, |X_2|\} \qquad (4.18)$$

and the cost of combining N signals is given by:

$$\text{UNION–COST } (X_1 \cup X_2 \cup \cdots \cup X_N) = \sum_{i=2}^{N} |X_i| \qquad (4.19)$$

where (as in the addition example) the $|X_i|$'s form a nondecreasing sequence. For a collection of greyscale signals, the union operator is replaced by a maximum operator and eq. (4.19) (for the two-input logic processor) becomes:

$$\text{MAXIMA–COST } (f_1 \vee f_2 \vee \cdots \vee f_N) = \sum_{i=2}^{N} |D_i| \qquad (4.20)$$

where it is again assumed that the $|D_i|$ form a nondecreasing sequence. For the case where the logic processor will accept M inputs, (4.19) becomes:

$$\text{UNION–COST } (X_1 \cup X_2 \cup \cdots \cup X_N) = \sum_{i=1}^{\lfloor \frac{N-1}{M} \rfloor} |X_{i \times M}| + |X_N| \qquad (4.21)$$

Sec. 4.5 An Example of Signal Property Analysis: Computational Costs

Using the definition of Minkowski addition, $X \oplus B = \bigcup_{b \in B} X_b$, and following the union cost definition, the union cost of the Minkowski addition of a signal X with a structuring element B is given by:

$$\text{UNION-COST } (X \oplus B) = |X|(|B| - 1) \tag{4.22}$$

This is consistent with the Minkowski addition operation defined in eq. (4.6), since there would be $|B| - 1$ translations and unions associated with the operation.

This method of computing the union cost allows the computation of nested expressions within an algorithm. For instance:

$$\text{UNION-COST } ((X \oplus B) \cup (Y \oplus C)) = |X|(|B| - 1) + |Y|(|C| - 1) \\ + \max\{|X \oplus B|, |Y \oplus C|\} \tag{4.23}$$

where the nested union cost of $X \oplus B$ and $Y \oplus C$ respectively generate the $|X|(|B| - 1)$ and $|Y|(|C| - 1)$ components while the remaining term follows from the definition of the union cost in eq. (4.18). For the case where X is a signal and B is a structuring element, the expression $|X \oplus B|$ reduces to $|X|$ (i.e., both the signal and its dilate have the same region of support), and the union cost in (4.23) becomes:

$$\text{UNION-COST } ((X \oplus B) \cup (Y \oplus C)) = |X|(|B| - 1) \\ + |Y|(|C| - 1) + \max\{|X|, |Y|\} \tag{4.24}$$

The above results form the basis for defining the intersection and minima costs. In particular, the cost of combining some number of binary (greyscale) signals with the intersection (minimum) operator is defined as the maximum of the number of points in the region of support of the signals. When both the signal and structuring elements are greyscale functions, the computational cost of these operators consists of both a maxima/minima and an addition component as seen from eqs. (4.13) and (4.14).

Signal Access Cost

Each morphological operation requires a signal to be accessed in order to be operated on. The signal access cost does not attempt to count the number of reads/writes associated with an operation, but rather it tries to measure the number of times a signal must be accessed, or used. The signal access cost for a fundamental operator such as Minkowski addition is defined as:

$$\text{SIGNAL-ACCESS-COST } (X \oplus B) = |X| + |B| \tag{4.25}$$

which indicates that one signal of $|X|$ points and one structuring element of $|B|$ points had to be accessed. By defining the access cost in terms of the number of points in the region of support, a high cost is associated with large signals. The signal access cost associated with the union of two signals is given as:

$$\text{SIGNAL-ACCESS-COST } (X_1 \cup X_2) = |X_1| + |X_2| \tag{4.26}$$

which states that two signals must be loaded into memory.

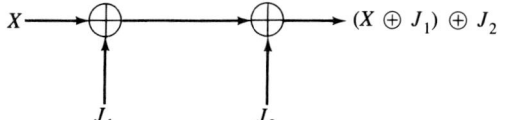

Figure 4.13 Serial implementation of Minkowski addition.

The signal access cost for a serial implementation of Minkowski addition (as shown in Figure 4.13) when B is a composite structuring element, i.e., $B = J_1 \oplus J_2$, may be defined as:

$$\text{SIGNAL-ACCESS-COST}\,((X \oplus B)$$
$$= ((X \oplus J_1) \oplus J_2)) = |X| + |J_1| + |X \oplus J_1| + |J_2| \quad (4.27)$$

If we were to compute this as $X \oplus (J_1 \oplus J_2)$ as shown in Figure 4.14, the signal access cost would become:

$$\text{SIGNAL-ACCESS-COST}\,((X \oplus B) = (X \oplus (J_1 \oplus J_2))) = |X| + |J_1 \oplus J_2| + |J_1| + |J_2| \quad (4.28)$$

Since $(X \oplus B) = X \oplus (J_1 \oplus J_2) = ((X \oplus J_1) \oplus J_2)$, the only difference between these forms is the intermediate signals that are created. In the latter case, fewer signal points were accessed because the intermediate signal, $(J_1 \oplus J_2)$, has fewer points than $(X \oplus J_1)$. Allowing the user to decide whether intermediate signal terms should be counted [such as $|X \oplus J_1|$ in (4.27) and $|J_1 \oplus J_2|$ in (4.28)] permits a coarse representation of both array-based and pipeline-based image processing architectures.

One of the primary motivations for the signal access cost is to differentiate between decomposition implementations such as (4.27) and the direct form of (4.28). The significance of this is that although (4.27) may require fewer union operations than (4.28), the signal access cost may be significantly higher. On a general-purpose architecture, the access cost may be one of the most expensive components of the implementation cost vector.

Figure 4.15 shows that there is a considerable difference in the relative cost of some of the forms generated in Figure 4.10. Because the signal, X, was defined to be greyscale, the morphological operations become maxima and minima over the region of support of the structuring element.

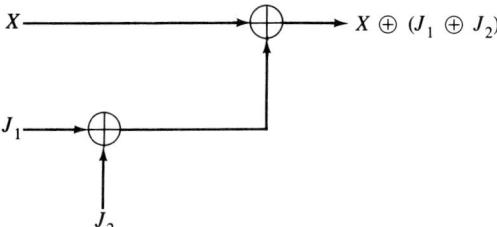

Figure 4.14 Direct implementation of Minkowski addition.

```
MORPH: (signal-cost (clos-open x circle))
----> The cost per output sample is:
            ADD-COST   =  0.00    INTERSECT-COST =  0.00
            MAX-COST   = 40.00    MIN-COST       = 40.00
            MULT-COST  =  0.00    UNION-COST     =  0.00
            ACCESS-COST=  4.0013

MORPH: (signal-cost (minkadd (erode (minkadd x circle)
            (homothetic circle 2)) circle))
----> The cost per output sample is:
            ADD-COST   =  0.00    INTERSECT-COST =  0.00
            MAX-COST   = 40.00    MIN-COST       = 68.00
            MULT-COST  =  0.00    UNION-COST     =  0.00
            ACCESS-COST=  3.0017

MORPH: (signal-cost (opening (closing x decomp-circle-3-1 decomp-circle-3-2
                                     decomp-circle-3-3 decomp-circle-3-4)
                   decomp-circle-3-1 decomp-circle-3-2
                   decomp-circle-3-3 decomp-circle-3-4))

----> The cost per output sample is:
            ADD-COST   =  0.00    INTERSECT-COST =  0.00
            MAX-COST   = 16.00    MIN-COST       = 16.00
            MULT-COST  =  0.00    UNION-COST     =  0.00
            ACCESS-COST= 16.0017
```

Figure 4.15 The cost of three of the CLOS-OPEN equivalent forms.

4.5.2 Cost Weighting

Because of the vector representation of the computational cost associated with an algorithm, it may not always be evident how to compare algorithms whose cost appears roughly equal, but distributed differently within the cost vector. One approach would be to use vector inequalities of the costs where $\vec{a} \leq \vec{b} \Leftrightarrow a_i \leq b_i \, \forall i$. Another approach is to map the cost vector with a user defined weighting function that reflects some particular architectural emphasis. By accumulating the elements of the mapped cost vector (thought of as a generalized inner product), a single figure of merit may be associated with a particular algorithm.

The weighting function provides some flexibility in representing different types of architectures. The reasoning behind this is that the discriminating feature behind image processing architectures (or signal processing architectures) is that some type of performance is usually optimized. Although the way that costs are generated with different architectures may be fundamentally different, the weighting functions provide some measure of control over the computational cost analysis.

4.6 SYSTEM PROPERTY ANALYSIS

The types of manipulation and analysis tasks considered up to this point have been characterized by well-behaved inferencing mechanisms. The only problem has been the combinatorial explosion of uninteresting forms generated during the equivalent form search. The analysis of signal properties, as mentioned in section 4.5, is nearly a linear search through the rule base. This is a consequence of the representation of individual systems and their effects on signals being well matched with a rule representation.

The analysis of system properties is more involved than the analysis of signal properties. To illustrate this, we will concentrate on the properties of idempotency and increasingness with respect to the input. As given in Table 4.2, the idempotency property means that repeated applications of a transformation have no effect after the first application. A system with the increasing property means that the output of the transformation monotonically increases as the input increases in extent.

From a set theoretic standpoint, the determination of these system properties is equivalent to proving that certain relationships hold. Idempotency can be shown using set inclusion relations of the form $\Psi(A) \subset \Psi(\Psi(A))$ and $\Psi(\Psi(A)) \subset \Psi(A)$, which is the definition of set equality. The increasing property may be shown by selecting any sets A and B where $A \subset B$ and showing that $\Psi(A) \subset \Psi(B)$ for the system Ψ.

By defining inclusion analysis as a method for determining whether an expression is included (in a set theoretic sense) within another expression, the application of inclusion analysis may be sufficient for the most general case of idempotency. Table 4.6 lists some typical inclusion relations. Although the use of inclusion analysis has broad applicability, this is just one approach to the system property analysis problem. There are several "shortcuts," or heuristics that may be used to show certain system properties. These shortcuts/heuristics are sufficient conditions that are associated with the systems. Although the sufficient conditions described in section 4.3.3 were conditions in terms of other property values, the shortcuts described here are associated with certain forms of the system. Examples of shortcuts for an anti-extensive system are: if Ψ is a cascade of anti-extensive systems, then Ψ is anti-extensive; if Ψ is the dual of an extensive system, then it is anti-extensive; if Ψ is a union/intersection of anti-extensive systems, then it is anti-extensive. The

TABLE 4.6 EXAMPLES OF MORPHOLOGICAL INCLUSION RULES THAT ARE ALWAYS APPLICABLE

Rule Name	Rule Equation
minkadd-within-mask	$(A \cap Z) \oplus B \subseteq (A \oplus B) \cap (Z \oplus B)$
minkadd-within-se-mask	$A \oplus (B_1 \cap B_2) \subseteq (A \oplus B_1) \cap (A \oplus B_2)$
minksub-within-se-mask	$(A \ominus B_1) \cup (A \ominus B_2) \subseteq A \ominus (B_1 \cap B_2)$
minksub-with-background	$(A \ominus B) \cap (Z \ominus B) \subseteq (A \cup Z) \ominus B$

Sec. 4.7   Conclusions

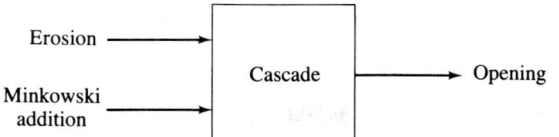

Figure 4.16   The use of the *cascade* meta-system.

name shortcut is used because these techniques allow the direct inclusion analysis effort to be avoided.

Just as systems produce output signals, new systems are generated through combinations of other systems. From a different perspective, the definition of the opening operation is the output of a *meta-system* whose inputs are not signals, but rather systems. Meta-systems may be viewed as objects that take systems for inputs and produce systems as outputs—exactly analogous to how systems take signals as inputs and produce signals as outputs. The meta-system in the opening example is called the *cascade* and is shown in Figure 4.16.

Since it is not only possible to represent signal properties but also how these properties are modified by systems, it is also possible to do this from the system/meta-system perspective. Since systems have observable properties that describe their characteristics, the effect of meta-systems on these properties is known. It has been convenient to associate the shortcuts described above with meta-systems. The most common meta-systems are the cascade (serial form) and summation (parallel form). Within morphological systems, the summation meta-system is most often presented as an intersection or union of systems. A summary of the current system property analysis technique is shown in Figure 4.17.

In Figure 4.17 it is seen that if the system property is part of the system definition, then the property holds, while if the property is excluded because of another property, the property does not hold. If the property is implied by other properties that hold, then the property holds. If the *need* for the property is above a lower threshold, then shortcuts can be found and applied. If none of the shortcuts are applicable and the *need* for the property is above an upper threshold, the system will resort to inclusion analysis using the functional definition of the property. The use of thresholds prevents nested property requests from consuming more resources than the original property request.

4.7 CONCLUSIONS

An approach for performing the symbolic manipulation and analysis of morphological expressions has been described. Morphological system theory differs from linear system theory in that the operations are based on maxima and minima, and do not satisfy an additive superposition property. The requirements for representing and manipulating such systems include:

- representation of system knowledge (facts about particular systems),

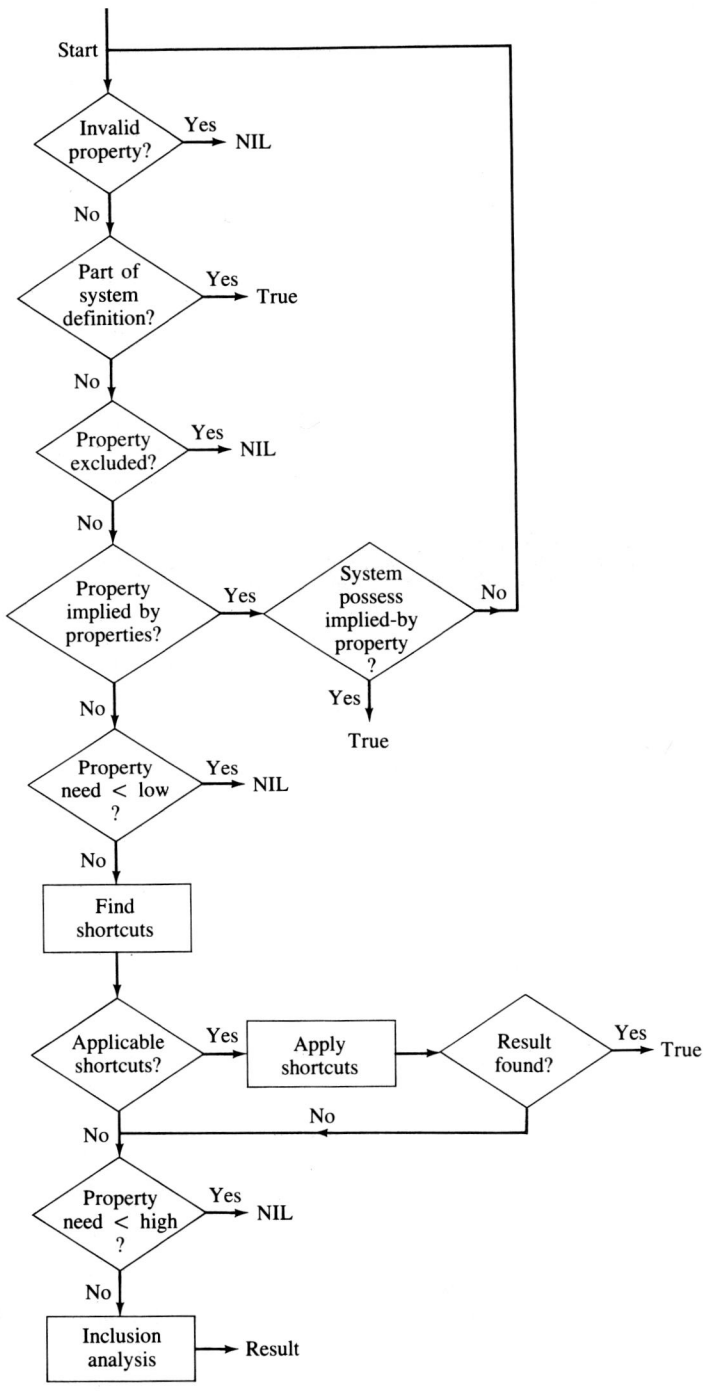

Figure 4.17 Typical steps involved in showing that a system possesses a system property.

Sec. 4.7 Conclusions

- representation of systems, signals, and structuring elements,
- rules to specify the interaction of signals and systems.

The MetaMorph system, a Common Lisp environment designed on these premises, has been introduced. Because of its importance, the representation and definition of the computational cost associated with morphological expressions was presented in detail. This cost can be represented as containing both an iterative component and a fixed component, and should include union, intersection, maxima and minima, addition, and multiplication cost elements. Generalizations have been introduced to allow a comparison of implementation cost on different types of architectures. In particular, the number of inputs to the arithmetic, multiplication, and logic units may be varied from a minimum of two to some user defined maximum, and a weighting function has been defined that may be applied to the cost vector to emphasize certain architectural features. By using the figure of merit together with the generation of equivalent forms, one may produce forms that are targeted for a particular implementation. For instance, assessing a large penalty on maxima and minima costs would attempt to generate implementations using binary unions and intersections.

The idea of introducing a computer at the symbolic stage of the algorithm development cycle is very attractive. By relegating the task of expression analysis and manipulation to the computer, the result is a design tool that is tireless and potentially more thorough in its analysis than we are capable of being.

The present MetaMorph environment has demonstrated that knowledge-based techniques can be used to create an effective analysis and design tool for morphological signal processing systems. However, further research should be done in the areas of knowledge verification, system property analysis, and iterative systems.

With any large aggregation of knowledge, it is necessary to ensure that the system has consistent information. The issues of knowledge validation are ones that we are just beginning to address. Currently, after new information is added to the MetaMorph system, a collection of manipulation and analysis examples that have been designed to test the system are run. The system output is compared to the expected results (which have been manually verified) and any differences are investigated. Unfortunately, as the size of the knowledge base increases, so does the need for more thorough validation and consistency testing.

The use of inclusion analysis with system properties is beginning to take on the appearance of theorem proving systems. More work is needed to see if a formal logic representation coupled with a resolution method (or one of its variants) is necessary to solve complex analysis problems.

The iterative cost component of systems does not provide any information about the convergence characteristics of the system. This makes it impossible to compare two iterative forms—one may have a high iterative cost with few iterations required, while the other has a low iterative cost but requires an infinite number of iterations.

Finally, a future area of research is how to add more high-level knowledge such as design techniques and heuristics. This addition would lead to a system where the environment could design the algorithm from a collection of specifications, and then optimize the algorithm using symbolic manipulations.

ACKNOWLEDGMENTS

The authors wish to thank the reviewers for thoughtful suggestions that have improved the chapter significantly. This work was supported in part by the Joint Services Electronics Program under Contract DAAL-03-90-C-0004 and in part by ESL.

REFERENCES

[1] G. Matheron, *Random Sets and Integral Geometry*. Wiley Series in Probability and Mathematical Statistics (New York: John Wiley and Sons, 1975).

[2] J. Serra, *Image Analysis and Mathematical Morphology* (London: Academic Press, 1982).

[3] P. Maragos, "A Unified Theory of Translation-Invariant Systems with Applications to Morphological Analysis and Coding of Images" (Ph.D. thesis, School of Electrical Engineering, Georgia Institute of Technology, July 1985).

[4] S. R. Sternberg, "Grayscale Morphology," *Computer Vision, Graphics, and Image Processing*, 35 (1986), 333-55.

[5] R. M. Haralick, S. R. Sternberg, and X. Zhuang, "Image Analysis Using Mathematical Morphology," *IEEE Trans. on Pattern Analysis and Machine Intelligence*, 9 (July 1987), 532-50.

[6] J. Serra, ed., *Image Analysis and Mathematical Morphology Volume 2: Theoretical Advances* (London: Academic Press, 1988).

[7] P. Maragos and R. W. Schafer, "Morphological Systems for Multidimensional Signal Processing," *Proceedings of the IEEE*, 78 (April 1990), 690-710.

[8] J. R. Slagle, "A Heuristic Program that Solves Symbolic Integration Problems in Freshman Calculus," *Journal of the Association for Computing Machinery*, 10 (1963), 507-20.

[9] D. E. Knuth and P. B. Bendix, "Simple Word Problems in Universal Algebras," in *Computational Problems in Abstract Algebra*, ed. J. Leech (New York: Pergamon Press, 1970), pp. 263-97.

[10] W. W. Bledsoe, "Non-Resolution Theorem Proving," *Artificial Intelligence*, 9 (1977), 1-35.

[11] D. A. Waterman, *A Guide to Expert Systems*. The Teknowledge Series in Knowledge Engineering (Reading, Mass.: Addison-Wesley, 1986).

[12] J. Moses, "Symbolic Integration: The Stormy Decade," *Communications of the Association for Computing Machinery*, 14 (August 1971), 548-60.

[13] G. Kopec, "The Signal Representation Language SRL," *IEEE Trans. on Acoustics, Speech, and Signal Processing*, 33 (August 1985), 921–34.

[14] W. P. Dove, "Knowledge-based Pitch Detection", RLE Technical Report 518, Cambridge, Mass.: MIT (1986).

[15] C. S. Myers, "Signal Representation for Symbolic and Numerical Processing", RLE Technical Report 521, Cambridge, Mass.: MIT (August 1986).

[16] C. S. Myers, "Symbolic Representation and Manipulation of Signals," *International Conference on Acoustics, Speech, and Signal Processing* (1987), 2400–03.

[17] M. M. Covell, "An Algorithm Design Environment for Signal Processing", RLE Technical Report 549, Cambridge, Mass.: MIT (December 1989).

[18] M. M. Covell, "An Algorithm Design Environment for Signal Processing," *International Conference on Acoustics, Speech, and Signal Processing*, 3 (1990), 1779–82.

[19] B. L. Evans, J. H. McClellan, and W. B. McClure, "Symbolic Z-Transforms Using DSP Knowledge Bases," *International Conference on Acoustics, Speech, and Signal Processing*, 3 (1990), 1775–78.

[20] M. Schmitt, "Mathematical Morphology and Artificial Intelligence: An Automatic Programming System," *Signal Processing*, 16 (April 1989), 389–401.

[21] C. H. Richardson, "The Symbolic Representation, Analysis, and Manipulation of Morphological Algorithms" (Ph.D. thesis, School of Electrical Engineering, Georgia Institute of Technology, December 1991).

[22] C. H. Richardson and R. W. Schafer, "An Environment for the Automatic Manipulation and Analysis of Morphological Expressions," *Proc. of SPIE, Image Algebra and Morphological Image Processing*, 1350 (July 1990), 262–73.

[23] F. Meyer, "Automatic Screening of Cytological Specimens," *Computer Vision, Graphics, and Image Processing*, 35 (1986), 356–69.

[24] C. H. Richardson and R. W. Schafer, "Application of Mathematical Morphology to FLIR Images," *Proc. of SPIE, Visual Communications and Image Processing II*, 845 (October 1987), 249–52.

[25] J. S. J. Lee, R. M. Haralick, and L. G. Shapiro, "Morphologic Edge Detection," *IEEE Journal of Robotics and Automation*, 3 (April 1987).

[26] S. Peleg and A. Rosenfeld, "A Min-Max Medial Axis Transformation," *IEEE Trans. on Pattern Analysis and Machine Intelligence*, 3 (March 1981), 208–10.

[27] P. Maragos and R. W. Schafer, "Morphological Skeleton Representation and Coding of Binary Images," *IEEE Trans. on Acoustics, Speech, and Signal Processing*, 34 (October 1986), 1228–44.

[28] P. Maragos, "A Representation Theory for Morphological Image and Signal Processing," *IEEE Trans. on Pattern Analysis and Machine Intelligence*, 11 (June 1989), 586–99.

[29] P. Maragos and R. W. Schafer, "Morphological Filters—Part I. Their Set-Theoretic Analysis and Relations to Linear Shift Invariant Filters," *IEEE Trans. on Acoustics, Speech, and Signal Processing*, 35 (August 1987), 1153–69.

[30] P. Maragos and R. W. Schafer, "Morphological Filters—Part II. Their Relations to Median, Order-Statistic and Stack Filters," *IEEE Trans. on Acoustics, Speech, and Signal Processing*, 35 (August 1987), 1170–85.

[31] X. Zhuang and R. M. Haralick, "Morphological Structuring Element Decomposition," *Computer Vision, Graphics, and Image Processing*, 35 (1986), 370–82.

[32] C. H. Richardson and R. W. Schafer, "A Lower Bound for Structuring Element Decompositions," *IEEE Transactions on Pattern Analysis and Machine Intelligence*, 13 (April 1991), 365–69.

[33] A. Barr and E. A. Feigenbaum, eds., *The Handbook of Artificial Intelligence*, vol. 1, 2, 3 (Los Altos, Calif.: William Kaufmann, Inc., 1981).

[34] E. Rich, *Artificial Intelligence*. McGraw-Hill Series in Artificial Intelligence (New York: McGraw-Hill, 1983).

[35] P. H. Winston, *Artificial Intelligence*, 2nd ed. Reading (Addison-Wesley, 1984).

[36] E. Charniak and D. McDermott, *Introduction to Artificial Intelligence* Reading (Addison-Wesley, 1985).

[37] K. Preston, "The Abingdon Cross Benchmark Survey," *Computer*, 22 (July 1989), 9–18.

[38] R. M. Lougheed, D. L. McCubbrey, and S. R. Sternberg, "Cytocomputers: Architectures for Parallel Image Processing," *Proc. Workshop on Picture Data Description and Management* (August 1980), 281–86.

5

INTERACTIVE SIGNAL PROCESSING DOCUMENTS

• Malcolm Slaney
Apple Computer, Inc.

5.1 INTRODUCTION

During the last few years, microcomputers have become powerful and inexpensive enough so that every technical person has access to significant computer power on his or her desk.[1] These machines have changed the way many facets of business and research are conducted because users can easily interact with the computer.

Spreadsheets are a common example of how a relatively simple computer program has enhanced personal productivity by allowing users to harness the available computer power. Spreadsheet programs allow the user to develop a model and ask many "what-if" questions, thus gaining a better understanding of the system being modeled. High-level languages and spreadsheets do not make the most efficient use of the machine's computational power, but their speed and ease of use make them valuable to users.

A similar revolution is now possible in the world of signal processing. Recently it has become possible to combine a powerful mathematical program with a word processor and thus create interactive scientific documents. An interactive document includes text and a computer model so that it is easier for readers to understand the

---
[1] This chapter is an expanded version of [1].

material. These documents can be read like a normal technical paper, but since the document includes computer models, the reader can ask it questions. In this way, the material is learned much more quickly. I believe that interactive documents will eventually change publishing as much as did Gutenberg's invention of movable type. This chapter will describe the important features of an interactive document and why it helps make research and learning more efficient.

An important characteristic of an interactive signal processing document is the ability of the reader to simulate, change, and inquire about properties of the system being described. This chapter describes how a symbolic manipulation program can be used to model a DSP system and present it to a reader in a highly interactive form. Many programs have been written to aid portions of a DSP problem, but symbolic math programs allow a system to be modeled at any number of levels. Symbolic manipulation programs use their mathematical knowledge to automate common mathematical operations such as manipulating polynomials, calculating integrals, and finding limits. The resulting interactive signal processing document, or notebook, can be a very efficient method to teach DSP concepts.

The ideas expressed in this chapter evolved as I wrote an electronic notebook describing a cochlear model developed by R. F. Lyon [2]. This technical report began as a modest notebook for my own use as I wrote a conventional paper. As the work evolved, I realized that the examples in the notebook would also be useful to the reader. The original notebook was not comprehensible to other readers, but by combining the original paper with electronic models and interactive examples, a new type of document was created.

This electronic notebook was created using *Mathematica*. As suggested in Chapter 3, *Mathematica* notebooks exemplify many of the desirable characteristics of a system for research in signal processing. The resulting document is a powerful tool for research and development and an effective tool for teaching. Not only is *Mathematica* an example of a tool for symbolic mathematics, but it includes elements of hypermedia, interactive modeling, and literate programming. The ability to perform symbolic manipulations is important for reasoning about the models, and its other characteristics make it easier for readers to learn the material. How *Mathematica* uses these ideas is discussed in Chapter 3 and section 5.2.

This chapter is not a review of a software product. That function has been ably covered in many articles [3–5]. Instead, *Mathematica* notebooks are used to illustrate many useful features and as a framework to describe additional functionality. Some of the needs of an interactive signal processing document are met by *Mathematica* and related software, while others are not.

Figure 5.1 shows a portion of the type of interactive document this chapter describes. In this case, an animation of wave motion in the cochlea is included as part of the document. The reader can see the animation by clicking on the button with a mouse. Most important, the animation is not just a pretty picture, or even just a pretty animation. The computational model is defined elsewhere in the document, but it is available to the user to change. The reader can modify the parameters of the model and see how the animation changes. If the reader has a

Sec. 5.1 Introduction

This animation shows a gray scale representation of the pressure in the cochlea due to a single tone at 1000Hz. Darker and lighter regions correspond to pressures above and below the average. The basilar membrane is shown along the horizontal center line.

```
In[42]:
          ShowCochlearGrayScaleAnimation[1000];
```

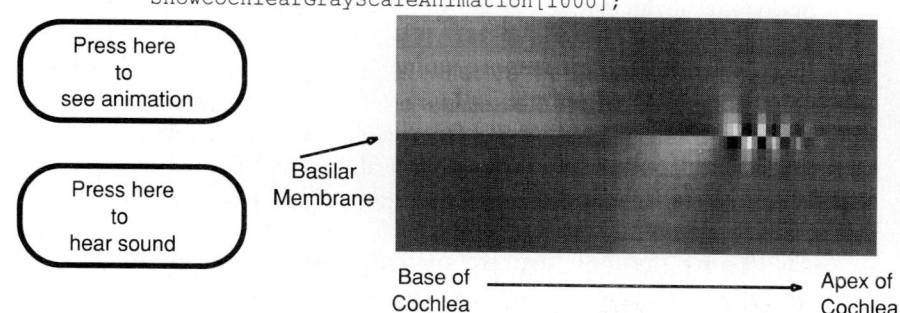

The change in propagation speed means that the long-wave, or one-dimensional, approximation is only valid in the early part of the wave's travel. Near the response peak the wavelength is short enough that energy can flow both down the cochlea and perpendicular to the membrane, so a two-dimensional model is needed.

> **Figure 5.1** An electronic notebook combines many features to make it easier for the reader. This example combines text, a mathematical model, graphics, and sound with a user interface that helps guide the reader. One frame of a real *Mathematica* animation computed by R. F. Lyon showing fluid pressure in the cochlea (inner ear) is shown here. The reader can press the button with the mouse and see an animated representation of the fluid flow. The user can modify the frequency of the tone or the parameters of the model and see how the animations change.

favorite model of the middle ear, it can be added to the model. It is important to realize that, except for the buttons, this example is possible with commercially available software. Examples of similar notebooks will be presented throughout this chapter, and I hope it will encourage other researchers to build the necessary software and prepare additional electronic notebooks.

Today, many signal processing problems are easily described using such an interactive scientific notebook. Recent work described in other chapters of this book demonstrates the ability to automatically reason at a high level about a signal processing problem. In addition, personal computers are now powerful enough to implement the digital signal processing algorithms with real data. The problem that remains is, how can authors describing signal processing results take advantage of these technologies to better interact with their readers? This question is the primary focus of this chapter.

There are two concerns in the design of an interactive signal processing document: the content and the form. Certainly the most important part of such a document is the intellectual content. This chapter describes the techniques an author can use to write an interactive document. The second concern, the form of an interactive document, is set by the available tools. This chapter describes the ideal tool and how the tools available today can be used to describe real signal processing problems.

First, the form of an interactive signal processing document is discussed. Section 5.2 describes some of the important properties of an electronic document and some of the characteristics of an ideal tool for researching and teaching signal processing. It is important that the system have powerful tools for modeling and also that this power is easily available to casual readers. Section 5.3 describes the state of DSP tools today. It describes conventional tools for DSP research and development and explains the use of *Mathematica* to create signal processing notebooks, and is illustrated with two *Mathematica* notebooks. The first example shows features of *Mathematica*; the second notebook shows how *Mathematica* can be used for signal processing research.

Given the tools that are currently available for writing an interactive signal processing document, how should the intellectual content be structured? Section 5.4 presents some of the design issues that must be faced by an author of an electronic document. All writing is difficult, and making it interactive adds another dimension. Some problems become easier to explain, but the additional flexibility can be difficult to manage. Finally, section 5.5 talks about research issues that should be addressed in the future.

5.2 PROPERTIES OF AN INTERACTIVE DOCUMENT

Communicating results is an important part of research. While developing the solution to a problem, it is often useful to share the results with colleagues. Later, when the results are polished, the report will be copied and more widely distributed.

An important characteristic of an interactive system for signal processing is that the work can be developed and documented in the same system. Results can be discovered and easily presented to colleagues. If changes are necessary, they can be made quickly without having to transfer the results between a symbolic math system and a word processor. Most important, the reader has the same tools available and can modify or extend the model as desired.

Signal processing is an ideal subject for an interactive document. It is hard to imagine using a computer to teach a carpenter, for example, how to hammer a nail, but computers are a necessary part of most signal processing work. Signal processing researchers already use computers to calculate and simulate algorithms. An interactive signal processing document can provide test data and an evaluation function so students can design a filter and automatically verify that it meets the design goals.

An interactive system is the most effective means to teach and disseminate signal processing results. Much is learned from passive documents, like books, but the learning process is more effective when the reader can ask questions. With interactive documents, readers effectively ask the document questions.

During the last ten years much effort has been expended to create interactive learning environments. The results have been mixed. Perhaps the biggest impediment to their success has been their goal to create an all-encompassing environment. This goal has brought with it the need for specialized and expensive computer systems, which are only available to a small number of users.

Interactive notebooks are an important intermediate step between paper and fully interactive environments. If written well a notebook can be printed on paper and read without any special technology. But if the reader does have the appropriate hardware and software, he or she will benefit from a much richer interaction.

This section describes the ideal properties of an interactive signal processing document; the current state of the art will be described in section 5.3. A well-designed interactive document will have several features. It will be an example of hypermedia; the user will be able to explore the document, study the sections that are new, and dig deeper into those that are at first not understood. It will be easy to refer to other parts of the document where references are first explained. In addition, computer models will allow the reader to ask questions and more easily integrate the new material with what is already known. Finally, the algorithm must be explained in a way that is easy for both a reader and a computer to understand. This is known as Literate Programming, or the creation of a document that is both a well-written program and a well-written paper.

The features of an interactive signal processing document as previously described will be discussed in the remainder of this section. But, these features are only part of a complete system. A researcher will also need drawing programs to create graphics to explain the work and spelling checkers to help get the descriptions right. These other components of a complete system are vital but are not discussed here.

5.2.1 Symbolic Manipulation

Perhaps the most important characteristic of an interactive signal processing document is the ability to reason symbolically about a system or a design. Certainly, many very important signal processing problems have been solved without symbolic manipulation programs, but their use allows much of the drudgery involved in the mathematics to be automated. This makes it easier for the casual reader to verify the results and to explore new ideas.

Many tools are good at numerical math. These tools might be used to calculate the eigenvalues of a matrix or to design a filter with a specified pass-band. The software might be supplied as a subroutine library that is used with a conventional language (such as IMSL [6]) or as a complete environment with a customized language and graphical output (such as MATLAB [7]).

There are many problems, however, where numerical answers do not provide much insight into the solution. Numerically integrating an equation to discover that the maximum transmission rate of a channel is 29 kbits/sec is useful, but an answer in terms of the bandwidth of the system and the antenna gain provides more insight. Symbolic math software allows the user to perform algebraic manipulations, do symbolic calculus, and find the solutions to equations. In addition, symbolic math packages include numerical routines to deal with problems that cannot be done symbolically. This makes symbolic math packages a superset of numerical software, but often this generality makes strictly numerical calculations slower.[2]

Although symbolic math tools can be used to perform numerical calculations similar to those done by conventional programming tools, their real power is in manipulating the algebraic expressions that describe the system's behavior. An introduction to *Mathematica* as it relates to our goals in this chapter, follows shortly.

A symbolic system allows a filter to be studied in many different forms. For example, an expression that makes clear the poles and zeros of a filter can be expanded into the direct-form polynomial often used in a digital filter. This is how it can be done with *Mathematica* (Note: *Mathematica* uses In and Out to represent what the user typed and *Mathematica*'s response):

$$\text{In}[1] := \text{Expand}[(z - 1/2 - 1/2I)(z - 1/2 + 1/2I)(z + 1/2)]$$

$$\text{Out}[1] = \frac{1}{4} - \frac{z^2}{2} + z^3$$

The filter's characteristics can then be shown graphically in any number of forms. This chapter shows several examples: the magnitude of the frequency response, pole-zero plots, and rubber sheet diagrams of the filter's z-domain response.

A symbolic math package allows a user to reason about a system in ways that are not possible with conventional tools. One question that came up when preparing this chapter is: where did the algorithm for designing a second-order digital filter with a given center frequency and bandwidth come from? A reader, unsure about the source of an algorithm in a paper, might want to do a bit of exploration. There will be false starts, but eventually the user should be able to find the correct answer. Note that a symbolic manipulation program does not answer the question automatically, but it does provide some of the necessary mathematical knowledge and expertise to allow the question to be answered.

Symbolic math programs do not normally contain domain-specific knowledge about fields such as DSP, but they can be easily extended. One such knowledge base for use with *Mathematica* is described in Chapter 3; their packages allow *Mathematica* to perform Fourier, Laplace, and z-transforms [8]. It is easy to add other packages containing, for example, knowledge about acoustics or speech synthesis.

Other types of common DSP operation that are difficult with symbolic math programs are the algorithm optimization built into ADE [9] and described in Chapter 2, and the morphological algorithm manipulation described by Richardson in Chapter 4. ADE allows a user to describe a signal processing algorithm in such a way that ADE can reason about the design, try alternate implementations of the algorithm, and show an equivalent design that is more efficient. Symbolic math programs will often rearrange an equation to put it in a standard form for display

---

[2] On the other hand, a symbolic package can evaluate a definite integral by first performing symbolic integration and then substituting the numerical limits. This might be faster than numerically evaluating the integrand and summing the results.

Sec. 5.2   Properties of an Interactive Document   **179**

to the user, but the standard form is defined more by mathematical convention than by computational complexity. Thus, a calculation might produce the result

$$ax + b + cx^2$$

which will be displayed to the user as

$$cx^2 + ax + b$$

yet a potentially more efficient computational form is

$$(cx + a)x + b$$

As far as I know, there is no symbolic math program that allows users to specify the relative costs of different representations, let alone suggest the changes that might be made to optimize the algorithm as is done in ADE.

5.2.2 Hypermedia

At one time, information was passed from generation to generation via the story teller. This was inherently an interactive process, but it also limited the speed at which information could be conveyed. By the middle ages, publishing was well established. This increased the rate at which information could be disseminated, but it lost its interactive nature.

Most papers and books are designed to be read in a linear fashion. (Dictionaries and encyclopedias are examples of works that are meant to be accessed randomly.) The order of presentation is determined by the author, and the reader is expected to make the best of it. Some browsing is possible, but the paper medium makes this inconvenient. The reader is a passive part of the learning process.

Readers have questions to be answered. Often readers do not have the same goals as the original author and might want to skip around in the material. Sometimes they will look in the index for the subject in which they are interested and then work backwards until they understand enough to solve their problem. This sometimes involves just paging back through a few pages of material, but at other times the necessary preliminary information spans many chapters of the book, often with intervening material the reader does not need to understand to solve the problem at hand.

Hypermedia is often described as a solution to this problem [10]. Using a computer, the reader can browse through a document and then quickly move to other sections based on what is read. Thus, if a reader encounters a new concept, it might be accompanied by a button on the screen that will display more detailed information. In this way, the reader can fashion his or her own path through the material.

Hypermedia has many forms, but they all represent enhancements of the paper world. There are four hypermedia features of an electronic notebook that will be considered here. The simplest is using multiple media: sound, animations, and other ways of presenting information that are hard to print on paper. Second, the notebook can be easily changed by the reader, to emphasize the points that are more interesting, or to add additional notes. Third, the most traditional form of hypermedia is represented by links, or the ability to move quickly from one topic to another in the paper. Finally, an electronic notebook has many functional links based on the mathematical definitions and algorithms in the paper. Each of these ideas will be considered in turn.

Multimedia

Hypermedia can mean the use of more than one type of media in a document. There are many examples of papers that would be much more effective by adding audio and visual demonstrations. Why should a paper on sound perception not include audio examples so readers can make their own judgments about what is heard? A paper on speech coding should have audio examples so the results are meaningful to readers not familiar with intelligibility scores. A paper on acoustics is more readable with simple animations showing the propagation modes. A discussion of video compression algorithms should include a sequence of images so the reader can modify the algorithms and see how their changes affect performance.

Modifiable

An electronic notebook can be easily changed. Users can rearrange the report to make the presentation more natural for their background. In addition, users can add their own material. Since there does not have to be any difference in appearance between the original text and the "margin" notes, the document becomes personalized for the reader.

Links

The most conventional form of hypermedia allows the reader to browse through a document in any desired order. In its ideal form, the reader should be able to follow ideas anywhere they might go within a hypermedia document. But this leads to a multidimensional web of links that is difficult to organize on paper.

Instead, an electronic notebook has a hierarchical organization which can then be flattened when rendered on paper. Text, equations, output, and graphs are grouped into sections, and the entire paper organized into a hierarchy. Most papers have a tree-like structure, but the electronic version of a paper's most detailed sections can be hidden from the user so as to make the presentation easier to follow. These hidden sections can be easily opened by the user and can be used to hide details of the presentation, attempts that did not work, or test code that is not strictly necessary for the presentation.

Most writing efforts include unpublished examples that were used to test and refine the material in the paper. This material is probably not interesting to most readers, but in some cases it is. Inquisitive readers might want to know where a derivation came from or would like to double check a result. For instance, an interactive signal processing document that I wrote describing the classic algorithms for filter design includes many examples that were used to test the algorithms in the paper. Some of these examples are interesting to the casual reader and are shown in the main body of the paper. Other examples were used to explore more difficult cases or to compare my solution to published designs. These examples are hidden, yet still available to the interested reader. The purpose of writing is to communicate a result to the reader. Allowing the reader to peek behind the scenes, so to speak, is beneficial to the communication process.

On top of this hierarchy, a separate set of links is represented by the functions and algorithms that are defined. This set of links might take the form of a help facility to make it easier for the user to understand the paper. Since the help facility includes information about any function defined in the paper, it is a very simple form of hypermedia. A user can select any function name in the paper, select help, and read a short description of the function. When more information is needed about a function, the system can take the reader to the point in the paper where the function is first defined.

Smart Links

The links defined by the functions in an interactive notebook are not just navigational aids. Instead, these functions have mathematical meaning, and their links often include dependencies on other mathematical definitions. For example, my report describing a cochlea model includes a relatively simple model of the outer and middle ears. This model is used when showing the neural firings due to a particular sound. But if a reader has a better model, it can be substituted in the report and the new graphs can be computed.

Thus, a scientific notebook extends the hypermedia concept because a function is not just a collection of symbols, but, more important, has a mathematical meaning. Using a function in a notebook not only implies a link back to the original definition, but its usage implies a specific mathematical operation. These "smart links" are an important part of an interactive signal processing notebook.

5.2.3 Interactive Models

An interactive signal processing document extends the hypermedia concept by casting each of the new results as an equation or computer model with which the user can interact. Since the system includes a computation engine, readers can change the model and see the effect. The results are shown graphically. By controlling the parameters of the model or system, the user can gain a better understanding of how it works.

This book describes several tools for signal processing research and development. Just as an interactive system for signal processing is a useful research tool, such a system can also be valuable to a reader trying to understand the results. A good instructor or piece of writing should guide the student to the same conclusions that were reached by the research. Hopefully, the learning process will be more efficient than the original research, but the same tools are useful in both cases.

An interactive signal processing document contains equations and computer models that the reader can manipulate. In some cases, the reader will be content to change the parameters of a model and see how the results change. Other readers might want to study the model from a different angle. Perhaps due to a different background, the reader will want to analyze the system response in the time domain instead of a frequency domain approach more natural to the original author. With a symbolic math package, readers can apply the appropriate transformation and study the result in their preferred domain.

Users should be able to interact with a signal processing document in two ways. First, a mathematical model can be modified to extend it into the reader's own domain. For example, a reader of a paper on reconstruction theory might want to try the algorithm using data from his or her own research problems. This makes the solution described in the notebook more realistic to the reader.

A second, more important, aid to learning is direct manipulation. In many systems, the behavior is controlled by a numerical parameter. A paper describing such a system will probably include an equation describing the system's behavior as a function of this parameter. But the user would have a much better feel for the behavior of the system if there was a knob (or slider) that could be manipulated with a mouse and would immediately vary the system's output. A simple example of this behavior in an interactive signal processing document is shown in Figure 5.2.

A prototype of a system like this is distributed with version 1.0 of the NeXT workstation software. Unfortunately, many problems cannot be recomputed at rates fast enough to be interactive. Designing a simulation that can be solved with easily accessible hardware is a problem that is addressed in section 5.4.

Most symbolic math packages are interactive. An electronic notebook is unique, though, because the writer can guide the reader by suggesting areas to

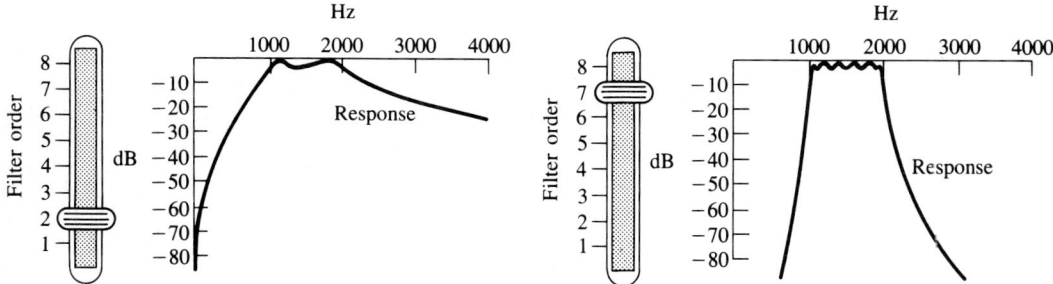

Figure 5.2 Animated Figures. Two samples are shown of a proposed scheme for allowing the reader to modify a figure in an interactive signal processing document. As the reader uses a mouse to change the position of the slider, the order of the band-pass filter is changed and a new response curve is calculated.

Sec. 5.2 Properties of an Interactive Document

explore. The notebook should encourage the reader to try new ideas. The symbolic math package can show the result as an equation, a figure, or even an animation.

It is unfortunate that a *Mathematica* notebook does not support direct graphical manipulation like that shown in Figure 5.2. Currently, notebooks have to be carefully written so that the reader does not have to understand much of *Mathematica* to know what parts of an expression to change. A graphical control, such as a slider, would make it possible for users, both novice and experienced users, to directly control an electronic notebook in a very intuitive fashion. Even the ability in a notebook to place a mousable button that would perform some action would be useful. An example of this capability is shown in Figure 5.1.

5.2.4 Literate Programming

Documenting a computer model or a signal processing algorithm is a useful way to transfer knowledge about a new result. Conventionally, this has been done using a simplified form of a language like Algol or Fortran. It is difficult within the constraints of these languages to eloquently express the ideas that went into the model.

An alternate style of expressing an algorithm is known as Literate Programming. Denning's introduction to Van Wyk's column on *Literate Programming* says [11]:

> A literate program contains not only the needed statements in a programming language, but also a precise problem statement, a summary of trade-offs between the running time and space, or between running time and programming time, and suggestions on how to modify the program. Program code segments are inserted in the text at points logical to the intellectual development of the algorithm. A literate program pays careful attention to lucidity of presentation and presents all arguments needed to understand why the program will actually work as intended.

The combination of an interactive computer model and literate programming is a very powerful tool for learning. Before a reader can intelligently change a model, the description must be read and understood. Expecting the reader to understand the program from just the embedded comments is not very practical. Instead, a literate program should include graphics, examples, and even interactive controls. All these techniques will help the reader to better understand the results.

Changing the focus of the programming effort can have a great benefit. A computer language is designed to make it easy to implement algorithms, not to explain material to another reader. Comments and sometimes even pictures are added to a program to describe an algorithm, but the text of the program is still one-dimensional. In a literate program the primary goal should be to describe an algorithm for a reader, but to do it in such a way that the computer can also understand it. This is especially true when describing highly mathematical material such as signal processing algorithms.

5.3 DSP TOOLS

Unfortunately, there is no software that possesses all the features needed to build the interactive signal processing document described in section 5.2. Most conventional tools for signal processing are libraries or environments that are designed for programmers. Although these tools are powerful, they are difficult for casual readers to use.

There are many programs that can perform symbolic manipulations, and one of them, *Mathematica*, has an electronic notebook interface. *Mathematica* and its notebooks come closest to the ideal interactive signal processing environment described here. Using a symbolic math program to do signal processing problems is not new (MACSYMA, the grandfather of all symbolic math programs, has built-in support for Fourier and Laplace transforms), but adding a notebook interface allows an author to make the DSP knowledge more accessible to a reader.

It is not possible to describe all the tools designed to help solve DSP problems. This section describes several of the conventional tools and then describes the use of *Mathematica* to research and teach signal processing ideas. The purpose of this chapter is to discuss the use of a symbolic manipulation program to describe in an interactive manner the solution of a signal processing problem. To provide some context, several conventional tools will first be reviewed.

5.3.1 Conventional Tools

Perhaps the most commonly used tools for signal processing are subroutine libraries. The two best-known libraries for signal processing are IMSL [6] and the IEEE Signal Processing Library [12]. These libraries include code to perform many common signal processing operations in a user's program. The user must still do much of the programming, but the difficult numerical work is handled by these subroutine libraries. Much research has gone into these algorithms and they represent a significant step in the use of structured programming techniques. The routines in these libraries, especially in IMSL, are highly optimized, and their numerical stability is well documented. Most of the work required to use these libraries consists of reading in the data and putting it in the proper form for the appropriate subroutine.

Subroutine libraries eventually led to complete programming environments for signal processing. SRL (the Signal Representation Language) [13] and SPLICE [14, 15] are two examples of specialized programming environments for signal processing. These systems are built on top of the Lisp programming language, and use object-oriented techniques to make it easy for researchers to extend the environment. Two extensions for computing sine wave signals in SRL are shown in Figure 5.3.

Systems such as SRL and SPLICE are very powerful, but their nonstandard programming methodology (object-oriented Lisp) requires much effort to learn. In a sense, these tools are programmer friendly but not necessarily user friendly. This has limited their success.

```
(defsigtype sine-wave-signal-type
  :a-kind-of basic-signal-type
  :parameters (ncycles length phase)
  :finder signal-sine-wave
  :init (setq-my dimensions (list length))
  :fetch ((i) (sin(*3.141592 2.0 ncycles (/ i length )))))

(defsigtype sine-wave-with-zero-phase-signal-type
  :a-kind-of sine-wave-signal-type
  :parameters (ncycles length)
  :finder signal-sine-wave-with-zero-phase
  :init (setq-my phase 0.0))
```

Figure 5.3 Two examples of signal definitions in Kopec's Signal Representation Language (SRL). The first example, `sine-wave-signal-type`, defines the scheme to calculate a sine wave with `ncycles` in `length` samples. The second example, `sine-wave-with-zero-phase-signal-type`, further refines this signal type to include a default `phase` of 0 degrees. The rest of the behavior of the `sine-wave-with-zero-phase` is inherited from the signal's parents, `sine-wave-signal-type` and `basic-signal-type`.

As the number of people wanting to solve signal processing problems has grown, the need for systems that do not require any programming has also increased. These environments include a large number of specialized signal processing algorithms that can be applied in a cookbook fashion to solve a problem. With a high-level tool the language becomes more specialized, making it easier to express some algorithms. A drawback of such specialization is that concepts that do not fit within the tool's model are much harder to program.

MATLAB is another example of a specialized signal processing environment [7]. MATLAB includes a large number of routines for linear algebra and signal processing which can be used interactively or combined by the user into new functions. MATLAB also includes functions to import data and to plot the results. A large number of common signal processing operations are part of MATLAB's libraries, and it is easy to add new functions. MATLAB provides a simple command line interface for the user and a single window for graphics.

An even more specialized tool that can be used for image processing is Adobe's *Photoshop* [16]. *Photoshop* is designed to make it easy to perform many common operations on images and see the results immediately. This tool is very powerful but is not programmable.

Other tools have been designed to deal with specific signal processing problems. For example, there are tools for speech analysis, all sorts of filter designs, and even VLSI layout for signal processing algorithms. The price paid for this power is the limited domain. For example, it would be difficult to design a filter using one of the filter design programs and then use the results in one of the speech analysis tools.

5.3.2 About *Mathematica*

Mathematica is an example of a system for creating interactive scientific documents [17]. It is first and foremost a program for doing mathematics on a computer. The program allows a user to pose both symbolic and numerical mathematics questions. Thus, a user can symbolically integrate an expression, and then find its numerical solution over any domain. *Mathematica* includes a programming language to allow more complicated models to be described and graphical functions to allow the user to visualize the results more easily. When mathematical definitions, graphics, and words are all combined, the resulting document is called a notebook. *Mathematica* is available for most of the popular scientific and personal computers, but the notebook feature is only currently available on the Apple Macintosh and the NeXT Machine. Notebook support for other machines has been promised.

Like its predecessors (MACSYMA, SMP, Maple, Reduce, etc.), *Mathematica* includes many facilities for doing symbolic mathematics and numerical calculations. The fact that these systems can easily work with polynomial equations makes them useful, for example, when designing filters. A family of filters can be designed and then analyzed for their behavior at DC. This section will talk about some of the features of *Mathematica* and how it can be used to create an interactive signal processing document.

The *Mathematica* system is divided into two halves. The user interacts with a front end while a back-end kernel provides the computational engine. The front end is unique for each type of machine, defines the behavior due to typed commands and mouse actions, and provides an interface to the host's window system. The back end is relatively machine independent and does all the calculations.

One useful feature of *Mathematica* is that the user interface (front end) and the kernel (back end) do not have to be on the same machine. Thus, a user can interact with a relatively inexpensive graphics machine on the desktop, while a more powerful shared back-end machine does all the calculations. The communications between the front end and the back end can be carried out over a serial line or a network connection.

On some machines, the *Mathematica* front end allows the user to create what is called a notebook. A *Mathematica* notebook is much like a scientist's notebook since it can contain data and thoughts about work in progress. However, these notebooks are unique in that they also contain computer models and even animations. The ability to create a live mathematical notebook is probably the feature that most distinguishes *Mathematica* from other systems.

Notebooks contain text, equations, and graphics. When a problem is first proposed, the notebook will reflect the steps actually used to carry out the calculation. It might contain false starts and notes understood only by the author. As the theory and calculations are refined, the notebook becomes more polished. Eventually, the notebook is cleaned up and the necessary text written so it can be distributed to colleagues. The finished notebook will contain explanatory text, symbolic and numerical models, and graphs and animations to explain the system and its solution.

Sec. 5.3 DSP Tools

Figures 5.4 and 5.5, my own technical report [2], and a recent book by Gray [18] are examples of notebooks in their polished form.

A notebook showing some of the capabilities of *Mathematica* is shown in Figure 5.4. *Mathematica* can be used as a calculator, even with arbitrary precision, but its real power comes from the symbolic functions. Equations can be integrated and differentiated, and algebraic manipulations can be performed to put the result into a simpler form. The results can be plotted to help understand how the system works. A second, more complete example showing the use of *Mathematica* in a signal processing application is shown in Figure 5.5.

The signal processing example in Figure 5.5 uses a special data structure to represent each filter. Ratios of polynomials are supported by *Mathematica*, but then any filter operation that needed the location of the poles and zeros would have to factor a polynomial with floating point coefficients. This can be done but is prone to errors and is time consuming.

Instead of representing filters as ratios of polynomials, the filter design functions shown here use a special structure that contains the gain, zero locations, and pole locations. This structure is called a GZP (Gain, Zeros, and Poles). The GZP structure is represented in *Mathematica* as a three-element list with the zeros and poles each being represented as a list of points in the complex plane. Choosing the appropriate data structure to model a system is important both for computational efficiency and for literate programming. One advantage a symbolic environment has over a purely numeric system such as MATLAB is that arrays and lists can be combined in arbitrary ways to represent the data at hand.

Both the GZP and ratio of polynomial representations of filters have their advantages. The GZP is used in Figure 5.5 because it is more accurate (roots of polynomials are already known) and it is easy to transform a filter from the GZP form to a ratio of polynomials. Consider the following filter design. The call to ChebychevLp returns an eighth-order filter with a 1 dB pass-band ripple. The gain of this filter has been adjusted so that it has unity gain at DC. Here, the GZP structure returned by the ChebychevLp function, the Laplace domain representation, and finally an expanded version are shown.

```
In[1]:=
      ChebychevLp[8,1]
Out[1] =
          .01720755
       {---------- , {} .
         10^{1/20}
       {-0.0350082 - 0.996451 I, -0.099695 - 0.844751 I,
        -0.149204 - 0.564444 I, -0.175998 - 0.198206 I,
        -0.175998 + 0.198206 I, -0.149204 + 0.564444 I,
        -0.099695 + 0.844751 I, -0.0350082 + 0.996451 I}}
```

Mathematica Introduction

This notebook is a short introduction to the features of *Mathematica*. In the two notebooks accompanying this chapter, the *Mathematica* input is shown in a `bold face Courier font` and the *Mathematica* output is shown in a normal `Courier font`. See Wolfram's *Mathematica* book for more examples.

Numerical Calculations

Mathematica can be used like a calculator to do numerical arithmetic. Here is Pi calculated to 50 decimal places.

`N [Pi, 50]`

3.1415926535897932384626433832795028841971693993751

Or *Mathematica* can be used to numerically integrate an expression which can't be integrated symbolically.

`NIntegrate [Sin [Sin [x]], {x, 0, 1.0}]`

0.43060610312069006045

Polynomial Manipulation

Manipulating algebraic expressions is easy for *Mathematica*. Here are some examples. First we multiply out the terms of an expression.

`Expand [(x + 1) (x + 2) (x + 3) ^3 (x + 4)]`

$216 + 594 x + 639 x^2 + 350 x^3 + 104 x^4 + 16 x^5 + x^6$

We can then factor this equation to find the original expression.

`Factor [216 + 594*x + 639*x^2 + 350*x^3 + 104*x^4 + 16*x^5 + x^6]`

$(1 + x) (2 + x) (3 + x)^3 (4 + x)$

`Solve [216 + 594*x + 639*x^2 + 350*x^3 + 104*x^4 + 16*x^5 + x^6 == 0, x]`

{{x -> -1}, {x -> -2}, {x -> -4}, {x -> -3}, {x -> -3}, {x -> -3}}

Here is a graph showing the behavior of this function as *x* varies between –5 and 0.

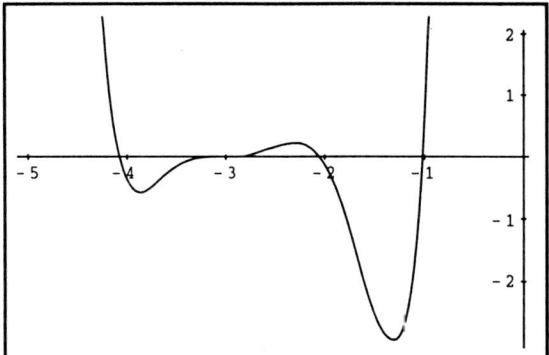

`Plot [(x + 1)(x + 2)(x + 3)^3(x + 4),{x,-5,0}];`

Calculus

Mathematica knows a lot about calculus. After reading in *Mathematica*'s integration rules, we can easily find the integral of $x/(1 - x^3)$.

```
<<IntegralTables.m;
Integrate [x/(1-x^3), x]
```

$$- \frac{\text{Sqrt}[3] \text{ArcTan}\left[\frac{1 + 2x}{\text{Sqrt}[3]}\right]}{3} - \frac{\text{Log}[1 - x]}{3} + \frac{\text{Log}[1 + x + x^2]}{6}$$

Now let's differentiate this result and see if we get the original expression.

```
Simplify [D [- (3^ (1/2) *ArcTan [(1 +
    2*x) / 3^ (1/2)]/
    3 - Log [1 - x]/3 +
    Log [1 + x + x^2]/6, x ]]
```

$$\frac{x}{1 - x^3}$$

We can also find the series expansion of an expression.

`Series [Exp [x] Cos [4x], {x, 0, 6} ]`

$$1 + x - \frac{15 x^2}{2} - \frac{47 x^3}{6} + \frac{161 x^4}{24} + \frac{1121 x^5}{120} - \frac{11 x^6}{16} + O[x]^7$$

Figure 5.4 An example of a *Mathematica* notebook showing elementary numerical, symbolic manipulation, graphing, analysis, and programming features.

Sec. 5.3 DSP Tools

Solving Equations

Mathematica can be used to solve simultaneous equations. Here is a simple example.

```
Solve [{x^3 + y^3 == 1, x + y == 2},{x, y}]
```

$$\left\{ \left\{ x \to \frac{6 - \sqrt{-6}}{6}, \quad y \to \frac{12 + 2\sqrt{-6}}{12} \right\}, \right.$$

$$\left. \left\{ x \to \frac{6 + \sqrt{-6}}{6}, \quad y \to \frac{12 - 2\sqrt{-6}}{12} \right\} \right\}$$

Graphics

Mathematica can graphically show you the results of your calculations. Here is a plot showing the previous two equations. Note: there are no solutions for real values of x and y.

```
Plot [{(1 - x^3) ^ (1/3), 2 - x},
      {x, -2, 2}]
```

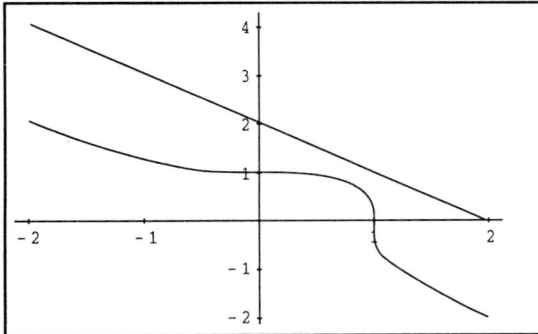

Data Analysis

Mathematica can be used to analyze the results of your experiments. Let's create a sample data set by adding random noise (uniform between 0 and 1) to a sine wave.

```
data = Table [N [Sin [1/25] +
       Random []] , {1, 200}];

ListPlot [data];
```

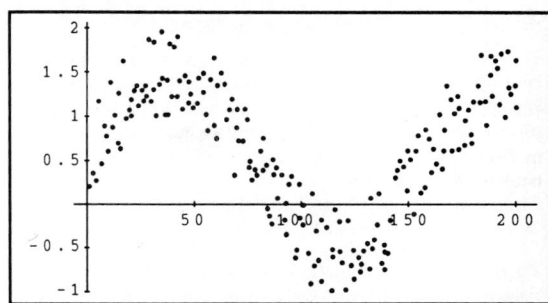

Now let's fit the data to a constant term (to get the mean of the random variable) and a sine and cosine of the appropriate frequency. Note that the term multiplying the cosine in the result is small compared to the factor multiplying the sine.

```
Fit [data,{1, Sin [x/25], Cos [x/25]},x]
```

$$0.499471 - 0.0565294 \; \text{Cos} \; \left[\frac{x}{25}\right] +$$

$$0.944494 \; \text{Sin} \; \left[\frac{x}{25}\right]$$

Programming by Example

Rules can be added to *Mathematica* to specialize it for your own problem domain. Here is an alternate definition of factorial. The first rule defines the stop condition. The second rule is the basic recursion to solve the problem.

```
Fact [0] = 1
Fact [x_] : = x Fact [x - 1]
Fact [6]
720
```

Here is an example of defining rules for simplifying logarithms in *Mathematica*.

```
log [a_ b_] : = log [a] log [b]
log [x y^2 z]
log[x] log[y^2] log[z]
```

Now we can tell *Mathematica* about powers

```
log [x_^n_] : = n log [x]
log [x y^2 z]
2 log[x] log [y] log [z]
log [x/y]
-(log[x] log[y])
```

Figure 5.4 (continued)

Signal Processing Example

This notebook is an example of using *Mathematica* to describe continuous filter design. This notebook is fully functional. It includes some *Mathematica* notation, but readers who are not *Mathematica* users should have no problem understanding the notebook by just reading the text. See [Wolfram88] for an explanation of the special notation used by *Mathematica*. Thanks to Ray DeCarlo at Purdue University for providing the original motivation to write this notebook.

There are a number of design techniques for high-order filter design. This notebook will show how to design a Chebychev low-pass filter, and then how to transform the original lowpass poles into a bandpass filter. Section 1 of this example defines a number of functions used to work with filter polynomials. Section 2 describes the techniques to design Chebychev lowpass filters with a corner frequency of 1 radian per second (rps), and Section 3 shows how to transform these generic lowpass filters into bandpass filters with arbitrary passbands.

1 Continuous Filter Functions

Continuous-time filters are described using polynomials of complex frequency s. A filter's response function is evaluated along the imaginary axis by making the substitution s->I w (or $j\omega$ in conventional EE notation.) The following function is used to evaluate the complex response of a filter radians. Additional functions compute the gain, magnitude, and phase response of the filter. The expression **filter** can be an arbitrary function of the complex frequency s.

```
FilterGain[filter_, w_] :=
    ReplaceAll[filter, s -> I w];

FilterMag[filter_, w_] :=
    Abs[FilterGain[filter,w]]

FilterPhase[filter_, w_] :=
    Arg[FilterGain[filter,w]]

FilterDb[filter_, w_] :=
    20 Log[10,FilterMag[filter,w]]
```

The following function is used to display the frequency response of a continuous filter. (The plot starts at 0.01 Hz to avoid any problems with filters that have a zero at DC.)

```
FreqResponse[filter_, maxf_,
    opts_:{}] :=
  Block[{response},
    response = N[FilterDb[filter, 2 Pi f]];
    Plot[response,{f,.01,maxf},
      AxesLabel->{" Hz","dB"},
      PlotLabel->"Response",
      opts]];
```

We define a similar function for displaying the frequency response of a filter as a function of radian frequency (ω or radians per second, rps).

```
FreqResponseRadians[filter_, maxw_,
    opts_:{}] :=
  Block[{response},
    response =
N[FilterDb[filter,w]];
    Plot [response,{w,.01,maxw},
      AxesLabel->{" RPS", "dB"},
      PlotLabel->"Response",
      opts]];
```

Note that for each of these functions there is a third optional argument that allows additional options to be set. We use this feature to pass special parameters to the **Plot** function. The frequency response of a fourth-order filter is shown below.

```
FreqResponseRadians[.197 s^2/
((0.09 - 1.3I + s)(0.09 + 1.3I + s)
  (.12-1.8I + s) (.12 + 1.8I + s)),4];
```

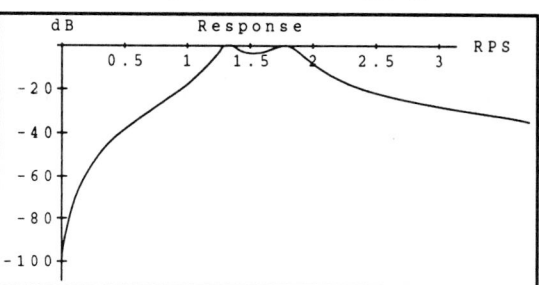

The **AdjustGain** function is used to modify a filter so that it has unity gain at any desired frequency.

```
AdjustGain[filter_,f_] :=
    filter/FilterMag[filter,f]
```

Higher order filters could be designed with *Mathematica* using either rational polynomials or lists of poles and zeros. Rational polynomials would be nice because all intermediate results would look like filters. Unfortunately, we sometimes need to talk about individual poles and zeros, for example when doing partial-fraction expansions. This is difficult if the filter is described as a polynomial. If a filter is described by its poles, zeros, and gain, we can always regenerate the polynomial.

A list of polynomial roots is turned into a polynomial in s using this *Mathematica* expression.

```
PolynomialFromRoots[roots_] :=
    If[Length[roots] == 0,
      1,
      First[Apply[Times,
        Map[{s-#}&,roots]]]]
PolynomialFromRoots[{4,2,1}]

(-4 + s) (-2 + s) (-1 + s)
```

Figure 5.5 A sample notebook shows the use of *Mathematica* to design and document a filter design paper.

Sec. 5.3 DSP Tools

In this notebook we use a list to keep track of the zeros, poles, and gain of a filter. Functions that transform filters will take as input a list of these three items and return a similar structure. We abbreviate the name of this structure to just GZP (Gain, Zeros, and Poles). The following function is then used to take one of these lists and transform it into a filter in the s domain. Note that we have used the pattern matching facilities of *Mathematica* to pick out the three elements of the input list.

```
FilterFromGZP[{gain_, zeros_,
        poles_}] :=
    gain*PolynomialFromRoots[zeros]/
        PolynomialFromRoots[poles]//N
FilterFromGZP[{2.4, {4,2,1},
                {12,10,7}}]

  2.4 (-4. + s) (-2.+ s) (-1. + s)
 ---------------------------------
    (-12. + s) (-10. + s) (-7. + s)
```

2 Chebychev Filters

The simplest high-order filters to design are the Butterworth and the Chebychev. The poles of a Butterworth low-pass filter are arrayed so that the filter's response is flat through most of its passband. As the frequency approaches the corner frequency, the gain quickly falls off. In some cases this characteristic is an advantage because the gain between DC and the corner frequency is nearly flat.

For a given stopband or transition band specification, filters with a much smaller variation in gain in the passband can be designed using the Chebychev polynomials. Chebychev filters do not have a flat response in the passband, but, as in Butterworth filters, the passband error can be made arbitrarily small.

The poles of the Chebychev polynomials are given by the following expression [Daryanani76]. This expression is a function of the desired order of the polynomial (**n**) and the maximum error (**amax**) in dB in the passband.

```
ChebychevPoles[n_,amax_] :=
  Block [{e},
    e = Sqrt[10^(amax/10)-1];
    Table[Sin[Pi/2(1 + 2k)/n]
        Sinh[ArcSinh[1/e]/n] +
    I Cos[Pi/2(1 + 2k)/n]
        Cosh[ArcSinh[1/e]/n],
        {k,n,2n - 1}]]
```

The **ChebychevPoles** function returns the location of the poles of a **n**-th order low-pass Chebychev filter with a cutoff frequency of 1 rps and a maximum pass-band error of **amax** dB.

```
ChebychevPoles[6,2]//N

{-0.0469732 - 0.981705 I,
 -0.128333 - 0.718658 I,
 -0.175306 - 0.263047 I,
 -0.175306 + 0.263047 I,
 -0.128333 + 0.718658 I,
 -0.0469732 + 0.981705 I}
```

As can be seen from the pole plot below, the roots of a Chebychev polynomial fall on an ellipse. This plot shows the roots as the maximum error in the passband is varied from 10^{-10} (the ones that look most like a circle) to a passband error of 1 dB (the rightmost arc).

```
PlotPoles[Flatten[Map[N[
    ChebychevPoles[16,#]]&,
    {10^-10,10^-4,.1,1}]]];
```

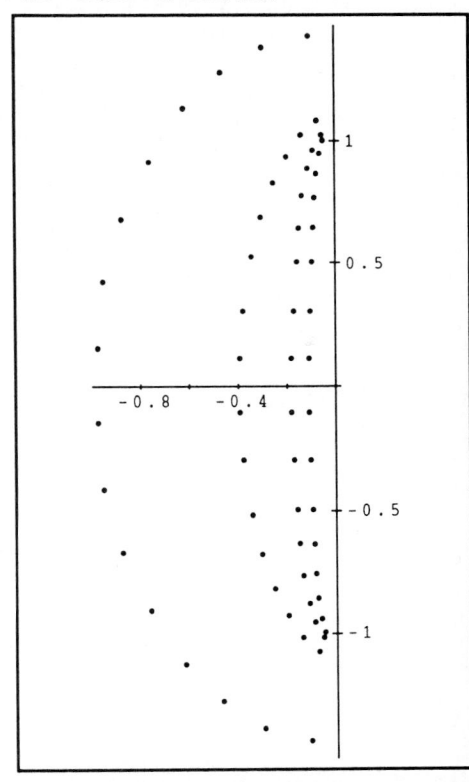

The next function computes a Chebychev low-pass filter and returns a list with the gain, zeros, and poles. Note that a Chebychev low-pass filter has only poles so the list of zeros is empty. The resulting GZP list can be passed to the filter transform routines to realize other types of filters (band-pass, band-reject, and high-pass). In this filter design function the gain at the corner frequency (1 rps) is adjusted so that it has a

Figure 5.5 (continued)

loss of **amax**. As will be seen in the plots to follow, this will set the maximum gain of the filter (at the peaks in the passband) to 0 dB.

```
ChebychevLp[n_,amax_] :=
  Block[{poles, gain},
    poles=ChebychevPoles[n,amax]// N;
    gain = FilterMag[
      PolynomialFromRoots[poles],
                   1/ (2 Pi)]// N;
    gain = 10^(-amax/20) * gain;
    Return[{gain, {}, poles}]]
```

The following plot shows the magnitude and phase response of an eighth-order Chebychev low-pass filter with a pass-band error of 1 dB.

```
ft=FilterFromGZP[ChebychevLp[8,1]];
FreqResponseRadians[ft,2,
             PlotRange->{-10,0}];
```

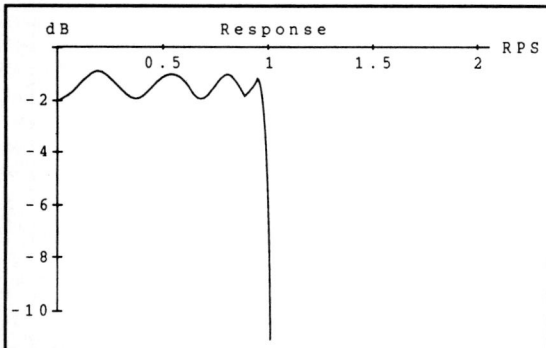

```
Plot[FilterPhase[ft,r],{r,0,2},
     AxesLabel->{" RPS", "Radians"},
     PlotLabel->"Phase Response"];
```

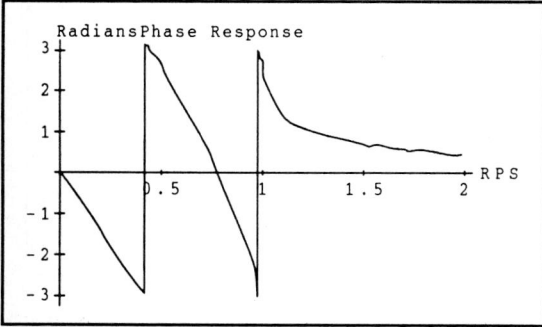

Chebychev filters can have an arbitrarily small error in the passband but this does not come for free. The following plot shows the gain at twice the corner frequency as a function of pass-band error. In each case an eighth-order Chebychev low-pass filter was designed. Note that if more error in the passband can be tolerated then a much sharper cutoff can be realized.

```
Plot[FilterDb[
  FilterFromGZP[ChebychevLp[8,e]],2],
    {e,.01,3},
    AxesLabel->{"Passband Error",
           "Gain at 2rps (dB)"}];
```

3 Band-pass Filters

Section 2 showed how to design a generic Chebychev low-pass filter. These low-pass filters can then be transformed into low-pass, high-pass, band-pass, and band-reject filters with arbitary cutoff frequencies. This section will show how to transform a low-pass filter into a band-pass filter. We use the gain, zero, pole structure to keep track of the filter parameters.

A low-pass filter is transformed into a band-pass by specifying the location of the two corner frequencies. We make this transform by substituting the following expression for **s** into the normalized low-pass filter [Daryanani76]:

$$s = \frac{s^2 + w0^2}{B\,s}$$

In these expressions **B** is the difference (in radians) between the two edges of the passband and **w0** is the geometric mean of the frequencies at the edges of the passband. The function **BpTransform** is used to transform a single root of the normalized filter into two new roots due to the substitution above. (The extra root at zero is ignored for now.)

```
BpTransform [roots_, w0_, B_] :=
  N[Flatten[Map[{B # / 2 +
    Sqrt [B^2#^2-4w0^2]/2, B # / 2 -
    Sqrt [B^2#^2-4w0^2]/2}&, roots]]]

BpTransform [ButterworthPoles [3],
        2Pi Sqrt[1000 2000],
        2Pi 1000]

{-1104.8 - 6450.39 I,
 -2036.79 + 11891.8 I,
 -3141.59 + 8311.87 I,
 -3141.59 - 8311.87 I,
 -1104.8 + 6450.39 I,
 -2036.79 - 11891.8 I}
```

Figure 5.5 (continued)

The function **LpToBp** transforms each of the poles and zeros in the original low-pass filter according to the **BpTransform** function. In addition, each zero in the original low-pass filter contributes a pole at zero, and, likewise, the original poles contribute a zero at DC. The difference between the number of poles and zeros tell us the number of roots at zero to add, and extra factors of **B** to add to the gain.

```
LpToBp [{gain_, zeros_, poles_},
        fp1_, fp2_] :=
  Block[{w0, B, RootDiff,
         ExcessPoles, ExcessZeros},
    w0 = 2 Pi Sqrt [fp1 fp2];
    B = 2 Pi (fp2 - fp1);
    RootDiff = Length[zeros] -
        Length[poles];
    If[RootDiff > 0,
      ExcessZeros = RootDiff;
      ExcessPoles = 0,
      ExcessPoles = -RootDiff;
      ExcessZeros = 0];
    {gain/B^RootDiff,
     Join[BpTransform[zeros, w0, B],
     Table[0,{ExcessPoles}]],
     Join[BpTransform[poles, w0, B],
     Table[0, {ExcessZeros}]]}]
```

This transform is applied to a third-order low-pass filter to determine a sixth-order band-pass filter with a passband between 1 kHz and 2 kHz and a maximum pass-band error of 3dB.

```
LpToBp [ChebychevLp [3, 3], 1000,2000]
                                //N
            10
{4.4443 10  , {0., 0., 0.},
    {-326.13 + 6478.28 I,
        -612.013 - 12157.1 I,
        -938.143 + 8836.1 I,
        -938.143 - 8836.1 I,
        -326.13 - 6478.28 I,
```

The frequency and phase response of this sixth-order band-pass is shown below.

```
flt = FilterFromGZP [LpToBp[
    ChebychevLp[3, 3], 1000, 2000]];
FreqResponse[flt, 4000]
```

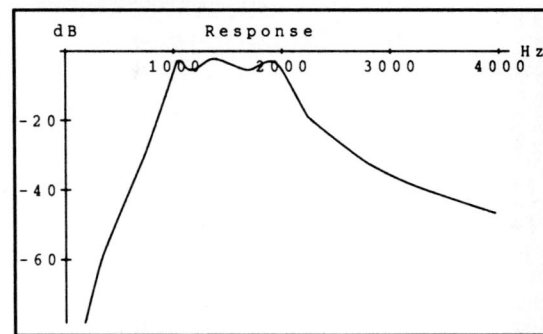

[Wolfram88] S. Wolfram, *Mathematica* (Redwood City, Calif.: Addison-Wesley, 1988).

[Daryanani76] G. Daryanani, *Principles of Active Network Synthesis and Design*, (New York: John Wiley and Sons, 1976).

Figure 5.5 (continued)

```
In[2]: =
    FilterFromGZP[ChebychevLp[8,1]]
Out[2] =
    .0153517 /
    ((0.350082 + 0.996451 I + s) (0.0350082 − 0.996451 I + s)
     (0.099695 − 0.844751 I + s) (0.099695 + 0.844751 I + s)
     (0.149204 − 0.564444 I + s) (0.149204 + 0.564444 I + s)
     (0.175998 + 0.198206 I + s) (0.175998 − 0.198206 I + s))
In[3]: =
    Chop[ExpandDenominator[FilterFromGZP[ChebychevLp[8,1]]]]
Out[3] =
    .0153517 /
    (0.0172267 + 0.107345 s + 0.447826 s² + 0.846824 s³ +
     1.8369 s⁴ + 1.65516 s⁵ + 2.42303 s⁶ + 0.919811 s⁷ + s⁸)
```

The Chop function is used here to drop the very small imaginary terms that are caused by roundoff error in the floating point calculations.

One of the more useful features of a symbolic math system is that it can be extended. Rules can be added or programs written to specialize the behavior of the system. In *Mathematica*, rules are defined using a pattern matching language much like Prolog. On the left-hand side of a rule the underscore character (_) indicates a wildcard position where any quantity can be substituted. Furthermore, the underscore character can be appended to a variable name to make a named wildcard variable. The pattern matching capability built into *Mathematica* is very powerful. The simple expression

```
foo[a_, b_, c_]
```

on the left side of a rule matches the function foo called with three arguments. The expression on the right-hand side of the rule will be used with the appropriate substitutions whenever the variables a, b, and c are used. A more complicated expression like

```
factorial[n_Integer]
```

matches any time the factorial function is called with an integer argument. The matching expression can include arbitrary *Mathematica* notation. For example

```
diff[a_ + b_]
```

can be used to pick apart a sum and define a new differentiation rule.

Mathematica notebooks also include the ability to display animations. This is useful as a way to show how simple parameter changes affect the solution, or to rotate a three-dimensional graph around its origin so the reader can more easily perceive its form. In my own work we have used *Mathematica* animations to show wave propagation solutions.

Readers of this chapter can get a sense of the readability of a notebook from the examples in Figures 5.4 and 5.5. A problem with *Mathematica* is that the language is new and probably unfamiliar to many readers. Consequently, notebooks should be written, much like mathematical papers, so that the general flow can be understood by skipping the equations [19]. A brief description of unusual syntax might be given the first time it is used. In this sense, a notebook is no different from a normal paper.

Mathematica includes elements of all the important characteristics of a system for creating and exploring interactive signal processing documents. First, *Mathematica* notebooks are organized hierarchically. Within a notebook, equations, paragraphs, mathematical results, and graphics are each cells that can be grouped into larger cells. Cells can be hidden (or closed) in such a way that only the first line of a cell is visible. From this information the reader can decide whether the rest of the cell needs further attention. The first line of large, grouped cells is typically used as a section title.

The help system in *Mathematica* is a simple example of hypermedia. A user can select a function in a notebook and ask for more information. A new window appears describing the usage of the function. These simple usage statements are handy, but the original definition will include a more complete description of the algorithm. The system would be even more useful if the user could ask about a function and immediately move to the part of the notebook where the function is first defined.

All of these features of a *Mathematica* notebook would not be interesting if notebooks could not be distributed. The essential information in a notebook is conventional ASCII text, which is easily moved through the email and computer networks. While only some computer systems support the complete notebook concept, all *Mathematica* systems can understand the data and the mathematics contained in a notebook. Thus, a reader with a version of *Mathematica* without the full notebook capability can study the printed version and still try the examples.

One disadvantage of systems like *Mathematica* is that strictly numerical calculations are not efficient. Symbolic manipulation programs are designed to work with any type of mathematical quantity. Thus, when performing a multiplication, the terms could be symbols, arbitrary precision integers (bignums), high-precision floating point numbers, or simple integers or floats in the machine's native format. Of these, only the native format calculations are fast. Still, a symbolic manipulation program must check for all of these possibilities each time it does an operation, and this type checking can be more expensive than the mathematical operation. Conventional programming languages do not pay this penalty because all types are known at compile time and the proper machine instructions generated ahead of time.

Finally, *Mathematica* is a commercial product that not everybody will be able to afford. Wolfram Research has put into the public domain a *Mathematica* notebook reader. This notebook reader does not have any of *Mathematica*'s mathematical ability, but it does allow people to view a notebook and play the animations on any Macintosh computer. My own cochlear notebook [2] has been published on paper and with a floppy disc containing the *Mathematica* notebook and the notebook reader to allow the material to have the widest possible distribution.

5.4 DESIGN ISSUES

Sections 5.2 and 5.3 have described the form an interactive document might use to describe a signal processing algorithm. How should the author structure his or her writing to make the best use of the technology available to describe a signal processing problem and its solution? There is no question that an interactive signal processing document requires new skills from an author. Some of these skills are the subject of this section.

The benefits of interactive signal processing documents described here do not come for free. Certainly the design and writing of such a document takes more thought and care than conventional papers do. When the interactive document is done, there is no easy way to disseminate its electronic form. Both of these problems should diminish as people become more familiar with this new medium for research and publishing.

This section describes some of the factors that make an interactive signal processing document a success. Certainly the biggest factor is writing the document so that it invites the reader to interact with the material. Fortunately, this is easily addressed by the author. Other factors, for example distribution and notation, are more difficult. Each of these difficulties are addressed in the remainder of this section.

5.4.1 How to Write an Interactive Signal Processing Document

Designing an interactive document is not easy, but the effort is worthwhile since writing an interactive mathematical document becomes as much a learning experience as reading it. Teaching new material is often the best way to learn it. By preparing an interactive document, one is forced to study the material as a reader, and by having a tool such as *Mathematica* it is possible to explore more of the subject area. In addition, when a common system is used to research and present a new result, the effort in creating an interactive document is minimized.

Making a document interactive adds another dimension to the writing task. In some ways this makes the task more difficult, but in other ways the task becomes simpler. Just as a picture is worth a thousand words, an interactive example showing how frequency response changes with pole location can be worth a thousand figures.

The interactive dimension might require extra work by the author of a signal processing document. One can always use the tools described in this chapter to write a conventional paper; it would not be any more or less efficient than using a word processor. Fortunately, some of the interactive features described here make the writing process easier. It is often easier to show a reader an interactive graphic than it is to explain it in words. Other parts of an interactive signal processing document, such as working models and simulations that would otherwise never leave the lab, require more effort to polish and make ready for publication. As always, it is up to the author to decide on the proper amount of effort to apply. A short note explaining a new algorithm for a research group probably doesn't need as much polish as an undergraduate signal processing text.

One might think that this extra work would slow the rate of research. This is not necessarily true. Preliminary signal processing ideas are already exchanged within one's research or development group as small code samples and in interactive discussions. The point of this chapter is to describe the benefits of allowing a reader to more easily benefit from this rich form of interaction.

There are four skills and practices that should be remembered when writing an interactive signal processing document. They are:

1. Use good writing and graphics.
2. Iterate the design with real readers.
3. Guide readers to interactions.
4. Limit interactions to fit the reader and available computer power.

That good writing and graphics are important for an interactive document should go without saying. It is especially important that the ideas expressed by Tufte [20] should be applied to the interactive display and models. The remaining techniques for designing an interactive signal processing document will be discussed next.

Adding an interactive component to a signal processing paper is not a panacea. Just as there are bad papers, there will be bad notebooks. Fortunately the technology encourages a closer collaboration between the author and the reader. A successful interactive document will often go through several iterations. When first writing an interactive document it is hard to know how much detail to include or what kinds of models are useful to any particular reader. At successive stages, observation of how readers interact with the document will help guide its evolution.

The key task is to design the interactive document so that the reader can profit from the information and the changes can be shown at interactive rates. The first part of this problem, inviting the reader to play with the model, is easily solved with words. Telling the reader, "Here's some equations, play around with them," will only help the most motivated readers. Instead, the introduction of an interactive document could guide the reader to those parts of the document that can be modified, such as is done in this example [2]:

> The best way to interact with this notebook is to read the description, study the examples, and then modify an example to see how different parameters give different results. For example, an appendix to this report describes digital filtering and provides

functions to design first and second order filters. Much can be learned about digital filtering by combining these filters and studying the resulting frequency response or pole-zero plots.

Readers might also want to modify this model to better fit their own experience or ideas. For example this notebook describes a relatively simple model of the effects of the outer and middle ears on the sound. A reader might be interested in providing a better model or removing the outer and middle ear filters completely and studying the change in response. As another example, this report describes a simple Automatic Gain Control (AGC) to compensate for the large range of sounds produced by humans. This notebook explores several variations on the basic AGC but readers might want to try their own.

Simple examples, spread liberally through the text, encourage the reader to "kick the tires." It is probably not important that every reader understand the details of a filter design algorithm. But labeling a simple figure showing a Butterworth filter response with the *Mathematica* text

```
PlotFilterResponse[DesignButterworthLowPass[8,1000]]
```

makes it clear to casual readers that this is a low-pass filter with a cutoff frequency of 1,000 Hz. If the `1000` is changed to `2000`, the cutoff frequency should change by an octave. The reader might not appreciate exactly what an eighth-order filter is but should see that the filter attenuates more quickly with higher order.

Readers also need to be guided toward those portions of the document that are interactive. Not every part of a document can be changed in a meaningful way, but those parts that can be changed or perform an action for the user (play a sound or display an animation) should be marked. If the reader changes the title of the paper it probably will not automatically change the contents. Highlighting the input text in a special font or with graphics tells the reader which parts of the document can be changed.

It is easy for both the reader and the computer to be overwhelmed by an interactive document. Without adequate guidance the reader might wander down paths where there is no hope that any meaningful conclusion can be reached. In addition, an all-encompassing simulation would model many details that are not interesting to the reader. Limiting the domain over which the reader can change the simulation helps control the amount of processing power that is needed to answer a reader's question. If the problem is well defined it might be possible to precompute all of the interesting results and then use interpolation to display the correct simulation result.

Figure 5.6(a) shows a simulation where the reader has too much freedom. In this example, the reader can place the poles of a filter at any place on the s-plane and see the resulting frequency response. This gives the reader too much freedom and it is unlikely, for example, that the special properties of the classical filter design techniques will be found. In addition, it will be hard to find an easily affordable

Sec. 5.4 Design Issues 199

document reader that will have the computational horsepower to keep up with the user's requests.

Figure 5.6(b) shows a modified version of this example where the reader is limited to studying the relationship between the Butterworth and Chebychev filters. By moving a knob that controls the eccentricity of the pole locations, the reader can see the effect on the pass-band ripple and the filter attenuation rolloff. This is now simple enough that even a dozen precomputed frequency responses would show the concept to the reader, without the reader knowing that the computations were done ahead of time. It would even be possible to precompute audio examples so that the reader can listen to the effect of each filter.

I do not mean to say that a paper including an example like that shown in Figure 5.6(a) is not useful. There will always be readers who will understand the basics of such an example and will want to explore the effect that quantization has on pole location or any number of ideas that never occurred to the original author. It is up to the author to carefully draw the line between guiding the reader and allowing the reader to become lost in the details.

5.4.2 Problems with Interactive Signal Processing Documents

Other parts of the problem are not as easy for the author to address. These problems include choosing a system, publishing the electronic document, and picking a notation. Each of these problems will be addressed in the remainder of this section. Other issues, such as version control or keeping track of what has changed between versions, and maintaining correctness in the face of changes by the reader, are secondary problems and are not discussed here.

Choosing a system for writing an interactive signal processing document is not easy; the ideal system does not exist yet. I have used *Mathematica* for writing several interactive signal processing documents. It has many of the desired characteristics but is lacking in other areas. *Mathematica* would be a much better environment for creating electronic DSP notebooks if it had better multimedia support, was more efficient at strictly numerical calculations, and if it had better text formatting and graphics support. Hopefully these and other problems will be addressed in future releases of the software.

On the other hand, it is hard to believe that any one system will solve everybody's problems. Large, all-encompassing tools tend to be unwieldy and not solve anybody's problem very well. Instead it will probably be best if a user can mix a tool for filter design from one vendor with a special-purpose accelerator from another to make the most efficient research and learning environment.

Publishing notebooks and other forms of electronic documents is not easy. Magazines and journals usually used to disseminate research results have evolved efficient mechanisms and designs to effectively communicate the printed word. But

200 Interactive Signal Processing Documents Chap. 5

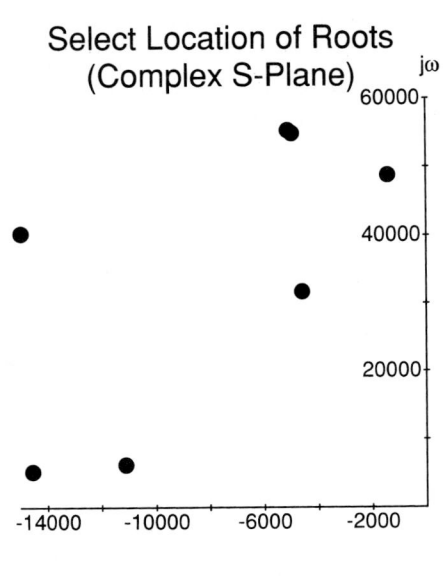
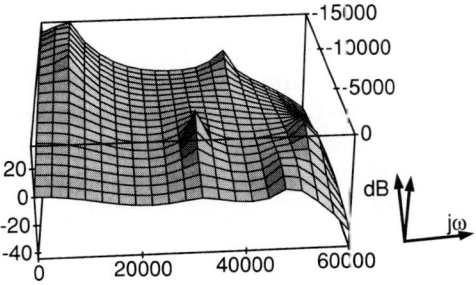
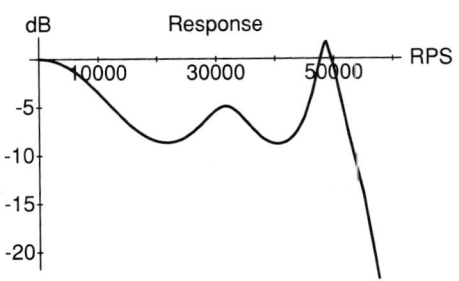

(a)

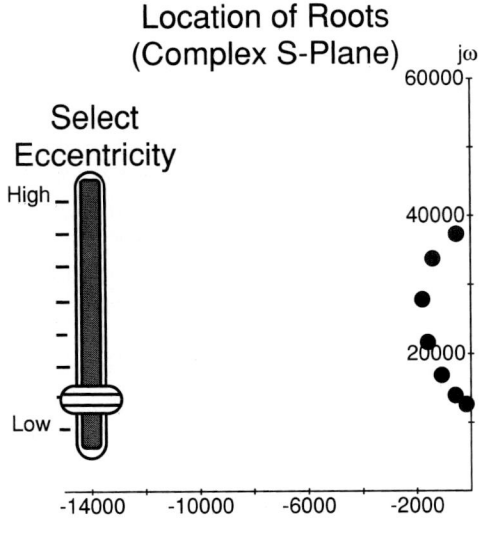
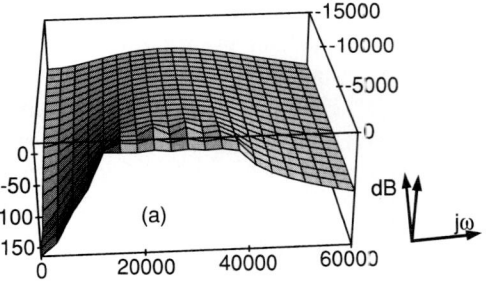
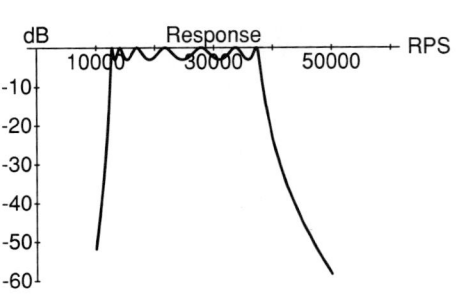

(b)

Sec. 5.4   Design Issues

a single floppy, or other form of electronic media, can cost as much as the magazine it accompanies.[3]

One of the more successful schemes for electronic publishing is based on the international computer networks. Dongarra at Argonne Labs maintains an electronic mail system for distributing many large numerical software packages [21]. Users can send electronic requests to a special address and receive more information or the software by return mail. Another scheme is to broadcast the software on one of the computer bulletin boards. This is commonly done, for example, on the Usenet bulletin board comp.sources [22].

Unfortunately, not everybody has access to the computer networks. Instead, electronic material is often made available on floppy disks that can be read on a user's own computer. For instance, my own report on the implementation of a cochlear model [2] was published as a technical report so it could be accompanied by a floppy disk containing the *Mathematica* notebook. A recent issue of the *Communications of the ACM* [10] included an advertisement for floppy disks containing hypermedia examples.

An additional problem is that an electronic document is not as convenient as a magazine or a book. Curling up with a computer will probably never have the same appeal as curling up with a good book, but the next generations of portable computers should make this easier. For example, my own notebook was designed so that it can be read as a normal paper but without all the benefits of an electronic document.

Finally, there is the issue of notation. Within any one technical area, for example, signal processing or high-energy physics, the notation is well established, but it often differs widely between areas. Even a concept as simple as an integral is written in many different ways with marks to indicate different flavors of integration.

Mathematica solves this problem by defining a new language based on the ASCII alphabet. Wolfram has exchanged the rich notation that scientists and math-

---

[3] There are more and more examples of printed works accompanied by electronic media. Often books on microcomputer programming include a floppy disc with programming examples [for example, *Programming with MacApp* by David Wilson]. *develop*, a magazine that Apple publishes for its software developers, includes a CD-ROM with every issue. Each CD-ROM contains the complete text for all the issues of the magazine published to date and source code. According to the editor, the magazine and the CD-ROM each cost approximately $2 in 1990. The *Mathematica Journal* distributes a floppy disk with source code with each issue, but the articles are written like a conventional paper accompanied by source code.

Figure 5.6 Two examples are shown of interactive signal processing models. The first example (a) allows the user to pick any location for the filter poles. It probably provides too much freedom for most users to gain any useful insight. The second example (b) allows the user to change the eccentricity of the pole locations and compare the Butterworth and Chebychev methods. Each figure shows the locations of the poles in the upper left quadrant of the s-plane, the magnitude of the response in the same quadrant, and the conventional frequency response.

ematicians have evolved through the ages for a very precise functional notation. For example, one writes

```
Laplace[ f[t], t, s]
```

to represent the Laplace transform of a function of t in terms of the complex variable s.

Other programs, such as *Milo*, by a company called Paracomp, use a more conventional mathematical notation, but their knowledge of mathematics is limited. A research system called *CaminoReal* [23] gives authors an interactive interface to a writing program and symbolic algebra programs. The result is a nicely formatted paper document without the hypermedia and interactive features that are part of an electronic notebook. Perhaps the best solution is to allow users of symbolic math programs to define graphical templates, which are used when the system wants to translate its internal representation into something to be displayed to the reader.

5.5 RESEARCH ISSUES

Electronic notebooks are possible today and the resulting document, for example [2], can successfully communicate a signal processing idea. Currently available software, however, limits the topics that can be covered readily. A notebook on filter design is relatively straightforward. Graphics and animations can show most of the ideas, but a notebook on audio compression would be frustrating without the ability to include high-quality audio examples in the notebook.

Some needs of future electronic notebooks go without saying. There will always be a need for more computational horsepower to enable more realistic models to be built. Higher quality audio and easier ways to integrate video will allow more signal processing topics to be described.

Other needs, such as providing instructional directions and tuning the human interface, are difficult issues. It is important that an electronic notebook encourage the user to interact with the material by making it easy for the reader to navigate through the notebook and ask reasonable questions.

Before concluding, two areas of future work are worth noting. These ideas have been alluded to in other parts of this chapter but they are worth repeating. First, more powerful symbolic tools will make it easier for researchers and readers to explore difficult signal processing problems. Second, there are many software engineering problems in designing a system that allow users to move between different types of simulations and include the necessary tools for a complete system.

Several improvements to the mathematical symbolic manipulation world would be nice to see. More powerful symbolic tools will allow more difficult DSP problems to be solved without resorting to brute force. Allowing the user to specify the cost of mathematical operations and then automatically optimize an algorithm as is done with ADE would make it easier to realize the solution to a DSP problem using the available hardware. Finally, better support for strictly numerical calculations will allow a system to be designed and the resulting algorithm applied to real data. All of these improvements would make it easier to write an interactive notebook to describe the solution to a signal processing problem.

The remaining problem is one of software engineering. Symbolically manipulating mathematical symbols is only part of the problem. An author of an interactive signal processing document will need other tools. Tools such as text formatters, spelling checkers, and drawing programs are needed but probably not within the domain of expertise of most people designing signal processing environments. A software component system is one solution to this problem.

ACKNOWLEDGMENTS

I would like to thank a number of people for help with this chapter. Theo Gray, Nancy Blachman, and Paul Abbott at Wolfram Research have been invaluable in helping me master *Mathematica*. They were very understanding when I tried to use *Mathematica* in ways that were never envisioned. I would also like to thank Richard F. Lyon, Robert Hon, Neenie Billawala, Monica Ertel, Pam Lau, and Nancy Tague for their help in producing the notebook [2] that led to this chapter. Finally, Richard Lyon, Michele Covell, Richard Fateman, Brian Evans, James McClellan, Norm Carter, and Dennis Arnon have all made many useful comments about the ideas expressed in this chapter.

REFERENCES

[1] M. Slaney, "Interactive Signal Processing Documents," *IEEE ASSP Magazine*, 7 (1990), 8–20.

[2] M. Slaney, "Lyon's Cochlear Model," *Apple Computer Technical Report #13*, Corporate Library, 20525 Mariani Avenue, Cupertino, Calif. (1988).

[3] "Enter *Mathematica*," *MacUser* (November 1988), 199–216.

[4] J. Barwise, "Computers and Mathematics," *Notices of the American Math Monthly*, 35 (1988), 1333–49.

[5] R. J. Fateman, "A Review of *Mathematica*," *Journal of Symbolic Computation*, to appear.

[6] *IMSL Library Reference Manual*, IMSL Inc., Houston, Tex. (1980).

[7] *MATLAB for Macintosh Computers*, The MathWorks, Inc., 24 Prime Park Way, South Natick, Mass. (1989).

[8] B. Evans and J. H. McClellan, "Symbolic Transforms with Application to Signal Processing," *Mathematica Journal*, 1 (Fall 1990), 70–80.

[9] M. Covell, "An Algorithm Design Environment for Signal Processing," RLE Technical Report 549, Cambridge, Mass.: MIT, 1989.

[10] *Special Issue on Hypertext*, in *Comm. of the ACM* (July 1988).

[11] P. J. Denning, "Announcing Literate Programming," *Comm. of the ACM*, 30 (1987), 593.

[12] *Programs for Digital Signal Processing* (New York: IEEE Press, 1979).

[13] G. Kopec, "The Signal Representation Language SRL," *IEEE Trans. on ASSP*, ASSP-33 (1985), 921–32.

[14] C. S. Myers, "Signal Representation for Symbolic and Numerical Processing," RLE Technical Report 521, Cambridge, Mass.: MIT, 1986.

[15] W. Dove, C. S. Myers, and E. E. Milios, "An Object-Oriented Signal Processing Environment: The Knowledge-Based Signal Processing Package," RLE Technical Report 502, Cambridge, Mass.: MIT, 1984.

[16] *Photoshop User's Guide*, Adobe Systems, Inc., Mountain View, Calif. (1990).

[17] S. Wolfram, *Mathematica: A System for Doing Mathematics by Computer* (Redwood City, Calif.: Addison-Wesley, 1988).

[18] T. W. Gray and J. Glynn, *Exploring Mathematics with Mathematica* (Redwood City, Calif.: Addison-Wesley, 1991).

[19] D. E. Knuth, T. Larrabee, and P. M. Roberts, "Mathematical Writing," MAA Notes No. 14, The Mathematical Association of America (1989).

[20] E. R. Tufte, *Envisioning Information* (Cheshire, Conn.: Graphics Press, 1990).

[21] J. J. Dongarra and E. Grosse, "Distribution of Mathematical Software via Electronic Mail," *Comm. of the ACM*, 30 (1987), 403–07.

[22] J. S. Quarterman and J. C. Hoskins, "Notable Computer Networks," *Comm. of the ACM*, 29 (1986), 932–71.

[23] D. Arnon, R. Beach, K. McIsaac, and C. Waldspurger, "CaminoReal: An Interactive Mathematical Notebook," in *Proceedings of the International Conference on Electronic Publishing, Document Manipulation and Typography*, ed. J. C. van Vliet (New York: Cambridge University Press, 1988).

6

BLACKBOARD SYSTEMS FOR KNOWLEDGE-BASED SIGNAL UNDERSTANDING

- Norman Carver
  University of Massachusetts
- Victor Lesser
  University of Massachusetts

A sophisticated signal understanding system requires knowledge-based interpretation techniques as well as sophisticated signal processing techniques. Blackboard systems are the most widely used artificial intelligence (AI) frameworks for understanding and interpretation problems. This chapter introduces the blackboard model of problem solving and discusses how blackboard systems can be used for understanding problems. The chapter does not specifically show how blackboard systems can be used for knowledge-based signal processing. Integration of signal processing and understanding via blackboard systems will be addressed in Chapter 7.

We use the term signal *understanding* rather than signal *processing* here, because our focus is on the development of "high-level" descriptions of data in terms of the "situations" it represents. Another way of saying this is that signal understanding involves the meaning of the data, while signal processing is generally limited to the extraction of important features—i.e., semantic versus syntactic descriptions. The notion of understanding has also been associated with enabling a system to *act* on the data in an appropriate manner. Thus, a sound understanding system must do more than just identify harmonically related acoustic signals; it must also determine their source—e.g., a ringing telephone (that may need to be answered). Likewise, a radar understanding system must do more than just identify echoes; it must also be able to explain the echoes in terms of the movements of certain vehicles, environmental disturbances, and so on.

Signal understanding problems cannot typically be solved by applying the kind of algorithmic data transformations used in traditional signal processing. This is because signal understanding applications often involve a great deal of underlying uncertainty: critical features cannot be accurately or precisely "extracted" or with the types of features that can be extracted the space of situations to be considered is still very large. For example, while there may be models of the output that individual sources/targets should ideally elicit from a sensor, it may not be possible to have exact models for the interactions of every possible combination of sources. Add to this the uncertainty over the effect that environmental conditions or other phenomena may have on the sensor output and the problem is very difficult to solve algorithmically.

Blackboard systems (and AI techniques in general) are appropriate for understanding problems that involve very large answer spaces as well as uncertain, errorful, or incomplete data and problem-solving knowledge. These characteristics can produce a combinatorial explosion in the number of situation models that must be considered (this issue is pursued further in section 6.1). Solving such problems requires an approach that is very different from the algorithmic approaches of traditional signal processing. Instead of attempting to directly determine the best complete answer, in blackboard-based problem solving partial possible solutions are incrementally constructed and refined as part of a *search* process. In such a process, the data (which may have already undergone algorithmic signal processing) is first examined to try to produce likely partial models. These preliminary potential models drive further processing, influencing decisions about the best models to pursue and the best ways to pursue them (they may also influence parameter settings for further algorithmic signal processing). In other words, instead of trying to develop a single strategy that is appropriate for all situations, a blackboard system is structured so that it is able to *adapt* to specific situations by *opportunistically* selecting among a large set of diverse problem-solving methods. This requires a system with a great deal of knowledge and flexibility. For instance, significant amounts of task-specific heuristic control knowledge must be encoded and applied in order to decide what to do next—to constrain the search and limit the portion of the answer space that must be examined.

The chapter begins with a section that identifies signal understanding as an example of a class of artificial intelligence problems called *sensor interpretation* problems. The next section introduces the *blackboard model of problem solving* and discusses why it is appropriate for signal understanding problems. The third section presents an example aircraft monitoring application to demonstrate how blackboard systems can be used for interpretation problems. Section 6.4 expands on the example by discussing the issues that affect the structuring of interpretation problems and problem-solving knowledge. Section 6.5 examines the critical issue of control in blackboard systems. It begins with an introduction to the major issues that are relevant to the control of blackboard systems and then examines several of the

Sec. 6.1 Signal Understanding as Interpretation

control schemes that have been developed for blackboard sytems. The chapter concludes with a summary of the major points in the chapter and a discussion of current blackboard systems research that is relevant to signal understanding.

6.1 SIGNAL UNDERSTANDING AS INTERPRETATION

Signal understanding is one example of a class of AI problems often referred to as *sensor interpretation* problems. Interpretation involves the determination of high-level, conceptual explanations of sensor and other observational data. The interpretation process is based on a specification of the relations between the data types and a set of abstraction types. These types form a hierarchy like the one for an example aircraft monitoring system shown in Figure 6.1.

An interpretation system analyzes the available data and creates hypotheses that provide possible explanations for subsets of the data. Each hypothesis represents a possible instance of its associated interpretation type. Higher-level hypotheses are said to explain lower-level hypotheses while, conversely, lower-level hypotheses are said to support higher-level hypotheses. Answers are typically in terms of top-level type hypotheses since these types represent the most abstract and most encompassing level of explanation for the data. The lowest levels (i.e., the

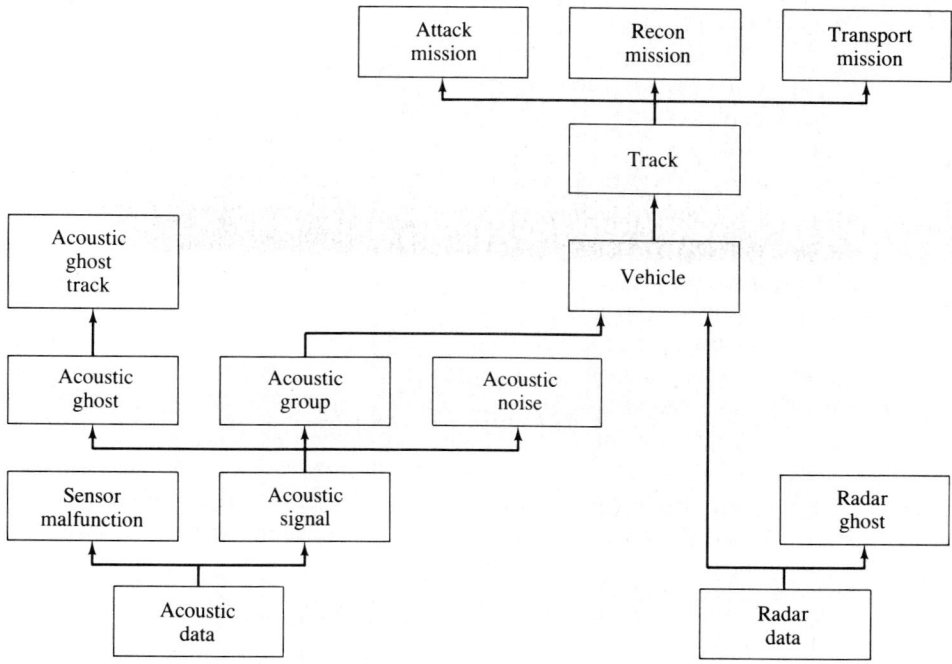

Figure 6.1 Interpretation type hierarchy for example aircraft monitoring application.

leaves) of the hierarchy represent the data upon which the interpretation process will be based.[1]

Each interpretation type has a set of "attributes" and hypotheses include corresponding *slots*. The relations between the types define certain constraints on the acceptable values of these slots. Thus, the creation of relations between hypotheses requires that these constraints be checked. In other words, the slots of lower-level hypotheses must meet certain constraints in order to create a particular type of higher-level hypothesis as a potential explanation of the lower-level hypotheses. The slot values of the lower-level hypotheses also define the slot values of consistent higher-level hypotheses.

In the aircraft monitoring example of Figure 6.1, data from acoustic and radar sensors (i.e., Acoustic Data and Radar Data) is abstracted and correlated to identify potential aircraft sightings (Vehicle hypotheses), movements (Track hypotheses), and goals (Mission hypotheses) that might explain the sensor data. Each abstraction level represents the application of additional constraints. For instance, Track hypotheses must have particular vehicle ID and movement slot values in order to be able to be consistent with each particular Mission explanation. A Track's slot values are defined/constrained by the slot values of its supporting Vehicle hypotheses which are in turn defined by the slot values in their supporting data or hypotheses. Besides the top-level Mission types, the hierarchy includes other top-level types—like Acoustic Noise—that are not of interest as answers. These "nonanswer" types are useful for interpretation because they can provide alternative explanations for the data. The blackboard system example in section 6.3 includes further explanations of the abstraction types in Figure 6.1.

Interpretation is based on the notion of *abduction*: if B types *cause* A types, then when an A is observed we might infer (hypothesize) a B as an *explanation* for the A. Likewise, given a possible B, in order to provide *support* for the B we can look for an A instance that the B should have caused. An interpretation type hierarchy like that in Figure 6.1 can be viewed as a *causal hierarchy*. For example, an aircraft on a certain Mission will cause the aircraft to travel a Track with certain attributes. This eventually causes acoustic signals with certain attributes, causing the acoustic sensor to produce corresponding data (the reasons why we do not want to go directly from the top-level types to the data are discussed in sections 6.3 and 6.4). The distinction between the explanation and causal views of the type relations is analogous to the difference between analytic and generative models [36].

Abductive inferences are uncertain; while plausible, they are not logically correct. Abductive inferences are uncertain because there can be multiple causes (and thus multiple possible explanations) for data and hypotheses. For example, it may be that both B and C types cause A types and that there is a C instance that

---

[1] Note that data types may feed into the hierarchy at different levels. Here Radar Data immediately provides support for Vehicles, while Acoustic Data must be abstracted to get to the Vehicle level. Likewise, it would be possible to have intelligence reports or other kinds of evidence that could provide direct evidence for Mission-types.

Sec. 6.1   Signal Understanding as Interpretation

is the true cause (explanation) for the A mentioned above. This potential for alternative explanations for data and hypotheses is the primary source of uncertainty in interpretation inferences (and thus interpretation hypotheses). However, besides this uncertainty inherent in abductive inferences, a number of other factors can contribute to uncertainty about the correctness of interpretation hypotheses. For example, because of sensor resolution limitations, data attributes like position and frequency class may be imprecise. As a result, the validity as well as the correctness of interpretation inferences may be uncertain (since it may not be possible to say the data conclusively meets the necessary interpretation constraints). Also, sensors may malfunction, causing events in the environment to be missed or data to be produced for nonexistent events.

Because abductive inferences are uncertain, interpretation hypotheses are uncertain and the primary goal of an interpretation system is to resolve its hypothesis uncertainty sufficiently to produce acceptable answers. There are two main techniques resolving hypothesis uncertainty: *incremental hypothesize and test* (also known as *evidence aggregation*) and *differential diagnosis*. Incremental hypothesize and test means that when a hypothesis has only partial support evidence, the system should attempt to find the complete support that the hypothesis would have if it were correct. Differential diagnosis means that the system should attempt to discount the possible alternative explanations for a hypothesis' supporting evidence.

Hypothesis correctness can only be guaranteed by doing complete differential diagnosis—i.e., discounting all of the possible explanations for the supporting data. Even if complete support can be found for a hypothesis, there may still be alternative explanations for all of this support. However, while complete support cannot guarantee correctness, the amount of supporting evidence is often a significant factor when evaluating the belief in a hypothesis (this is the basis of hypothesize and test). For example, once a Track hypothesis is supported by correlated sensor data from a "reasonable" number of individual positions (Vehicle hypotheses), the belief in the Track will be fairly high regardless of whether alternative explanations for its supporting data are still possible. Blackboard systems typically have been limited to hypothesize and test strategies because of the characteristics of their evidence and control frameworks (and because interpretation problems require constructive problem solving—see below). This issue will be discussed further in section 6.5.5.

In order to determine whether hypotheses are sufficiently certain and to make control decisions about the hypotheses to work on next, an interpretation system must be able to represent the level of uncertainty in its hypotheses. Hypothesis uncertainty is typically represented by attaching numeric "ratings" to the hypotheses when they are created or modified. However, the rating schemes used in most blackboard-based interpretation systems have been ad hoc—i.e., they have not used formal numeric representations of uncertainty (like probabilities) and they have not accurately treated interpretation inferences as abductive inferences. The development of techniques for reasoning about uncertainty is a major area of research in AI (see for example [8,41]) and is beyond the scope of this chapter.

While hypothesis uncertainty is an important issue, the key reason why interpretation problems require AI techniques is that solving them generally means a *search* process. The reasons for this were discussed in the introduction to this chapter. However, in order to better understand what makes these problems special, it is necessary to understand the differences between interpretation problems and similar problems that can be solved with *classification* techniques. Clancey [6] distinguishes between two classes of problem solving: classification problem solving and constructive problem solving. In classification problem solving, the problem solver selects "the answer" from among a pre-enumerated set of possible solutions.

"Classification problems" may sometimes be solved through pattern matching: comparing an object (or a processed representation of the object) against a set of templates and picking the "best match." This sort of approach has been used for isolated word recognition [43]. For "classification problems" in which evidence must be aggregated to make a selection, there are well-developed numeric evidential reasoning techniques like the Dempster-Shafer calculus and Bayesian networks [41] that can be used. For example, simple medical diagnosis problems [1,42] can be solved by entering evidence into a belief network (of findings and diseases) until the answer (the diagnosis) can be identified with sufficient certainty [41,42] (though determining the best evidence to gather may be critical to acceptable performance in such systems).

In constructive problem solving, both "the answer" and the set of possible solutions are determined as part of problem solving. Constructive problem solving is required when the set of possible solutions cannot be enumerated prior to problem solving. Interpretation problems generally require constructive problem solving because of the combinatorics of their answer spaces [2]. For example, in aircraft monitoring, an (effectively) infinite number of different Track hypotheses (possible aircraft movements) must be considered for each potential aircraft and an indeterminate number of Track hypotheses may be correct since the number of aircraft that will be monitored is unknown. The possibility of multiple aircraft also produces a combinatorial number of data combinations that may have to be considered due to "correlation ambiguity" [28].

Formal numeric techniques like the Dempster-Shafer calculus and Bayesian networks are often not directly applicable to constructive problem solving because they require a complete "frame of reference." Thus, in at least one sense, interpretation problems are more difficult than classification problems. In fact, constructive problem solving requires numerous capabilities that are not required for classification problem solving. For instance, constructive problem-solving systems must be able to represent incomplete and imprecise hypotheses, incrementally create and extend such hypotheses, maintain large numbers of these hypotheses, and apply significant amounts of knowledge to focus the construction process (i.e., control the search for the answers). This last point is critical because a combinatorial number of hypotheses may be able to be created to explain the data, creating each hypothesis may be computationally expensive, and a variety of methods may be used to resolve the uncertainty in each of them.

The distinction between classification and constructive problem solving is also

Sec. 6.2   Blackboard Systems

relevant to understanding why blackboard systems have typically relied on hypothesize and test strategies. Differential diagnosis requires an understanding of the possible alternative interpretations for the supporting data of a potential answer hypothesis. This is relatively straightforward in "classification problems" since the set of alternatives can be fully enumerated. In order to apply differential diagnosis techniques to "constructive problems," more sophisticated methods for representing and evaluating alternatives is required (see section 6.5.5) because the alternatives cannot be fully enumerated.

6.2 BLACKBOARD SYSTEMS

Blackboard systems are the most used artificial intelligence architectures for sensor interpretation and related problems (see, for example, [4,15,28,30,32,35,37,39,44]). This section introduces the blackboard model of problem solving and examines the major elements of a basic blackboard architecture. Section 6.3 uses an aircraft monitoring example to make the concepts presented here more concrete. Sections 6.4 and 6.5 expand on the issues involved in implementing an interpretation problem using a blackboard framework.

The concept of a blackboard system originated with the Hearsay-II speech understanding project [18,30,31,40]. Understanding connected/continuous speech is a difficult problem. It involves the kinds of uncertainties and combinatorics that were discussed in the chapter introduction and section 6.1: variability in the speaking process of different people, environmental noise and distortions in the sensing process, incomplete/imprecise models of speech, combinatorial explosion of model interactions (e.g., co-articulation phenomena result in words "looking" different depending on the adjacent words), and so on. This means that the speech understanding problem cannot be solved using classification techniques like template matching (though this can succeed for the recognition of isolated words). Understanding connected speech requires a search process that generates and evaluates numerous uncertain, partial solutions.

The Hearsay-II group came up with a system based on the following idealized model of problem solving: a group of experts sits watching solutions being developed on a blackboard and whenever an expert feels that he can make a contribution toward finding the correct solution, he goes to the blackboard and makes appropriate changes or additions. The first thing to note about this idealized model is that it is very different from the highly structured and deterministic model that underlies standard computer programming languages. In the blackboard model, there is no predetermined order in which the experts apply their knowledge and no predetermined order in which data/hypotheses are pursued.

Among the key ideas behind this idealized model are that problem solving should be both *incremental* and *opportunistic*. That is, solutions should be constructed piece by piece and at different levels of abstraction, basing the selection of the methods to be used on the data that is available and on the intermediate state of the possible solutions. Solutions are constructed using an "island driving" strategy: hypotheses that are judged to be likely ("islands of certainty") are incrementally abstracted and extended to produce a complete final solution. This opportunistic strategy allows the system to work where it can make the best progress instead of being limited to a fixed, predetermined strategy that may not be best for every problem situation. In speech understanding, for example, well-sensed words can constrain the search for poorly sensed words—as compared with a fixed beginning to end strategy.

Other key ideas behind the model are that problem-solving knowledge should be partitioned into a set of independent modules and that alternative solution hypotheses should be contained in an integrated, global database. In conventional algorithmic problem solving, a single algorithm/program encodes all of the knowledge needed to solve the problem. This makes it difficult to change the control strategies and the methods for solving (particular aspects of) the problem. Partitioning the problem-solving knowledge into a set of independent modules (the "experts") provides the system with the flexibility to experiment with alternative control strategies and alternative problem-solving methods. The ability do this kind of experimentation is crucial for the successful development of knowledge-based systems. An integrated, global database permits module independence because it assures that modules can recognize when they should be invoked without requiring the modules to directly invoke each other. In addition, because there is a single integrated representation al all the developing solution hypotheses, it is possible to have multiple, alternative solution paths being pursued in parallel and to be able to recognize the relationships among the paths.

We will now look in more detail at the architecture of blackboard systems. Any problem-solving architecture must prescribe methods for structuring and using three major components: domain knowledge, a database, and control knowledge. Domain knowledge consists of the task-specific knowledge and procedures that are used to actually solve the problem. The database holds the initial data, intermediate results, and the answer. Control knowledge includes both task-specific and task-independent knowledge about how/when to use the domain knowledge to best solve the problem. The basic blackboard problem-solving model prescribes a database structure and domain knowledge interaction protocols, but says little about control knowledge (control is discussed further in section 6.5).

A blackboard system is composed of two main components: the blackboard and a set of knowledge sources (or KSs). Figure 6.2 shows the relationships between these components. The blackboard is a shared, global database that contains the data and the developing hypotheses. The knowledge sources contain the domain knowledge of the system (they are the "experts" of the idealized model described above). KSs examine the blackboard and construct new hypotheses or modify existing hypotheses when appropriate.

The knowledge sources are intended to be totally independent (modular): their execution should not explicitly depend on the execution of other KSs and they should not communicate directly with each other. In the blackboard model, sequencing of

Sec. 6.2 Blackboard Systems   213

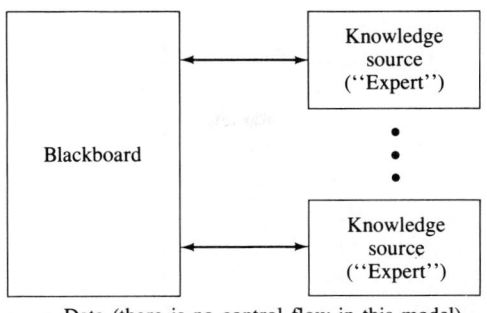

Figure 6.2 The idealized blackboard model.

KSs and communication between KSs occur only via the creation and modification of hypotheses on the blackboard. For example, a knowledge source that created an Acoustic Signal hypothesis from a piece of Acoustic Data (see Figure 6.1) would not directly call the knowledge source that abstracts this hypothesis to create an Acoustic Group hypothesis. Instead, the Acoustic Signal to Acoustic Group KS would see the creation of an Acoustic Signal hypothesis on the blackboard and recognize that it could now apply its knowledge.

In addition to being a global database, the blackboard model prescribes that the blackboard be structured as a hierarchy of *spaces* (also often referred to as levels). Each space should represent a different level of abstraction or a different point of view on the data and problem solution. Only particular types of hypotheses are allowed in each space. For interpretation problems, the set of blackboard spaces would correspond to the levels in the interpretation type hierarchy (Figure 6.1). Decomposing the blackboard into spaces (with corresponding hypothesis types) serves two related functions: it facilitates the incremental construction of solutions and it helps to partition the domain knowledge into KSs. Section 6.4 discusses the issues involved in structuring problems when using blackboard systems.

Blackboard spaces are themselves structured in terms of a set of *dimensions*. Each space has its own particular set of dimensions that depend on the slots in the types of hypotheses that can be placed into the space. Typically, each space in a blackboard will include one or more dimensions that are common to all of the spaces. For example, in the aircraft monitoring example of Figure 6.1 (see also section 6.3), all of the spaces include time, X, and Y dimensions. Additional dimensions may be specialized to the particular space—e.g., group-class in the Acoustic Group space of an aircraft monitoring system.

Dimensions are used to define the "location" of a hypothesis within a space in order to provide efficient *associative retrieval* of hypotheses. Event-directed or data-directed KS invocation means that KSs must evaluate hypothesis characteristics such as whether there are hypotheses that are "close" to a triggered hypothesis or are within a certain "region" of a space. For example, a KS for extending a Track hypothesis by adding another Vehicle support (see the example in section 6.3) would have to check if a Vehicle hypothesis exists for the right time and the right region of space to determine if it is applicable. Retrieval efficiency is an important issue for blackboard systems because of the large numbers of hypotheses they must often deal with and because of the data-directed invocation of knowledge sources.

The representation of hypotheses on the blackboard is termed "integrated" not only because of the global nature of the blackboard, but also because each hypothesis is kept linked to the hierarchy of hypotheses and data that represent the support and possible explanations for the hypothesis. This hierarchical structure can be used to determine the relationships between any two hypotheses. For instance, it can be used to determine that two hypotheses are competing alternatives and can show exactly why this is so (in terms of shared supporting data or hypotheses). An example of this sort of structure can be seen in Figure 6.10 in section 6.3. This integrated approach to representing hypotheses can be contrasted with the "possible words" approaches of Truth Maintenance Systems [3] which make it difficult to understand in detail the relationships between alternative hypotheses.

The idealized blackboard model does not prescribe any particular control framework: whenever a KS can contribute to the developing solutions, it immediately does so. Thus, knowledge sources must not only know how to create or modify hypotheses as part of constructing a solution, they must also be able to recognize when they can take these actions. For this reason each KS consists of two functional components: the *precondition* and the *action*. The precondition portion examines the state of the blackboard in order to determine if the action portion is applicable (as was discussed earlier in the context of associate retrieval of hypotheses). If it is, the precondition portion can then create an instantiation of the action portion in terms of the relevant blackboard objects.

Despite the appeal of not having to specify control strategies, there are two problems with the idealized blackboard model. The first problem is that most computers have had a single processor. Having a single processor means that KSs can actually only be executed one at a time (in effect, only one "expert" can be at the blackboard at any time). The second problem with the lack of control in the idealized blackboard model is that blackboards are typically applied to combinatorially explosive problems (like signal understanding and other interpretation problems). In such problems, computation time considerations make it undesirable or even impossible to execute all of the KS actions that are possible before an acceptable answer is found. Executing actions that create and pursue unlikely hypotheses "distracts" the system by forcing useful actions to compete with nonuseful actions for resources. This can greatly delay the construction of the final answers.

These problems suggest that a control mechanism that can help to focus the search for the answers must be added to the blackboard model presented so far. In effect, the idealized model must be modified so that the experts raise their hands when they can contribute something and someone else then decides whether their contributions are likely to be useful or not. This allows the system to focus its attention on the most promising partial solutions in order to make the best use of its limited resources and produce answers more quickly. *Focus-of-attention* is a critical issue for most blackboard applications and it has been the subject of much

Sec. 6.2 Blackboard Systems

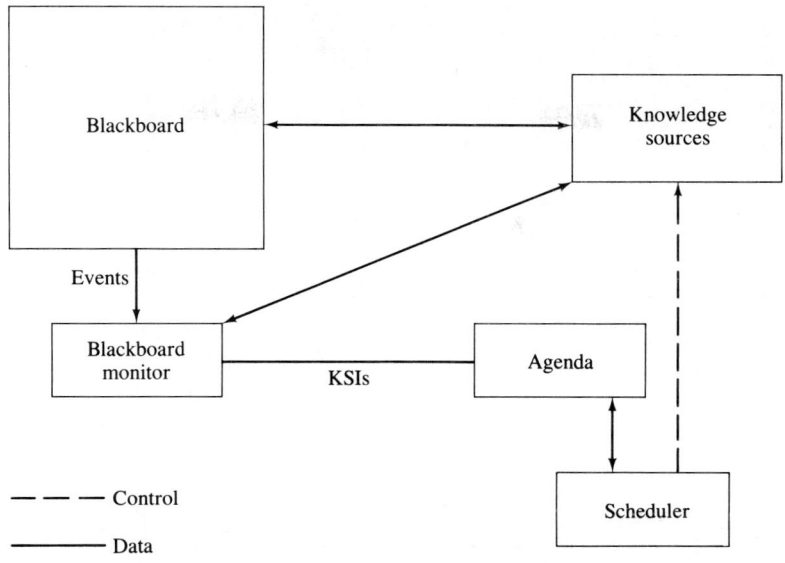

Figure 6.3 The data-directed, agenda-based blackboard architecture.

of the research on blackboard frameworks. Control issues will be pursued further in section 6.5.

To handle the control problems just discussed, Hearsay-II used an agenda mechanism: all potential actions are placed onto a scheduling "queue" where they are rated and selected for execution by the scheduler. The basic Hearsay-II architecture is shown in Figure 6.3. This agenda-based control approach has been followed in most subsequent blackboard systems. Thus, the Hearsay-II architecture as shown is the prototypical blackboard architecture. When reference is made to a standard, data-directed blackboard system, a system with this basic Hearsay-II architecture is meant.

Having just referred to Hearsay-II as a data-directed blackboard system, it is important to point out that complex interpretation problems require the integration of both data-directed and goal-directed control factors. Our description of Hearsay-II as a data-directed system is a simplifying abstraction of the Hearsay-II implementation that makes clear the key elements of a data/event-directed blackboard system. Hearsay-II actually made use of a number of additional features that allowed it to integrate goal-directed factors into the control decisions.[2] These features will be noted and discussed as appropriate in the following sections. Section 6.5 in particular will show how many of these features have been formalized or generalized in later blackboard frameworks.

Another problem with the idealized blackboard model is that checking the

---

[2] For example: prediction KSs, policy KSs, and focus-of-attention data.

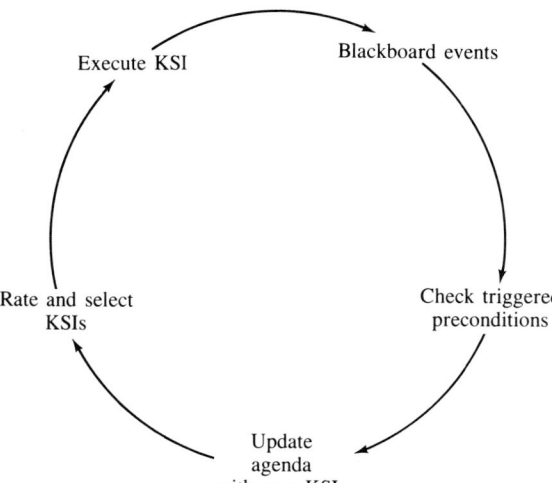

Figure 6.4 The basic control loop for the data-directed blackboard architecture.

preconditions of every KS after each action is taken could become very expensive. Because of this, the blackboard model of Figure 6.3 includes a mechanism for limiting the checking of KS preconditions. All changes to the blackboard are categorized in terms of a set of *blackboard event* types. Every time a KS action is executed, the changes to the blackboard are described in terms of the blackboard event types. These events descriptions are passed to a *blackboard monitor* that uses a table of the KSs that are interested in each event type to identify those KSs whose preconditions should be checked (i.e., that should be "triggered").

Figure 6.4 summarizes the main control loop in a data-directed blackboard system. As we have just said, every time data is inserted onto the blackboard or hypotheses are created or modified, a description of these changes in terms of blackboard event types is passed to the blackboard monitor. The blackboard monitor identifies those KSs that might be interested in the particular blackboard events and causes the preconditions of these KSs to be checked.[3] Instead of executing a KS action as soon as it is found to be applicable, a successful precondition produces a knowledge source instantiation (KSI)[4] and places it onto the agenda (the scheduling "queue"). Each KSI represents an action that could potentially be taken by the system and includes the hypothesis bindings and other information necessary to execute the KS action. Thus, at any given time, the set of KSIs on the agenda

---

[3] In early versions of Hearsay-II, once a KS was triggered, the precondition procedure was placed onto the agenda instead of immediately being executed. This made it possible for the execution of preconditions to be controlled by the scheduler. However, this capability came at the cost of a much more complicated scheduler and was not ultimately useful given the level of the knowledge available to Hearsay-II to control its usage. BB1 uses a similar approach to implement precondition-action backchaining (see section 6.5.4).

[4] Knowledge source instantiations are also sometimes referred to as knowledge source activation records (or KSARs).

represent the actions that the system could take next. Once the triggered preconditions have all been checked and any new KSIs placed on the agenda, the scheduler selects the KSI to be executed next, removes it from the agenda, and executes it.[5] Execution of a KSI can result in the creation or modification of hypotheses bringing the system back to the beginning of the control loop.

Since it is the scheduler that selects the next action to take, it is the scheduler that determines the focus-of-attention in a blackboard system. The scheduler selects the next KSI to execute by calculating a rating for every KSI on the agenda and choosing the highest rated KSI. Thus, the scheduler's KSI rating function embodies the strategy (control) knowledge of a blackboard system. KSI ratings are typically based on factors like the ratings of the hypotheses that triggered the creation of the KSI and the "desirability" of creating the hypotheses which the KSI should create. We will look at ratings further in section 6.3 when we look at an example blackboard-based aircraft monitoring system. The blackboard control cycle repeats until acceptable answers have been found or else there are no KSIs on the agenda. Determining whether acceptable answers have been found is known as the *termination problem*. Termination can be difficult and it will be examined in section 6.5.

6.3 AN EXAMPLE: AIRCRAFT MONITORING

In this section we present a short example to show how a blackboard system may be used to implement the understanding portion of an aircraft monitoring application. Further details about this and related vehicle monitoring applications can be found in [3, 10, 32]. We will concentrate here on the basic, data-directed blackboard framework described in the previous section. The example will also be discussed in the succeeding sections of this chapter in the context of more detailed explanations of the structuring and the control of blackboard problem solvers.

As we have said, the interpretation process involves the incremental construction and extension of hypotheses representing possible partial solutions. The interpretation hierarchy for our example aircraft monitoring application was shown in Figure 6.1. In the aircraft monitoring hierarchy of Figure 6.1, there are two types of data (Acoustic (sensor) Data and Radar Data) and three top-level Mission types that represent the possible purpose of aircraft movements. However, for this short example, only Acoustic Data will be used and answers will be restricted to the Track level.

---

[5] Actually, before a KSI can be executed, the system must verify that it is still applicable. Because there can be a delay between when a KS's precondition is checked and when the resulting KSI is executed, the situation on the blackboard may have changed so that the KSI is no longer applicable. Reverifying KSIs has been handled in several ways, ranging from simply re-executing the precondition procedure of the KS to specifying "obviation conditions" [22] with each KS. Obviation conditions allow the system to recognize when KSIs are no longer applicable so that they can be removed from the agenda as soon as possible.

Acoustic Data and Radar Data are correlated at the Vehicle level. Vehicle hypotheses represent potential "sightings" of aircraft. Vehicle hypotheses include a position slot that is the point in space and time where the aircraft is hypothesized to be and an ID slot that identifies the type of aircraft. Track hypotheses represent the movements of aircraft over time. They must be supported by sequences of Vehicle hypotheses that could represent the movements of the same aircraft. Vehicle hypotheses must meet certain constraints to be able to support the same Track hypothesis: they must have consistent ID slot values and their positions must conform to the movement parameters of the particular type of aircraft represented by ID (e.g., maximum velocity). The Mission-level represents a further specialization of Track hypotheses in terms of the goal or purpose of aircraft movements. Mission-level hypotheses impose constraints on the ID of Tracks and on the permissible movements represented by the Track. The hierarchy may in fact be expanded further to the Scenario level (sometimes called the pattern level) in which Mission-level hypotheses from multiple aircraft are correlated to explain the overall goal of the individual aircraft Missions.

Acoustic Data is interpreted through the Acoustic Signal and Acoustic Group levels to support Vehicle hypotheses. An Acoustic Signal hypothesis represents a single sensed acoustic signal in the environment and includes position and frequency-class slots. An Acoustic Group hypothesis represents a set of harmonically related acoustic signals that all emanate from a single mechanism on an aircraft. Thus, an Acoustic Group hypothesis will be supported by a set of Acoustic Signal hypotheses whose positions overlap and whose frequency-class slot values can be part of the same harmonic group. Acoustic Group hypotheses each include a position slot and a group-class slot that are defined by the values of the position and frequency-class slots of their supporting Acoustic Signal hypotheses. Since each aircraft typically has a number of mechanisms that generate acoustic signals, Vehicle hypotheses typically will be supported by several Acoustic Group hypotheses. In other words, the acoustic frequency spectrum created by an aircraft is not directly interpreted; the spectrum is first divided into appropriate Acoustic Groups.

For this example we are assuming that only acoustic sensor data is available. The data is represented in Figure 6.5. The dots in the figure show the spatial orientation of clusters (spatially close groupings) of acoustic sensor data points. Associated with each dot is a label whose numeric portion denotes the time that the data was sensed—e.g., 1a data is from time 1. We are also assuming that the data point clusters can be used for indexing the data—having been created by automatic low-level routines. Being able to index the acoustic data via clusters is useful for focusing the search process since the data resulting from a single source tends to get clustered together (data points resulting from a single source may not end up with identical position attributes due to sensor imprecision and environmental disturbances). In this example we will make use of the clusters in certain "large-grained" KSs and for rating the data. Clustering can also be used as a type of *approximate processing* technique—see section 6.5.3.

Figure 6.6 gives the ratings that are associated with each data cluster (the relative size of the dots in Figure 6.5 reflects these ratings). Data is rated as it is placed

Sec. 6.3 An Example: Aircraft Monitoring

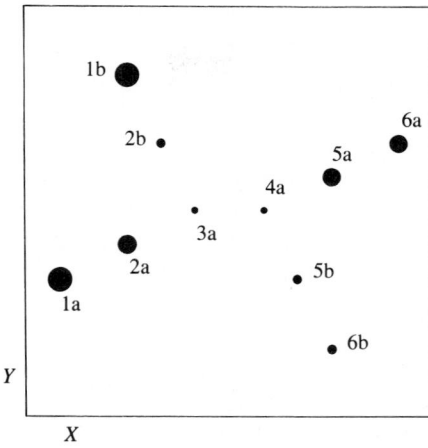

Figure 6.5 *X-Y* map of the clusters of acoustic data points for the example scenario.

on the blackboard. The data ratings are based on the loudness of the sensed frequencies and on the number of frequencies in the cluster. Since the data ratings influence the rating of data abstraction (synthesis) KSIs, the data rating scheme is intended to try to limit the processing of "noise" and to promote the use of "well-sensed" data from actual aircraft. The rating scheme used here is based on the following assumptions: noise is typically quiet relative to aircraft signals, noise typically has a sparse frequency spectrum relative to aircraft signals, and when the signals are louder they tend to be sensed with more precision (have less frequency-class and position uncertainty). Thus, the ratings are based on the average loudness of and the number of data points in the cluster.

The situation that the data represents is that the "a" data has resulted from an actual aircraft, while the "b" data is "ghost" data that has resulted from environmental reflections. Ideally, data from an actual vehicle would be rated higher than ghost data. However, due to environmental disturbances and the limitations of sensors, the relative ratings of individual noise or aircraft data clusters can vary (the

> **cluster 1a** Rating of 0.6—real data, loud, complete spectrum (6 data points)
> **cluster 2a** Rating of 0.5—real data, loud, partial spectrum (5 data points)
> **cluster 3a** Rating of 0.2—real data, weak, partial spectrum (2 data points)
> **cluster 4a** Rating of 0.2—real data, weak, partial spectrum (2 data points)
> **cluster 5a** Rating of 0.5—real data, loud, partial spectrum (5 data points)
> **cluster 6a** Rating of 0.5—real data, loud, partial spectrum (5 data points)
> **cluster 1b** Rating of 0.7—ghost and noise data, loud, partial spectrum plus inconsistent data (7 data points)
> **cluster 2b** Rating of 0.3—ghost data, weak, partial spectrum (3 data points)
> **cluster 5b** Rating of 0.3—ghost data, weak, partial spectrum (3 data points)
> **cluster 6b** Rating of 0.3—ghost data, weak, partial spectrum (3 data points)

Figure 6.6 The ratings and brief descriptions of the data for the example scenario.

ratings on Track hypotheses that are composites of aircraft data will be higher than Tracks that are composites of ghost data—given enough data). The ratings for the example correspond to 1a data that has been fairly well sensed for real vehicle data, but is not rated as highly as the 1b ghost data. This situation results from the 1b cluster including some additional random noise frequencies that boost its data point count and thus its rating. Because the 1b ghost data is highly rated, it will tend to distract the system, causing it to pursue this data until it can gather evidence from higher-level, more global interpretations to be able to discount it. The other source of difficulty in this example is that the 3a and 4a data, which is necessary to complete the "a" Track, has been very poorly sensed (environmental or sensor problems have resulted in only a portion of the aircraft's spectrum being sensed and with low signal loudness). Here again, there is a need for global information to direct the control so that this data is processed in a timely fashion.

Figure 6.7 describes the different knowledge sources that are used in this

> **synth:ADs → Ss**
>> **trigger:** Addition of Acoustic Data points to the blackboard.
>> **precondition:** Identify *clusters* of the trigger Acoustic Data points and create KSIs based on each such cluster.
>> **action:** Synthesize a Signal hypothesis from each of the Acoustic Data points in the cluster associated with the KSI.
>
> **synth:Ss → Gs**
>> **trigger:** Addition of Signal hypotheses to the blackboard.
>> **precondition:** Create a KSI based on the set of trigger signal hypotheses (derived from an Acoustic Data cluster).
>> **action:** Synthesize appropriate Group hypotheses from the set of Signal hypotheses associated with the KSI.
>
> **synth:Gs → V**
>> **trigger:** Addition of Group hypotheses to the blackboard.
>> **precondition:** Create a KSI based on the set of trigger Group hypotheses (derived from an Acoustic Data cluster).
>> **action:** Synthesize a Vehicle hypothesis from the set of Group hypotheses associated with the KSI.
>
> **synth:V → T**
>> **trigger:** Addition of a Vehicle hypothesis to the blackboard.
>> **precondition:** Create a KSI based on the trigger Vehicle hypothesis.
>> **action:** Synthesize a Track hypothesis from the Vehicle hypothesis associated with the KSI.
>
> **ext:V + T → T**
>> **trigger:** Addition of a Vehicle or Track hypothesis or extension of a Track hypothesis on the blackboard.
>> **precondition:** If Vehicle trigger, check for any Track hypotheses that are

adjacent in time to the Vehicle hypothesis and that could be extended using the Vehicle hypothesis—creating a KSI based on each such Track hypothesis. If Track trigger, check for any Vehicle hypotheses that are adjacent in time to the "new ends" of the Track hypothesis and that could be used to extend the Track hypothesis—creating a KSI based on each such Vehicle hypothesis.

action: Extend the Track hypothesis associated with the KSI by adding the Vehicle hypothesis associated with the KSI to the Track as further support.

merge: $T + T \rightarrow T$

trigger: Creation or extension of a Track hypothesis on the blackboard.

precondition: Check for any Track hypotheses that abut or overlap the trigger Track and are consistent with the Tracks being the same aircraft (e.g., if the Track hypotheses overlap in time, they must share supporting Vehicle hypotheses for these times). Create a KSI based on each such pair of Tracks.

action: Merge the two Track hypotheses associated with the KSI to create a new Track hypothesis.

predict: $T \rightarrow V_p$

trigger: Creation or extension of a Track hypothesis on the blackboard.

precondition: At each of the trigger Track's "new ends," check for any Vehicle hypotheses that could extend the Track. Create a KSI based on each new end extension time-region for which there are *not* any such Vehicle hypotheses.

action: Create a Vehicle prediction (a Vehicle hypothesis that is marked as a prediction since it has no supporting data) representing the time-region associated with the KSI (where a Vehicle hypothesis could be used to extend the trigger Track).

synth: $ADs + V_p \rightarrow Ss$

trigger: Addition of Acoustic Data points or creation of Vehicle prediction on the blackboard.

precondition: If Acoustic Data trigger, check for any Vehicle predictions that overlap clusters of the data and create a KSI based on each such cluster plus the Vehicle prediction. If Vehicle prediction trigger, check for any clusters of Acoustic Data points that overlap the Vehicle prediction (and have not been used to synthesize Signal hypotheses) and create a KSI based on each such cluster plus the Vehicle prediction.

action: Synthesize Signal hypotheses from each of the Acoustic Data points in the cluster associated with the KSI and delete the associated Vehicle prediction.

Figure 6.7 A description of the knowledge sources being used in the example.

222 Knowledge-Based Signal Understanding Chap. 6

example and Figure 6.8 shows how they relate to the blackboard spaces and hypothesis types. There are KSs here to synthesize new Track hypotheses by successively abstracting acoustic sensor data, extend existing Track hypotheses, merge existing Track hypotheses, and predict from existing Tracks (these KS categories and the components of a KS are explained further in section 6.4). The basic (incremental) control strategy is the following: start with data that looks promising and use it to create a likely preliminary (incomplete) Track hypothesis, make predictions of

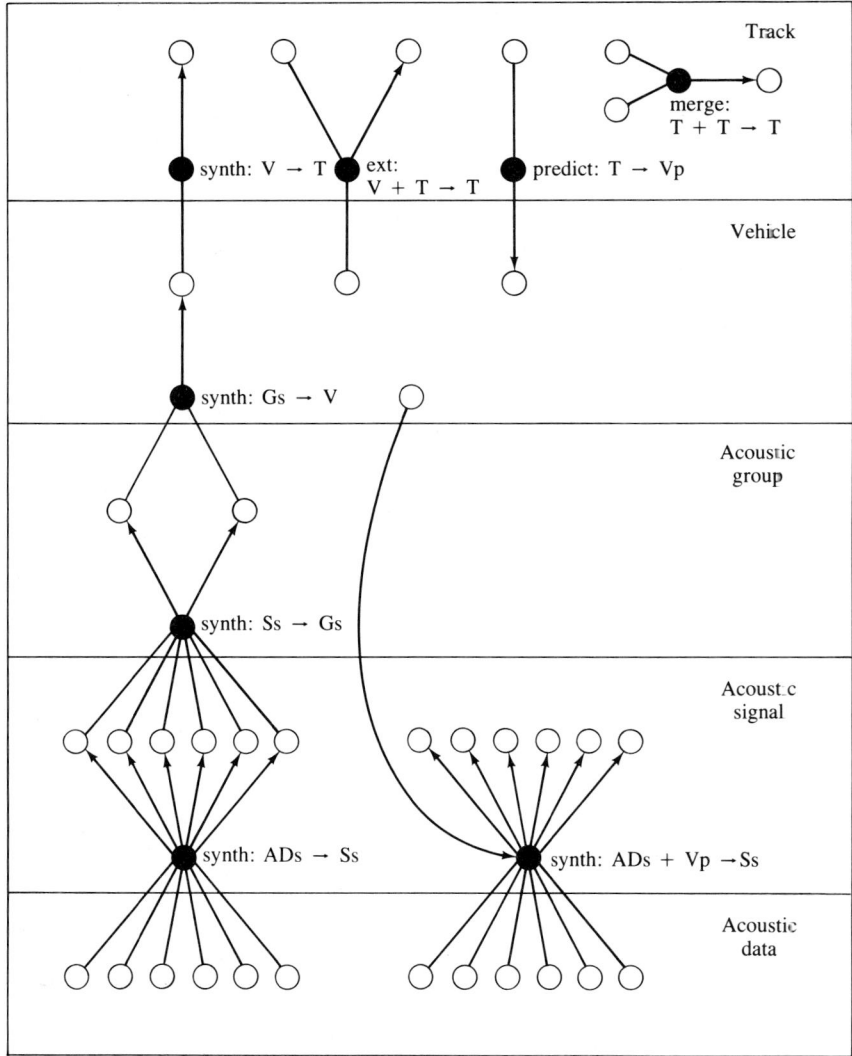

Figure 6.8 The relations between the example knowledge sources and the blackboard spaces.

extension Vehicle hypotheses from the Track hypothesis to provide some goal-directed control (discussed later) to select further data, abstract data to create the predicted Vehicle hypotheses, use the new Vehicle hypotheses to extend the Track, and repeat this process until a complete Track hypothesis is created. Note that many of the knowledge sources operate on a *set* of hypotheses that have been created from a single data cluster. The reasons for using such "large-grained" KSs will be explored in section 6.4. It is also important to note that the example assumes that hypothesis belief ratings are computed as part KS actions that create or modify hypotheses (see section 6.4 for more detail).

Figure 6.9 lists the actions taken by the system to solve the example problem. When the acoustic data is initially put onto the blackboard, *each* data cluster triggers the creation of a synth:ADs→Ss KSI that is placed onto the agenda (note that for simplicity, we have assumed batch mode data acquisition here rather than real-time data acquisition). On the first control cycle, the scheduler selects the KSI associated with cluster 1b because this KSI is the most highly rated as a result of the rating of the 1b cluster data. This results in the creation of a set of Acoustic Signal hypotheses based on the 1b data. The creation of these Acoustic Signal hypotheses on the blackboard triggers a synth:Ss→Gs KSI which can further abstract the newly created Acoustic Signal hypotheses to be placed onto the agenda. At this point, the Acoustic Signal hypotheses from the 1b data are relatively highly rated, resulting in a high rating for the associated synth:Ss→Gs KSI—which is executed next. Abstraction of the 1b cluster data continues through the Vehicle level because the relevant synthesis KSIs continue to have the highest ratings (the ratings result from both the ratings of the hypotheses that are based on the 1b data ratings and on the weight given to pursuing existing higher-level hypotheses rather than starting with more data). However, the Vehicle hypothesis that is created from the 1b cluster is not very highly rated because at the Vehicle level of constraints it is possible to recognize that the data represents an incomplete aircraft spectrum and that there is inconsistent data within the cluster.

As a result of this low rating, the synth:V→T KSI triggered by the creation of the 1b Vehicle hypothesis is not highly rated and is not selected for execution next. Instead, the synth:ADs→Ss KSI associated with the 1a cluster (which has been on the agenda since the data was first inserted) is now the most highly rated and is selected next. This change of focus shows what is meant by opportunistic control: the system shifts its focus because it is continually pursuing those actions that it believes will be most useful (as represented by the KSI ratings). Here the 1b data that initially looked promising has been found to be less promising after being partially processed and achieving a more comprehensive/global viewpoint. This leads the system to pursue actions not directly connected with the 1b cluster focus. It is important to note though, that whether this happens depends on the scheduler's rating function (and its use of data and hypothesis ratings). The importance of this issue for agenda-based blackboard systems will be pursued in section 6.5 and an alternative approach that is relevant to this scenario will be examined in section 6.5.3.

cycle 1 synth:AD→Ss for cluster 1b, producing a set of Acoustic Signal hypotheses.
cycle 2 synth:Ss→Gs for cluster 1b, producing a pair of Acoustic Group hypotheses.
cycle 3 synth:Gs→V for cluster 1b, producing Vehicle hypothesis V_{1b}.
cycle 4 synth:AD→Ss for cluster 1a, producing a set of Acoustic Signal hypotheses.
cycle 5 synth:Ss→Gs for cluster 1a, producing a pair of Acoustic Group hypotheses.
cycle 6 synth:Gs→V for cluster 1a, producing Vehicle hypothesis V_{1a}.
cycle 7 synth:V→T for cluster 1a, producing Track hypothesis T_{1a}.
cycle 8 predict:T→V_p from T_{1a} to time 2, producing a Vehicle prediction.
cycle 9 synth:ADs + V_p→Ss for cluster 2a, producing a set of Acoustic Signal hypotheses.
cycle 10 synth:Ss→Gs for cluster 2a, producing a pair of Acoustic Group hypotheses.
cycle 11 synth:Gs→V for cluster 2a, producing Vehicle hypothesis V_{2a}.
cycle 12 ext:V + T→T for cluster 2a and T_{1a}, producing Track hypothesis T_{1a2a}.
cycle 13 predict:T→V_p from T_{1a2a} to time 3, producing a Vehicle prediction.
cycle 14 synth:AD→Ss for cluster 6a, producing a set of Acoustic Signal hypotheses.

...

cycle 17 synth:V→T for cluster 6a, producing Track hypothesis T_{6a}.
cycle 18 predict:T→V_p from T_{6a} to time 5, producing a Vehicle prediction.
cycle 19 synth:ADs + V_p→Ss for cluster 5a, producing a set of Acoustic Signal hypotheses.

...

cycle 22 ext:V + T→T for cluster 5a and T_{6a}, producing Track hypothesis T_{5a6a}.
cycle 23 predict:T→V_p from T_{5a6a} to time 4, producing a Vehicle prediction.
cycle 24 synth:ADs + V_p→Ss from cluster 4a, producing a set of Acoustic Signal hypotheses.

...

cycle 33 merge:T + T→T with T_{1a2a} and $T_{3a4a5a6a}$, producing a complete Track hypothesis.

Figure 6.9 Actions taken by the system to solve the example scenario.

The 1a data is pursued through several KSI executions until a Track hypothesis (T_{1a}) is created. The creation of this hypothesis triggers a predict:$T \rightarrow V_p$ KSI being placed on the agenda. This KS effectively allows the data-directed agenda control scheme to integrate some goal-directed control (this is an important issue for blackboard systems and is discussed in section 6.5). The rating of this KSI is based on the rating of the Track hypothesis and on the recognition that goal-directed control is very useful for focusing the aircraft tracking process. As a result, the system selects this KSI next and creates a Vehicle prediction hypothesis for time 2. This prediction hypothesis specifies the region into which the Track hypothesis might be extended via its slot values. Since the extension region overlaps with the 2a data cluster (note the associative retrieval required here), this triggers the creation of a synth:ADs + $V_p \rightarrow$ Ss KSI for cluster 2a. While this KSI will do exactly the same work as the 2a synth:ADs $\rightarrow$ Ss KSI that is already on the agenda, it will be rated more highly because of its use of predicted (goal-directed) information.

The execution of the goal-directed synthesis KSI in cycle 9 eventually leads to the creation of a Vehicle hypothesis based on the 2a data in cycle 11. This now triggers the creation of two KSIs that are placed on the agenda: synth:$V \rightarrow T$ and ext:$V + T \rightarrow T$. The extension KSI is rated more highly since its rating is not only based on the 2a Vehicle hypothesis, but also on the existing Track hypothesis. Execution of this KSI in cycle 12 results in the Track hypothesis T_{1a2a}. The structure of this hypothesis is shown in Figure 6.10.

Extension of the Track again triggers the creation of a predict:$T \rightarrow V_p$ KSI and this KSI is executed next. However, due to the extremely low rating of the 3a data, the synth:ADs + $V_p \rightarrow$ Ss KSI that is created for cluster 3a is not rated highly enough to be executed next. Instead, the system selects the 6a synth:ADs $\rightarrow$ Ss KSI and eventually creates a second Track, T_{6a}, in cycle 17. This Track is then extended through 5a as described above. At this point, the synth:ADs + $V_p \rightarrow$ Ss for cluster 4a is sufficiently highly rated relative to the applicable synth:ADs $\rightarrow$ Ss KSIs on the agenda (for clusters 3a, 4a, 2b, 5b, and 6b) that it is executed next (note that in the example we have assumed that "obviation conditions" have been included in the KSs so that no longer applicable KSIs are removed from the agenda—e.g., synth:ADs $\rightarrow$ Ss KSIs for clusters processed by synth:ADs + $V_p \rightarrow$ Ss KSIs). The second Track is thus extended through clusters 4a and 3a. This triggers the creation of a merge:$T + T \rightarrow T$ KSI because the Track hypotheses T_{1a2a} and $T_{3a4a5a6a}$ cover adjacent time ranges. When this KSI is executed next (cycle 33) it results in the creation of the complete "a" Track.

6.4 STRUCTURING INTERPRETATION PROBLEMS FOR BLACKBOARD SYSTEMS

In this section we will discuss the factors that influence decisions about how to structure interpretation problems when using a blackboard system. In particular, we will provide criteria for partitioning the interpretation process into abstraction levels

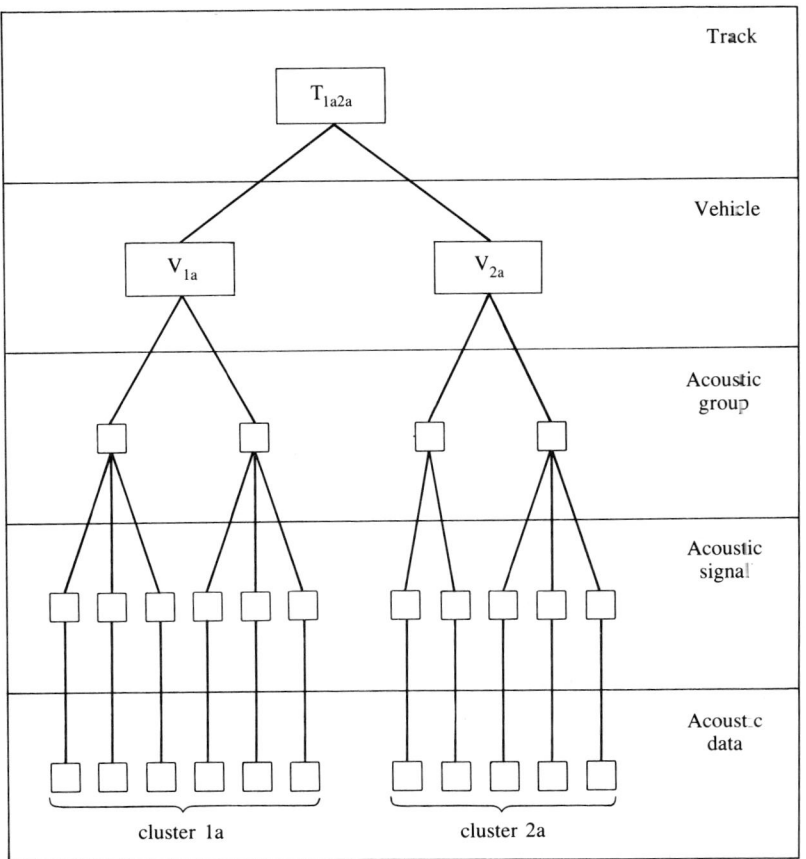

Figure 6.10 Hypotheses on the blackboard at the end of cycle 12.

and knowledge sources. Control issues will be touched on only briefly here—see section 6.5 for an examination of blackboard control issues and a summary of several important blackboard frameworks.

The first task in implementing a blackboard application is to decide on the structure of the blackboard. This requires the creation of an appropriate hierarchy of spaces. The space hierarchy determines how the problem-solving knowledge can be structured as well as what explanations of the data can be produced. For interpretation problems, the space hierarchy corresponds to the hierarchy of types. The interpretation type hierarchy for our example aircraft monitoring application was shown in Figure 6.1. The lowest levels of a hierarchy correspond to the sources of data that can be available. The highest levels of a hierarchy represent abstractions of the data that are suitable to be provided as answers by the system. In our aircraft monitoring example, there are three top-level types. These Mission types represent the possible purpose of aircraft movements.

Note that neither acoustic sensor nor radar data are directly abstracted to the Mission level. Instead, interpretation typically occurs through *intermediate types*. Several factors influence the selection of appropriate intermediate levels. As we stated in section 6.2, decomposing the blackboard into spaces serves two related functions: it facilitates the incremental construction of solutions and it helps to partition the domain knowledge into KSs. Whereas top-level hypotheses represent the application of a complete set of constraints and typically encompass a significant amount of data, intermediate-level hypotheses represent the application of only a portion of the constraints and data that must be aggregated into a final solution. Thus, having intermediate spaces serves to limit the amount of domain knowledge that must be applied in each step and localizes the effects of a KS. For example, having a Vehicle level as well as a Track level allows an aircraft monitoring system to reason about the spectral characteristics of sensor data being consistent with a possible aircraft sighting without having to simultaneously consider whether this data is consistent with a preliminary and uncertain hypothesis about the aircraft's movements. This also allows several KSs to be working in parallel on potential sightings of the aircraft since they would be working on separate Vehicle hypotheses instead of a single Track hypothesis.

Pursuing top-level answers via intermediate-level hypotheses also allows constraints to be incrementally aggregated. This can result in more efficient problem solving. Although intermediate-level hypotheses are still uncertain since they do not represent the application of all the constraints on answers, applying these partial constraints can be significantly cheaper than applying the full constraints and may eliminate large numbers of hypotheses from consideration. In other words, the ability to create intermediate-level hypotheses allows the application of reasonable chunks of constraint knowledge that can significantly reduce the search space for the higher-level interpretations without incurring the cost of applying the full, top-level constraints.

This approach also helps to control the creation of hypotheses since intermediate-level hypotheses implicitly represent uncertainty about the correct higher-level explanations—without having to directly create large numbers of alternative top-level hypotheses. For example, it would be unreasonable to try to directly interpret the aircraft monitoring data in terms of Missions because there would be too much uncertainty about possible interpretations of the data. An individual piece of acoustic sensor data can have many possible interpretations: it could be due to a number of different aircraft on a number of different missions or it could just be noise or a ghost signal. Thus, interpreting small amounts of acoustic data directly to top-level types would mean that the system would initially have a great deal of uncertainty about the correct Missions-level interpretations—which it would have to represent by creating a number of top-level hypotheses.

Another reason for defining an abstraction level is to be able to produce preliminary, partial explanations of data that are still conceptually useful; in other words, so that hypotheses below the level of the top-level answer types are meaningful to system users. This helps users to understand the interpretation process during system engineering and can provide useful results even if resource limitations prevent the system from fully pursuing top-level explanations. For instance, Track hypotheses are useful even without abstraction to the Mission level. They may even be acceptable answers depending on the particular system goals.

Defining the blackboard space hierarchy also requires that the dimensions of the spaces and the slots in the hypotheses be defined. The dimensions of each space must be based on the slots in the types of hypotheses that can be placed into the space. As we stated in section 6.2, space dimensions are used to define the "location" of a hypothesis within the space for associative retrieval. The "position" of a hypothesis along each dimension will be computed from the values of particular hypothesis' slot values. Space dimensions may directly correspond to hypothesis slots—e.g., frequency-class, X, Y, and so on. However, this need not be the case—e.g., the location of a hypothesis along the X and Y dimensions may be derived from the value of a single *position* slot. Typically, some space dimensions will be common to all spaces in the blackboard—e.g., X, Y, and time dimensions for the aircraft monitoring problem.

Another important issue for the representation of spaces and hypotheses is the issue of imprecise or uncertain hypothesis slot values. Because of sensor limitations, environmental conditions, and incomplete models of phenomena, hypotheses may not have precise values for each slot. For example, acoustic sensors will have some imprecision in the frequency-class attributes of the data that they produce, and not all of the frequency components of an aircraft spectrum may be sensed depending on the environmental conditions and position of the aircraft. As a result, there may be uncertainty about the ID (aircraft type) of the aircraft being sensed which must be represented as uncertainty in the ID slot value of the corresponding Vehicle or Track hypotheses. Thus, hypotheses must be able to use appropriate representations of uncertainty (like ranges, sets, or partially specified values).

Once the blackboard space and hypothesis hierarchy have been defined, the knowledge sources necessary to carry out actions must be defined. The exact form for the knowledge sources depends on the particular blackboard framework that is being used. The description here will be general enough to apply to all of the agenda-based control schemes in section 6.5 (but not to the RESUN framework of section 6.5.5). A knowledge source can be viewed as having four basic conceptual components: a set of triggers, a precondition, an action, and a rating function. *Triggers* are used to identify KSs that should have their preconditions checked following changes to the blackboard. They provide a quick, coarse filter on the KSs that are potentially applicable. Triggers are often specified nonprocedurally in terms of blackboard events. The *precondition* of a KS is what actually determines whether the KS is capable of being executed given the current state of the blackboard. Preconditions typically have been implemented as chunks of procedural code. Besides returning a signal of whether the precondition succeeds or fails, the precondition function must also return any information necessary to create a KSI. Typically this includes a record of the stimulus and response frames (the hypotheses that triggered the KS and the types of hypotheses likely to be produced) and any binding information needed to be able to carry out the action (action components should not have to duplicate precondition processing).

The *action* component of a KS is the portion that creates and/or modifies hypotheses on the blackboard. Knowledge sources proceduralize the constraints that exist between hypothesis types. KSs must both check the validity of the constraints in terms of the slot values of the hypotheses and compute newly constrained slot values if hypotheses are consistent. Thus, KSs must be able to deal with uncertainty in hypothesis slot values. For example, suppose that a certain Vehicle hypothesis represents a confirmed sighting of, say, aircraft type *aircraft-1*. Then, using this hypothesis as support for a Track hypothesis whose ID slot value is either *aircraft-1* or *aircraft-2* will constrain the Track's slot value to *aircraft-1* (it will also constrain the slot values for its "sibling" Vehicle hypotheses—i.e., those that are supporting the same Track hypothesis). Actions have been implemented in many different ways, from procedural code to sets of rules. This is one of the advantages of the modularity of blackboard KSs: each KS can be implemented using the methods and data structures that are most efficient for its particular task. There are no constraints on the internal form of KSs because all interaction between the KSs takes place through the single, consistent blackboard (hypothesis) interface.

Whenever hypotheses are created or modified by an action, a *rating function* must be invoked in order to provide updated ratings for the affected hypotheses (note that we are discussing hypothesis belief ratings, not KSI ratings). KS rating functions may be embedded in the action code or they may be separate, general routines that are called by the action components. Since the belief ratings must effectively combine the overall effect of multiple KS actions on the hypothesis, it makes more sense to use general rating routines that examine the support and explanation hypotheses instead of dealing with the KSs that created these relations. In other words, the relationships between the hypotheses should be treated as evidential relations (see section 6.5.5).

KSs typically can be assigned to one of the following four categories: synthesis, extension, merge, and prediction. *Synthesis* KSs abstract one or more hypotheses at some blackboard level to create a new hypothesis at a higher level. *Extension* KSs add support to an existing hypothesis using one or more existing lower-level hypotheses. *Merge* KSs combine overlapping hypotheses at the same level (i.e., hypotheses that share support and are consistent with each other). *Prediction* KSs create one or more new hypotheses on a blackboard level based on a hypothesis at a higher level.[6] Note that each category of knowledge source need not be provided for each abstraction level or pair of levels. The KSs that are provided will constrain the problem-solving strategies that can be used. For example, processing will always be bottom-up between particular levels if synthesis but no prediction KSs are provided for these hypothesis levels. Note also that KSs need not be limited to operating between adjacent spaces.

---

[6] In the goal-directed blackboard framework (see section 6.5.2), prediction KSs are not used; *goals* posted to a *goal blackboard* take the place of predicted hypotheses.

Blackboard implementations sometimes include special-purpose KSs that do not fall into any of these categories or that span several categories. For example, Hearsay-II used the special KSs *predict* and *verify* that were always scheduled together. *Predict* made predictions about words that could possibly extend a phrase. However, instead of creating (prediction) hypotheses on the blackboard for each of the possible extension words, the list of possible words was passed directly to *verify*. This was done because *predict* would typically create a large number of words, most of which could be relatively easily discounted. *Verify* checked each of the words against lower-level hypotheses to see which were possible and created hypotheses only for these possible words. Hearsay-II also made use of special "generator" and "policy" KSs to integrate goal-directed control factors and control the combinatorial explosion of hypotheses that might be created. Generator KSs were large-grained synthesis KSs that were capable of creating all possible explanations for hypotheses at some level. However, instead of always creating hypotheses representing *all* of the possible explanations, generator KSs could be controlled so that only a portion of these hypotheses would be created at once. This control was provided by corresponding policy KSs that specified how many hypotheses a generator KS should create and where in the space. Should processing "stagnate," policy KSs may be triggered to change the criteria for generator KSs and retrigger these KSs to extend or redirect the search process by creating additional hypotheses. Policy KSs provide a mechanism for implementing a more global search than would be possible with a strictly data-directed scheduler.

Another factor to be considered when defining KSs is *grain size*—i.e., how much work is done by a single KS execution and how many hypotheses are affected by the execution. Grain size involves a trade-off between control flexibility and control overhead. In the example in section 6.3, we used a number of large-grained KSs that processed sets of data or hypotheses from a single cluster. These KSs used fixed strategies to decide which hypotheses to create from the data in a cluster instead of posting a large number of KSIs and letting the scheduler select the best actions. One reason for doing this is that it can be more efficient since it eliminates the scheduler overhead from a sequence of actions. Another reason is that control decisions can be made using complex (numeric) algorithms that cannot easily be implemented through the KSI ratings and scheduling mechanism. However, poor quality data may lead to highly uncertain choices that would be better reasoned about via the scheduler.

One important aspect of the aircraft monitoring hierarchy shown in Figure 6.1 is that it explicitly specifies "nonanswer" types. In other words, while answers will only be given in terms of Track hypotheses or Mission hypotheses, we have included models of phenomena other than aircraft that may explain some of the data. This has not typically been done in blackboard-based interpretation because the representation of hypotheses has not enabled systems to understand the complex evidential relations between alternative explanations. However, recent work that we have done on the RESUN framework (see section 6.5.5) has extended the representation of hypotheses so that these alternatives may be explicitly pursued. This makes it possible to use the "nonanswer" abstraction types to directly resolve interpretation (possible alternative explanations) uncertainty through a differential diagnosis process.

Our models of the nonanswer phenomena are based on reasonable assumptions about how the data for these types should differ from that for aircraft. Typically, though, there will be much more uncertainty involved in models of such phenomena. For example, acoustic ghosts and noise both tend to result in clusters of data that are incomplete with respect to the set of acoustic frequency classes that result from aircraft. However, since incomplete spectra may also be detected from actual aircraft, this is not conclusive evidence. Ghost and noise hypotheses must also have limited "tracks." This means that for ghost and noise hypotheses, "tracking" has the opposite effect from what it has with Track hypotheses—i.e., the longer the sequence of ghost or noise hypotheses that are able to be correlated, the less likely it is that the data is actually due to ghosting or that it is noise. Again, however, even data from actual aircraft may be incomplete due to environmental disturbances and sensor malfunctions.

6.5 CONTROL IN BLACKBOARD SYSTEMS

This section expands on the critical issue of control in blackboard systems. The section begins with an introduction to the major issues involved in the control of blackboard systems: integration of data/event-directed and goal-directed control factors, opportunism, and coordination of sequences of actions. We then examine a number of different control schemes that have been developed as alternatives to the data-directed, agenda-based control of Hearsay-II (section 6.2): Lesser and Corkill's goal-directed blackboards, Durfee and Lesser's data abstraction-based planning, Hayes-Roth's control blackboards, and recent work that we have done on the RESUN framework which uses planning in conjunction with a symbolic representation of interpretation uncertainty.

6.5.1 Issues in Control

Although the idealized blackboard model described in section 6.2 was quite clear about the representation of the domain knowledge and the database, it did not specify any particular control scheme. In fact, one of the great advantages of the blackboard model of problem solving is that it allows the use of many different styles of control reasoning. By contrast, many AI problem-solving paradigms try to force all reasoning into a single framework—e.g., rule-based systems force all reasoning to be either forward chaining or backward chaining.

The agenda-based blackboard model described at the end of section 6.2 essentially only supports data/event-directed control.[7] One of the limitations of this blackboard model is that it makes it difficult to take goal-directed control factors into account in making decisions. A data-directed model of control implicitly assumes that if hypotheses are useful, then they will eventually be generated. Whether this is a good model or not depends on how well "good data" can be identified from a local perspective and how serious incorrect decisions are (i.e., how quickly the system can recover from erroneous decisions). Much of the work on sophisticated, goal-directed blackboard control has been in response to the need to limit poorly constrained problem solving and help minimize the effects of incorrect decisions. The example in section 6.3 demonstrates the problem: without the prediction KS, generation of the answer would be delayed because Vehicle hypotheses based on the low-rated 3a and 4a clusters would not be created until after all the ghost ("b") data was processed. In addition, we saw that the ghost 1b data was pursued first since the data rating scheme was "fooled" by the presence of noise. While these examples suggest the need for a better rating scheme, it should be noted that from the local perspective of the data there is very little context for judging "goodness" (in fact, we actually made use of the more global context of the clusters—ratings based on individual data point characteristics would generally give even worse results).

Intelligent control in domains with large answer spaces requires the integration of both data-directed and goal-directed control factors. Data-directed control factors tell the system what it is best able to do given the data it has. Goal-directed control factors tell the system what it would most like to do in order to solve the problem. Without goal-directed control, a system may waste time working on data and hypotheses that are not important for generating a solution. Without data-directed control, systems may spend time working on goals that cannot be (easily) satisfied or may get stuck working on goals that were good at the point where they were selected, but that have become suboptimal due to changed conditions.

One of the key reasons why Hearsay-II relied so heavily on data-directed control was to provide opportunistic control: the ability to rapidly shift the focus-of-attention of the system based on developing solution quality, the appearance of new data, and the like. Opportunism relies on the control decisions having a strong data-directed component. However, purely data-directed control decisions do not take into account the global context of possible actions. Goal-directed control is necessary for many types of control reasoning: scheduling an entire sequence of actions needed to satisfy some goal (planning), identifying low-level actions that are relevant to high-level goals (subgoaling), and identifying actions that can enable other actions that are necessary to solve the problem (precondition-action back-chaining). Planning is important in keeping a system focused on long-term goals and making sure it follows through on goals it begins to pursue—e.g., making sure that

---

[7] As we have already stated, the model described in section 6.2 is an abstract and idealized description of the Hearsay-II architecture. Hearsay-II was actually more complicated than what was described in the model because it used a number of mechanisms that allowed it to integrate goal-directed control factors into its decisions.

all the KSs necessary to abstract Acoustic Data to the Vehicle level to extend a Track when trying to extend a Track. Subgoaling focuses low-level processing so that hypotheses will be created that can extend existing high-level hypotheses. Precondition-action backchaining is useful when a certain action is necessary to meet a goal, but the action is not yet executable—i.e., its precondition is not yet fully satisfied.

One approach that has been used to provide a goal-directed component to blackboard control is planning: "deciding on a course of action before acting" [7]. Planning is advantageous because it coordinates sequences of actions and gives a system the ability to understand the purpose of each action in terms of its role in meeting goals. It is important to note that most AI planning research (i.e., "classical planning" research) is not applicable to blackboard control because it has been concerned with *strategic planning*. Strategic planning involves the determination of a complete sequence of actions that can solve the problem—prior to taking any actions. Strategic planning is seldom useful for blackboard control because the knowledge in blackboard applications is typically too uncertain to permit actions to be completely planned out (the outcome of actions is uncertain) and because dynamic applications involve constantly changing situations. Recently there has been more work in AI on *reactive planning*. In the following subsections we will examine three different reactive planning-based control schemes for blackboard systems.

Another set of blackboard control issues result from the way that agenda-based control has typically been implemented. In conventional agenda-based control schemes, a single numeric rating function (the scheduler) is used to make control decisions throughout the run of the system (as described in section 6.2). This function may have to be very complex and it may rely in turn on the ratings of the hypotheses, goals, and KSIs that are produced by the KSs (or other rating mechanisms). Thus, though KSs may be independent in principle, the ratings functions associated with each KS must be consistent if effective control is to result. The development of the various rating functions is one of the major engineering aspects of a conventional blackboard system. In effect, much of the reasoning about control decisions is really done during the engineering of the system rather than during the running of the system. As a result, the ease with which experimentation can be used to facilitate system development is critical for the successful development of blackboard systems (this is true for all knowledge-based systems).

The use of monolithic rating functions can also make the system development process more difficult because the system's reasoning is not made explicit. System decisions result from reasoning that is merely implicit in the resulting ratings. Working only with these final ratings, it can be very difficult for a system engineer to understand why actions were or were not taken and why data was or was not used. This has led to some research into diagnosing blackboard system activities [25]. It has also resulted in much research into alternative blackboard control mechanisms (as we will see in the remainder of this section). For example, both the BB1 and RESUN frameworks (sections 6.5.4 and 6.5.5) make control reasoning more explicit than in a conventional blackboard system. As a result, it is easier for humans to understand why applications built with these systems select the actions that they do. This makes it is easier to recognize when the encoded control knowledge is incorrect or incomplete.

Another problem with an agenda-based control scheme is that the rating process can be very inefficient if not carefully implemented. For example, suppose that only a small percentage of the possible actions that are triggered ever get executed and that all KSIs are rerated on each control cycle. In this case, a large number of KSIs may constantly remain on the agenda and may be repeatedly rerated. Interpretation applications in which large amounts of data are being automatically produced by sensors are especially susceptible to this problem: each piece of sensor data inserted onto the blackboard will trigger one or more KSs even if only a fraction of the data will ever be able to be processed (see [3]). However, it is typically possible to avoid continually rerating KSIs by including information with KSs that identifies the conditions under which KSIs need to be rerated—e.g., a change in the ratings of the hypotheses that stimulated their creation. In addition, in some domains the overall overhead of rerating may not be of concern because it ends up being only a fraction of the cost of executing the KSIs. We will examine this issue again in section 6.5.4 in the context of BB1's dynamically determined rating functions.

Blackboard systems must also deal with the termination problem: has the system done enough work? Termination can be a difficult issue for data-directed systems with large search spaces because the existence of a possible answer hypothesis (i.e., an answer-level hypothesis that is "sufficiently" complete and has a "reasonably high" belief rating) does not exclude the possibility that a better alternative has simply not yet been produced. Search strategies that are able to guarantee that the first answer that is produced is the most likely are termed *admissible* [18]. However, the cost of such strategies may not be worth the benefit. Since heuristic search strategies are not guaranteed to find the best answer first, they require a more global, goal-directed reasoning process to decide whether to continue to look for other answers. For example, Hearsay-II dealt with termination through a special *stop* KS which was triggered by the creation of a highly rated, complete phrase-level hypothesis. The *stop* KS then examined the existing alternative hypotheses and "pruned" those that were unlikely to be able to produce higher-rated answer hypotheses (it did this by looking at the ratings of word hypotheses not covered by the alternative phrases and determined whether these words could possibly improve the alternatives so that they would be better than the current phrase) [31]. *Stop* halts processing when all possible alternatives have been removed.

Termination is a particularly difficult issue for interpretation systems. One reason is that in many interpretation problems, "the answer" may actually consist of a set of hypotheses that explain the data. In the Hearsay-II speech understanding problem, "the answer" was just a single complete sentence spanning the sampled speech. However, in domains like aircraft monitoring, "the answer" can be a set of Track or Mission-level hypotheses each of which represents a different monitored aircraft. Thus, the system must satisfy itself that it has identified all possible aircraft before it can terminate. Another reason for interpretation system termination difficulties is that interpretation must often deal with large amounts of noisy and uncertain data. This means that decisions about whether to pursue additional answers may have to be made with only a portion of the data having been processed and only a portion of the possible explanations constructed.

6.5.2 Goal-Directed Blackboards

One response to the problem of integrating goal-directed factors with data-directed, agenda-based control was the development of the *goal-directed blackboard system* [10, 33]. The framework was first implemented in the Distributed Vehicle Monitoring Testbed (DVMT) [32]. The goal-directed blackboard architecture extends the Hearsay-II architecture by adding a *goal blackboard* and a *goal processor*. The goal processor instantiates *goals* on the goal blackboard whose structure mirrors that of the domain blackboard. Each goal explicitly represents the system's intention to create a hypothesis of a particular type with particular attributes. Goals are used to focus problem solving and still maintain the advantages of opportunistic, data-directed, agenda-based control. While the data-directed blackboard model allows the system to understand what results can be derived from the data, the addition of the goals allows the system to also understand what the immediate consequences of actions are—i.e., to connect the effects of a KS action with the existing hypotheses so the control decisions can take these relationships into account. The basic architecture of a goal-directed blackboard system is shown in Figure 6.11 and the main control loop is shown in Figure 6.12.

Goals are created on the goal blackboard by the goal processor in response to the changes on the domain blackboard—i.e., the creation or modification of hypotheses. The goal processor is driven by three tables: a hypothesis to goal table, a goal to KS table, and a goal to subgoal table. The insertion of a goal on the goal blackboard causes knowledge source instantiations (KSIs) which might achieve the goal to be added to the agenda. KSIs are created by executing the precondition procedures of the knowledge sources (KSs) to estimate whether the KS is likely to generate hypotheses to satisfy the desired goal.

Goals, KSIs, and hypotheses are all assigned numeric ratings as they are created. Goal ratings are based on the ratings of the hypotheses that stimulated the creation of the goal and on the ratings of their supergoals (goals that have the goal being rates as a subgoal). KSI ratings are then computed as a weighted sum of the ratings of the goals that stimulated the creation of the KSI and of the ratings of hypotheses in the KSI's stimulus frame. Thus, the KSI ratings reflect both data-directed and goal-directed factors. The relative role of each type of factor can be adjusted by simply changing the rating weights.

When using a goal-directed blackboard framework, prediction KSs are not used. Instead, the goals that are produced by the subgoaling process take the place of prediction hypotheses. There are both advantages and disadvantages to this approach. One problem with using prediction hypotheses in a conventional black-

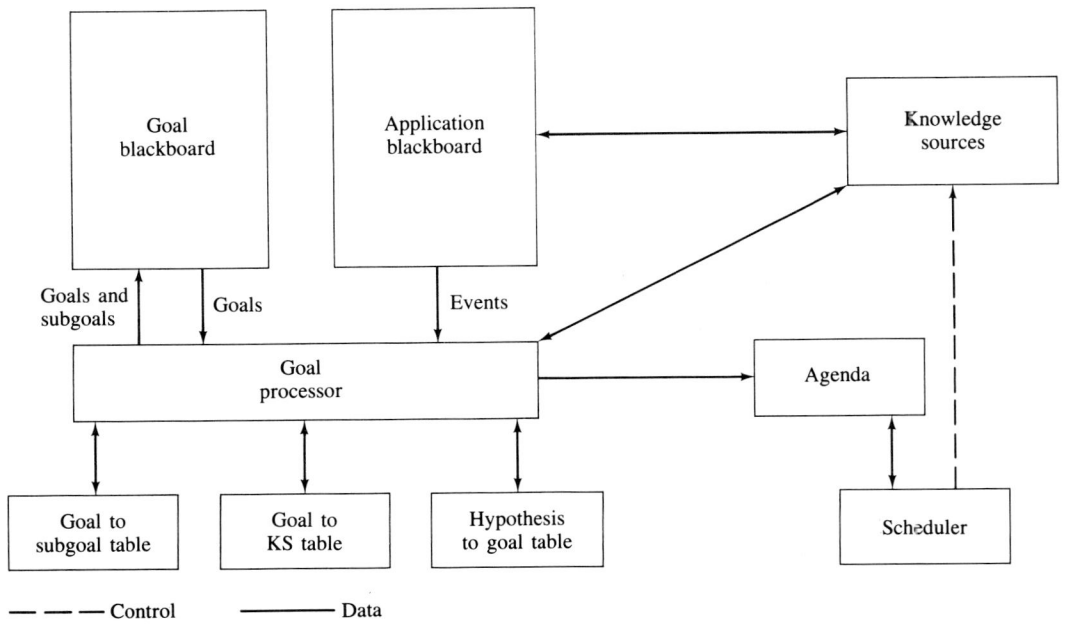

Figure 6.11 The goal-directed blackboard model.

board system is that they must be able to be recognized as such and must be treated specially. Simple single number hypothesis rating schemes do not allow the system to distinguish between hypotheses that have support and those that do not (i.e., predictions). Thus, it may be possible for predicted hypotheses to be used to provide (invalid) support for hypotheses that are alternatives to the original predicting hypothesis [10]. Since the purpose of making predictions is to guide control decisions, placing the "predictions" on a separate goal blackboard makes their role in control clear. In addition, it is possible to create more appropriate abstractions of desired hypotheses when using goals. For example, in the example in section 6.3,

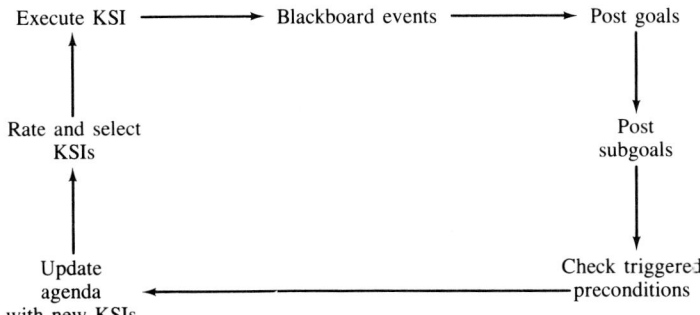

Figure 6.12 The basic control loop for the goal-directed blackboard model.

the prediction KS only created a single Vehicle prediction which then triggered the creation of a KSI to create a *set* of Acoustic Signals. The prediction process itself was not pursued down to the Acoustic Signal level because there is no way to use a single prediction hypothesis to represent a *set* of hypotheses that are desired. Goals can be used to represent such a set of hypotheses.

On the negative side for the goal representation, making predictions (i.e., subgoaling) may require complex computations that cannot be handled using a goal-to-subgoal *table*. Instead, these computations must be handled through what are effectively prediction KSs. However, when the prediction computations are complex and expensive, subgoaling needs to be controlled just like any other blackboard action. This can be done in a limited way in the basic goal-directed blackboard framework through the use of goal-rating thresholds and subgoaling from a limited number of abstraction levels. Work has also been done to extend this basic architecture through the addition of goal and hypothesis filters that control creation of goals [12]. Section 6.5.3 contains an alternative approach to controlling the creation of goals via a planning mechanism.

Another limitation of the goal-directed blackboard model is that its goals can only provide an understanding of the immediate consequences of actions. In other words, the goals can only represent the desire to carry out relatively simple actions like synthesizing a hypothesis of a particular type or extending an existing hypothesis with one more support. The goal-directed blackboard goals cannot represent complex, long-term goals like "resolve the uncertainty in a particular hypothesis." More general goals such as these require sequences of actions to be accomplished. In the next subsection, we will examine an extension to the goal-directed blackboard model which addresses this limitation by providing long-term goals via planning.

6.5.3 Data Abstraction-Based Planning

Work by Durfee and Lesser [15, 16] on incremental planning for control in blackboard systems is another approach for adding goal-directed control to focus blackboard activities. The mechanisms were developed in order to address some of the limitations of the goal-directed blackboard framework (discussed in section 6.5.2). This approach is able to provide a more long-term, global view of system goals which makes it possible to coordinate sequences of actions. The system was implemented in the Distributed Vehicle Monitoring Testbed (DVMT) and is built on top of the DVMT's goal-directed blackboard framework. The basic architecture of this planner-based blackboard system is shown in Figure 6.13.

The planner has two components: a clustering mechanism that creates an abstract model of the data and a planning mechanism that uses the data model to develop plans for working toward "long-term goals" of developing vehicle tracks. The clustering mechanism forms high-level models of the data by examining several types of relationships among the data: temporal relations, spatial relations, event-class relations, and so on. It provides the system with information about the possible vehicle tracks and identifies those that may be alternatives because they may require

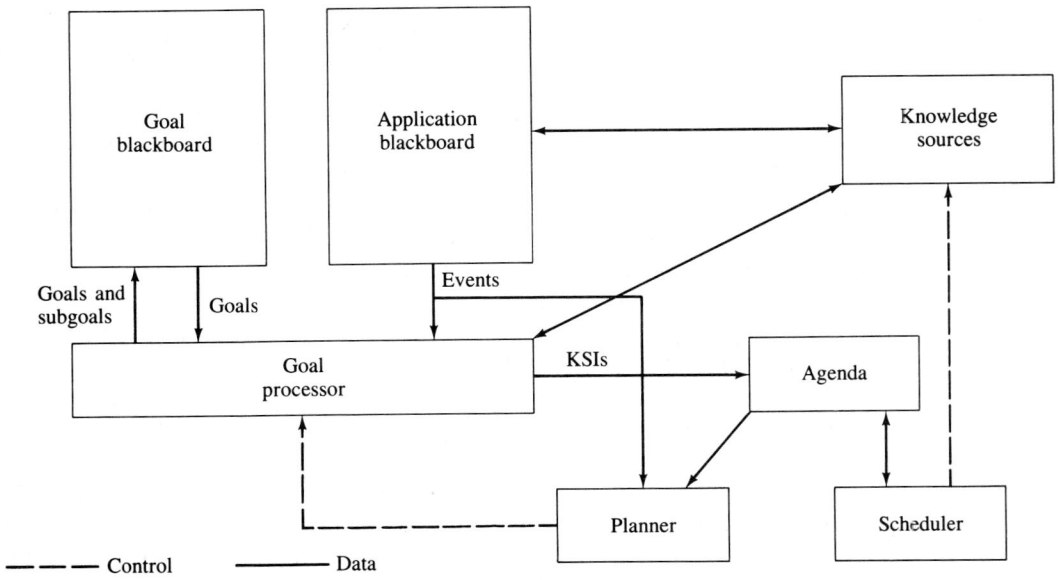

Figure 6.13 The data abstraction-based planning blackboard model.

the same supporting data. The modeling process effectively represents a rough pass at solving the problem and can be viewed as a kind of *approximate processing* technique [14].

The planner makes use of this model to create "long-term goals" for creating the potential tracks and resolving uncertainty about the possible alternatives. It does this by ordering the "intermediate goals" which represent the needs for the system to construct particular types of hypotheses to create track hypotheses and extend them in time. These goals are ordered using domain-independent heuristics. After the intermediate goals have been ordered, the planner determines how to achieve them by identifying appropriate sequences of KSIs on the agenda. This detailed level of planning is done incrementally as each intermediate goal actually has to be achieved. KSIs to achieve the intermediate goals are identified and ordered based on models of the KSs that provide rough estimates of their costs and the characteristics of their output. The planner controls the subgoaling process in the underlying goal-directed blackboard system. Only subgoals necessary to carry out the plan will be created.

The plans provide a global perspective on the data that can deal with some of the limitations of data-directed control that were identified and discussed in sections 6.3 and 6.5.1. For example, from the example scenario data of Figure 6.5, the abstraction-based planner can determine that the data in the 3a and 4a clusters is critical because it is shared by all of the potential tracks in the data. Thus, the planner would begin to work on this data and be able to take advantage of its constraints in limiting the ghost data that is processed. It would avoid the distraction of the 1b data and would not have to go through a merging process to form a complete Track hypothesis. In addition, once it has found a complete Track hypothesis it could immediately terminate problem solving because it would know that better alternatives could not be created (since it knows approximately what those alternatives could be).

The planning system developed for the DVMT is not general-purpose. Both the data modeling process and the planning process are highly specialized to vehicle monitoring and vehicle tracking. This can be seen by the fact that the planner has a very limited notion of goals: goals only refer to creating and extending the *possible tracks* identified by the modeling process. A major reason for this lack of generality may be the simplistic evidence model which was being used in the DVMT. The representation of evidence effectively limited the system to resolving its uncertainty by evidence aggregation—i.e., extending vehicle tracks. In fact, a key reason for the performance advantages of the planner-based system over the basic DVMT is that it provides the ability to do some differential diagnosis. This is because the modeling process identifies the possible alternative tracks and, thus, the possible alternative vehicles. Providing differential diagnosis in this way has one major drawback: all of the reasoning about alternatives is done by the control component rather than the evidential reasoning component. In other words, there is no representation within the hypothesis structures of the relationships between alternative hypotheses and thus there is no way to calculate the resulting effect on the belief of the hypotheses.

6.5.4 Control Blackboards

Hayes-Roth's *control blackboard* paradigm [22, 23] (now usually referred to as *BB1*—after the name of a particular implementation) extends the data-directed, agenda-based control of Hearsay-II through the addition of a "control planning" mechanism to help focus processing. The BB1 framework includes control knowledge sources and a control blackboard in addition to the standard domain KSs and blackboard of Hearsay-II. These mechanisms can be used to *dynamically* change the evaluation functions being used by the scheduler. This makes it easy to adjust the control decision criteria in response to changing problem-solving circumstances.[8] The architecture of the BB1 system is shown in Figure 6.14.

---

[8] Hearsay-II's *policy* KSs were an early precursor of KS-based control. Policy KSs were triggered not by blackboard events like standard KSs, but rather by the state of the control process—e.g., when there are no reasonably highly rated KSIs on the agenda. However, unlike the control plans of BB1, policy KSs could only provide a "single shot" at changing control parameters to respond to the situation and were not driven by explicit goals. It should also be pointed out that the concept of treating the blackboard control problem as a task that should be solved using the same blackboard model being used to solve the domain problem actually originated with Hearsay-III [19]. Hearsay-III was one of the first attempts to generalize and extend the Hearsay-II architecture. Scheduling in Hearsay-III could be based on very complex schemes because the scheduling functions could be changed by *scheduling* KSs that could also record control information on a scheduling blackboard. BB1 provides more structured methods for influencing control decisions and has been much more widely used than Hearsay-III.

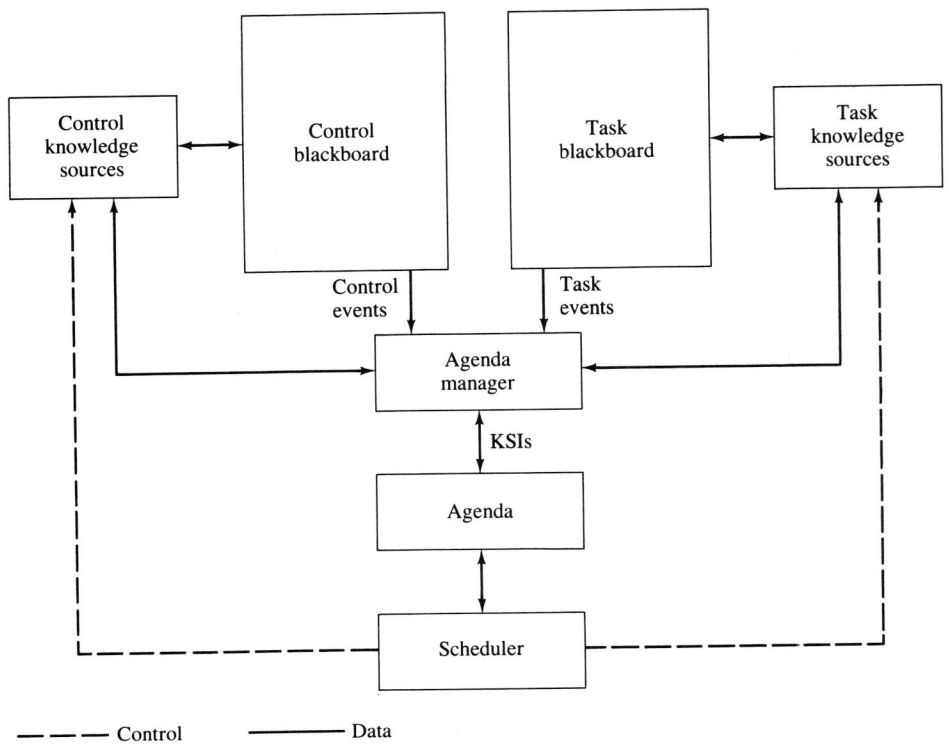

Figure 6.14 The BB1 blackboard model.

Control KSs are treated just like other KSs in that their KSIs—which represent potential (control) actions—are placed on the same agenda as the normal domain KSIs and are selected via the same scheduler rating functions (KSIs are actually referred to as *knowledge source activation records*, or *KSARs* in BB1 terminology). In other words, there is no special control planning loop in the BB1 model; control actions must compete with normal domain actions for resources. This makes it possible for the system to decide whether to reason about how to make control decisions (by executing control KSIs) or whether to just act based on its current strategies. When executed, control KSIs construct *control plans* on the control blackboard (we will not discuss the details of the structure of control plans). These control plans affect control decisions because the refinement of the control plans results in the selection of heuristic rating functions that are used by the scheduler. What BB1 effectively does is to eliminate the need to try to come up with a single complex rating/scheduling function that works properly for the entire problem-solving process. Instead, the planning mechanism allows the user to define several ratings functions that are appropriate for different stages of processing. Each "step" in a BB1 control plan has a corresponding rating function (actually it may have a set of these functions—called "heuristics"). As each plan "step" is completed, new rating functions are activated.

The BB1 model extends the data-directed blackboard model through the addition of the goal-directed planning mechanism. It is important to note that BB1 maintains the opportunism of the classic blackboard model because the identification of possible actions is still done in a data-directed fashion: all possible actions (KSIs/KSARs) are placed on an agenda as they are triggered. Since both domain and control actions are treated in this manner, both actions and plans (or goals) can be opportunistically instantiated. In fact, goals need not be instantiated through the hierarchical planning mechanism, but may be directly instantiated by appropriate control KSs. Because of this and because BB1 maintains its agenda in two parts (triggered "KSIs" and executable KSIs/KSARs), a BB1 system can implement and integrate other types of goal-directed reasoning including precondition-action backchaining and subgoaling [27].

When describing BB1's planning capabilities, it is important to understand that the notion of planning in BB1 is somewhat different from that of a typical AI "planner." BB1 control plans implement their problem-solving strategies through rating functions that select KSIs from the agenda, rather than directly identifying the actions needed to solve the problem as in AI planners. In BB1, the lowest level control "goals" are typically relatively general goals that will influence the selection of several actions—e.g., "work in x region" or "prefer actions of type y." This approach allows for highly opportunistic goal-directed control—e.g., when there are multiple, competing system goals and decisions must depend on the specific data or hypotheses that are available. However, the framework would make it difficult to specify detailed plans should this be desirable; there would be very substantial overhead in using the agenda and ratings functions to accomplish detailed planning.

The control plans provide a more explicit representation of strategy knowledge than is provided by a standard blackboard scheduler's complex rating function. Of course, while BB1 creates explicit control plan structure that denotes the strategies and goals to be pursued, it still relies on rating functions to actually select the next action to be executed. Thus, some control reasoning in BB1 is still implicit. In fact, the BB1 model makes it possible to have multiple active strategies and goals so there can be multiple active rating functions. This forces the BB1 scheduler to use "combining functions" to integrate ratings. Though control KSs can change combining functions according to the combinations of strategies or goals that are active, there is no support for identifying when this is necessary. Another control framework that facilitates explicit control reasoning is provided by the RESUN system (see section 6.5.5) which replaces the problem of making complex control decisions with a search process.

BB1's agenda/scheduler mechanism has two characteristics that have the potential for making BB1 very inefficient (see the discussion of agenda overhead in section 6.5.1). The first is that the ability to dynamically change ratings functions makes it more likely that KSIs will have to be repeatedly rerated as compared with a classic blackboard framework. The other is that both triggered and executable KSIs

(KSARs) must have their preconditions and obviation conditions repeatedly rechecked. This has not been a problem until recently, since BB1 has not been used for applications in which large amounts of data trigger numerous KSIs that remain on the agenda throughout problem solving—as in many interpretation problems. Recent work within the BB1 framework has addressed these efficiency issues as well as the need for timely/responsive control in real-time applications. In [9] the standard "best-next" BB1 control cycle is replaced with a "satisficing" control cycle which can respond to deadlines by not considering/rating all of the possible actions. In [24], not only is this satisficing control cycle used, but intelligent preprocessing of sensor data is used to "shield the reasoning system from data overload."

6.5.5 Planning to Resolve Sources of Uncertainty

We have developed an interpretation framework called RESUN [2, 3] that addresses many of the deficiencies of other blackboard frameworks. RESUN makes it possible to explicitly define complex strategies for interpretation problems. In particular, it makes it possible to use *differential diagnosis* techniques to resolve interpretation uncertainty rather than being limited to *incremental hypothesis and test*, as is the case for most blackboard-based interpretation systems. The designers of the Hearsay-II architecture believed that blackboard systems would have the capability to do differential diagnosis because of their integrated representation of alternative, competing hypotheses [31]. However, explicit differential diagnosis techniques have not been exploited by blackboard-based interpretation systems due to the limitations of their evidence and control frameworks.[9] In adding significant differential diagnosis capability, we have had to extend the representation of hypotheses and abandon the agenda-based control of standard blackboard systems. While our new control scheme does not maintain the modularity between knowledge sources and control as in a classic blackboard system, we feel that its advantages justify the change. The architecture of the RESUN system is shown in Figure 6.15.

RESUN models interpretation as an incremental process of gathering evidence to resolve *particular sources of uncertainty* in the interpretation hypotheses. There

---

[9] Hearsay-II did include a technique for implicitly doing some limited differential diagnosis through "KSI clustering." This involved pursuing sets of similarly rated hypotheses together. Though the similarly rated hypotheses were not necessarily competing alternative hypotheses, the technique was very useful when they were and caused no harm when they weren't. KSI clustering was developed because when similarly rated hypotheses were competing alternatives, Hearsay's "island driving" strategy would cause whichever hypothesis was first extended to then be pursued to the exclusion of the other alternatives. This is because the system had no way to recognize that one alternative had become more highly rated than another simply because more evidence had been gathered for it—not because there was any evidence against its alternative. Even if such evidential relations had been represented, they would have been difficult to exploit. In conventional agenda management, the rating of each KSI is done independently of rating the other KSIs. This local evaluation procedure limits the ability of the control system to understand the relationships among potential actions. Where explicit relationships among KSIs have been exploited [33], additional stages of processing beyond the local evaluation have been required.

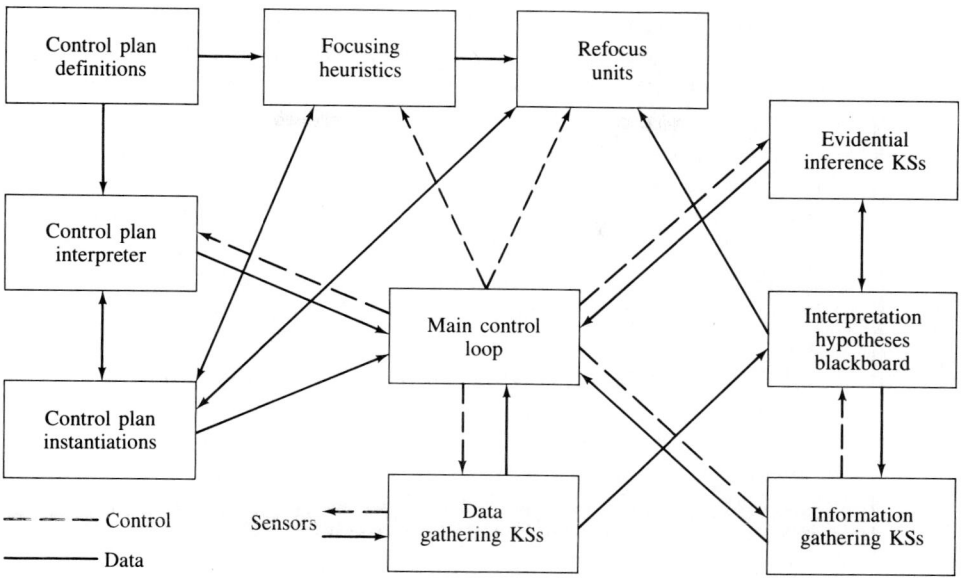

Figure 6.15 The RESUN architecture.

are two major contributions of this work: the evidential representation system and the control planner. Our representation of hypotheses maintains detailed information about the reasons hypotheses are uncertain and about the evidential relations between alternative hypotheses. The key feature of the evidential representation system is its use of explicit, symbolic statements of the *sources of uncertainty* (SOUs) in the evidence for interpretation hypotheses. For example, a Track hypothesis in an aircraft monitoring system may be uncertain because its supporting sensor data might have alternative explanations as a ghost, a malfunction, or a signal from a different aircraft; it may be uncertain because its evidence is incomplete and it may be uncertain because its mission-level explanation is uncertain. As interpretation inferences are made in our system, symbolic statements are attached to the hypotheses to represent their current sources of uncertainty. Thus, the Track hypothesis mentioned above might include statements that its supporting evidence is uncertain because of alternative possible explanations for the data, that its evidence is incomplete, and that there are alternative possible explanations for the Track. The symbolic SOUs provide more information than belief ratings. They allow the system to understand the reasons why its hypotheses are uncertain so that this uncertainty can be resolved directly (rather than indirectly through an incremental hypothesize and test approach).

Control in RESUN is provided by a planning mechanism that was designed to maintain the benefits of planners (such as detailed context information and long-term coordination of actions) while being sufficiently reactive and opportunistic for interpretation. The key obstacle to using planning for the control of problem solvers is that control planners must be able to deal with uncertainty and dynamically changing situations. We addressed these problems in several ways: our planner is *script-based* (i.e., it uses plan schemas), it interleaves planning and execution (i.e., planning is *incremental*), plans can invoke explicit *information gathering actions* to examine the current state of problem solving, and context-specific heuristic focusing knowledge is applied to control the planner's search. The basic control planning loop is shown in Figure 6.16. Unlike BB1 (section 6.5.4), the RESUN planner does detailed planning to directly identify the actions to be carried out. Actions are represented as *primitive* control plans and a RESUN application would include primitive control plan schemas that correspond to the knowledge sources in a conventional blackboard system.

It is interesting to note that while the other blackboard frameworks described in this section started with an agenda and then built goal-directed control mechanisms on top of the agenda, in RESUN we have abandoned the agenda. The primary innovation of the RESUN planner is its *refocusing mechanism*. It is this mechanism that allows the goal-directed RESUN planner to have the opportunistic control capabilities of an agenda-based blackboard system. Refocusing makes it possible for the planner to postpone focusing decisions (until additional information about the current situation can be accumulated) and to dynamically shift the system's focus-of-attention. Planning with refocusing results in a view of the control process as both a search for the correct problem solutions (interpretations) and a search for the best methods to use to determine these solutions. Thus, what we have effectively done is to extend the (explicit) search paradigm from just finding the answers to also finding the control decisions (finding how to find the answers). As a result, instead of trying to engineer complex ratings functions that directly select the best KSI (action) to execute next, our planning and refocusing mechanism allows us to make a series of less complex *search* decisions to select the next action.

One of the key motivations for using a planning-based control framework is that the goal/plan/subgoal hierarchy that is instantiated by a planner provides detailed and explicit context information for control decisions. In other words, control decisions result from planner focusing decisions and when focusing decisions are made, it is clear from the hierarchy exactly what the context of the decision is: what the purpose of the decision is in terms of the goals and subgoals to which it pertains and what the relationships among the various decision alternatives are. Having detailed and explicit context for each decision facilitates the implementation of sophisticated control strategies. This is because context information can be used to structure the control knowledge into *modular,* context-specific heuristics. This makes it much easier to encode and modify control knowledge than it is in systems based on global focusing schemes (e.g., conventional blackboard scheduler functions). Because planning-based control is highly goal-directed and creates detailed system subgoals, it provides other capabilities that most agenda-based blackboard frameworks lack. For example, a planning-based system can control the amount of data that undergoes (any) processing and can actively direct data gathering. These are important capabilities for interpretation problems which may involve passive

```
Initialize current-focus-points to the top-level control plan instance.
repeat: repeat: Pursue-Focus on each element of current-focus-points
               until null(current-focus-points)
        set current-focus-points to next-focus-points
        until null(next-focus-points)

Pursue-Focus(focus)
case on type-of(focus):
   plan instance  Focus on multiple-valued variable bindings to select plan instances.
                  Expand selected plan instances to next subgoals.
                  Focus on subgoals to select subgoals.
                  Match subgoals to control plans.
                  Focus on matching plans to select new plan instances for next-focus-points.
   primitive      Execute function associated with primitive to get status and results.
                  Update plans to select new focus element for next-focus-points:
                    propagate status and results of primitive to matching subgoal
                    and then up the control plan hierarchy to in-progress plan instance.
```

Figure 6.16 The basic RESUN control loop.

sensors that continuously generate large amounts of data and active sensors whose operation may be controlled by the interpretation system.

6.5.6 Blackboard Control: Still Evolving

In this section we have given the reader a taste of the difficult issues involved in the control of blackboard systems and other complex AI systems. We have looked at the evolution of blackboard control architectures from the basic, data-directed Hearsay-II model through highly goal-directed systems involving planning. Research on blackboard control architectures continues and it is important to state that there is no single best framework. The most appropriate architecture will depend on the application. Flexibility and sophistication typically increase both complexity and overhead cost.

6.6 CONCLUSIONS

In this chapter we have examined the notion of signal understanding as distinguished from traditional, algorithmic signal processing. We have shown how signal understanding can be viewed as an AI sensor interpretation problem and have examined blackboard systems as the most powerful AI approach for dealing with such problems. The chapter introduction as well as sections 6.1 and 6.2 discussed the reasons for using AI techniques like the blackboard model. Interpretation problems typically have large answer spaces and much uncertainty in the data and in the knowledge available to create explanations of the data.

Solving problems with these characteristics requires a search process in which potential solutions are incrementally constructed and the search is constrained by heuristic control knowledge. The blackboard model is ideally suited to this type of problem solving because it supports the incremental construction of hypotheses, the pursuit of multiple search paths (alternative explanations), opportunistic control (dynamic selection of problem-solving strategies), and so on. In addition, the modularity of the knowledge sources allows flexibility for the experimentation that is a necessary part of the development of any knowledge-based system. Of course, the flexibility of the blackboard is not without cost. Such a framework would not be appropriate for problems that do not involve large search spaces or data and problem-solving knowledge uncertainty; such problems should be solvable either through algorithmic or classification approaches.

Blackboard systems continue to be the subject of research. The current research directions in blackboard systems that are most relevant to signal understanding are: blackboard systems for real-time problem solving, parallel and distributed blackboard architectures and blackboard development environments. A collection of fairly recent papers that cover all of these research topics can be found in [26]. Other relevant papers include, for real-time systems [14,24], for parallel and distributed blackboard architectures [5,11,13,20], and for blackboard development environments [21,23,29,38].

Blackboard systems are well suited to the sort of dynamic control that is required for real-time problems. The key to effective real-time performance of blackboard systems is intelligent control. Research has centered on control mechanisms that can: use more detailed models of system goals, reason about the level of detail of their own reasoning, integrate more sophisticated models of uncertainty, adjust system goals to use approximate processing methods, and permit more asynchronous triggering and execution of knowledge sources. Real-time problem solving is relevant to many signal understanding applications because they involve real-time acquisition of data and must produce acceptable solutions within specified time limits.

The blackboard model was developed with concurrent execution of knowledge sources in mind. There are two ways in which concurrency may be obtained using the blackboard model: parallel blackboards or distributed blackboards. In the parallel blackboard model, there are multiple processors that can execute KSs in parallel, but the processors share a single blackboard database (and agenda, where applicable). The major problem that must be overcome is asynchronous modifications to the blackboard by KSs executing in parallel. In the distributed blackboard model, computation occurs throughout a set of fully separate blackboard systems. The major problem that must be dealt with is the need for the blackboard systems to communicate in order for the systems to cooperate: what should be communicated to whom and when it should be sent. Distributed blackboard systems are particularly well suited to interpretation problems that involve the fusion of data from a distributed network of sensors; distributing problem solving can reduce communication costs and allow the overall system to deal with hardware failures.

As there has been more interest in using the blackboard architecture to implement "real" applications, there has been more interest in the development of generic blackboard frameworks. The key issues for such frameworks are efficiency and flexibility/generality. These systems have to implement efficient retrieval and creation of hypotheses to provide efficient triggering and execution of KSs, but cannot do this by sacrificing generality or the ability to apply sophisticated control.

REFERENCES

[1] B. Buchanan and E. Shortliffe, eds., *Rule-Based Expert Systems* (Reading, Mass.: Addison-Wesley, 1984).

[2] N. Carver and V. Lesser, *Control for Interpretation: Planning to Resolve Sources of Uncertainty*, Technical Report 90-53, Computer and Information Science Department, University of Massachusetts (1990).

[3] N. Carver, "Sophisticated Control for Interpretation: Planning to Resolve Sources of Uncertainty" (Ph.D. thesis, University of Massachusetts, 1990).

[4] N. Carver and V. Lesser, "A New Framework for Sensor Interpretation: Planning to Resolve Sources of Uncertainty," *Proceedings of AAAI-91* (1991), 724–31.

[5] N. Carver, Z. Cvetanovic, and V. Lesser, "Sophisticated Cooperation in FA/C Distributed Problem Solving Systems," *Proceedings of AAAI-91* (1991) 191–98.

[6] W. Clancey, "Heuristic Classification," *Artificial Intelligence*, 27 (1985), 289–350.

[7] P. Cohen and E. Feigenbaum, eds., *The Handbook of Artificial Intelligence, Volume 3* (Los Altos, Calif.: William Kaufmann, 1982).

[8] P. Cohen, *Heuristic Reasoning about Uncertainty: An Artificial Intelligence Approach* (Boston, Mass.: Pitman Publishing, 1985).

[9] A. Collinot and B. Hayes-Roth, *Real-Time Control of Reasoning: Experiments with Two Control Models*, Technical Report KSL 90-17, Knowledge Systems Laboratory, Stanford University (1990).

[10] D. Corkill, "A Framework for Organizational Self-Design in Distributed Problem Solving Networks" (Ph.D. thesis and Technical Report 82-33, Department of Computer and Information Science, University of Massachusetts, 1982).

[11] D. Corkill, "Design Alternatives for Parallel and Distributed Blackboard Systems," in *Blackboard Architectures and Applications*, eds. V. Jagannathan, Rajendra Dodhiawala, and Lawrence Baum (New York: Academic Press, 1989).

[12] K. Decker, V. Lesser, and R. Whitehair, "Extending a Blackboard Architecture for Approximate Processing," *The Journal of Real-Time Systems,* 2 (1990), 47–79. (Also available as Technical Report 89-115, Computer and Information Science Department, University of Massachusetts.)

[13] K. Decker, A. Garvey, M. Humphrey, and V. Lesser, "Effects of Parallelism on Blackboard System Scheduling," *Proceedings of IJCAI-91* (1991), 15–21.

[14] K. Decker, A. Garvey, M. Humphrey, and V. Lesser, "Real-Time Control of Approximate Reasoning," to appear in *Proceedings of the 25th Hawaii International Conference on Systems Sciences* (1992). (Also available as Technical Report 91-50, Computer and Information Science Department, University of Massachusetts.)

[15] E. Durfee and V. Lesser, "Incremental Planning to Control a Blackboard-Based Problem Solver," *Proceedings of AAAI-86* (1986), 58–64.

[16] E. Durfee, "A Unified Approach to Dynamic Coordination: Planning Actions and Interactions in a Distributed Problem Solving Network" (Ph.D. thesis and Technical Report 87-84, Department of Computer and Information Science, University of Massachusetts, 1987).

[17] E. Durfee and V. Lesser, "Partial Global Planning: A Coordination Framework for Distributed Hypothesis Formation," *IEEE Trans. on Systems, Man and Cybernetics* (1991).

[18] L. Erman, F. Hayes-Roth, V. Lesser, and D. Raj Reddy, "The Hearsay-II Speech-Understanding System: Integrating Knowledge to Resolve Uncertainty," in *Blackboard Systems*, eds. R. Engelmore and T. Morgan (Reading, Mass.: Addison-Wesley, 1988).

[19] L. Erman, P. London, and S. Fickas, "The Design and an Example Use of Hearsay-III," in *Blackboard Systems*, eds. R. Engelmore and T. Morgan (Reading, Mass.: Addison-Wesley, 1988).

[20] R. Fennell and V. Lesser, "Parallelism in Artificial Intelligence Problem Solving: A Case Study of Hearsay-II," *IEEE Transactions on Computers*, 26 (1977), 98–111.

[21] K. Gallagher, D. Corkill, and P. Johnson, *GBB Reference Manual*, Technical Report 88-66, Computer and Information Science Department, University of Massachusetts (1988).

[22] B. Hayes-Roth, "A Blackboard Architecture for Control," *Artificial Intelligence*, 26 (1985), 251–321.

[23] B. Hayes-Roth and M. Hewett, "BB1: An Implementation of the Blackboard Control Architecture," in *Blackboard Systems*, eds. R. Engelmore and T. Morgan (Reading, Mass.: Addison-Wesley, 1988).

[24] B. Hayes-Roth, R. Washington, R. Hewett, and M. Hewett, "Intelligent Monitoring and Control," *Proceedings of IJCAI-89* (1989), 243–49.

[25] E. Hudlicka and V. Lesser, "Meta-Level Control Through Fault Detection and Diagnosis," *Proceedings of AAAI-84* (1984), 153–61.

[26] V. Jagannathan, R. Dodhiawala, and L. Baum, eds., *Blackboard Architectures and Applications* (New York: Academic Press, 1989).

[27] M. Vaughn Johnson, Jr. and B. Hayes-Roth, "Simultaneous Dynamic Integration of Diverse Reasoning Methods," in *Blackboard Architectures and Applications*, eds. V. Jagannathan, R. Dodhiawala, and L. Baum (New York: Academic Press, 1989).

[28] W. L. Lakin, J. A. H. Miles, and C. D. Byrne, "Intelligent Data Fusion for Naval Command and Control," in *Blackboard Systems*, eds. R. Engelmore and T. Morgan (Reading, Mass.: Addison-Wesley, 1988), 325–26.

[29] H. Laasri and B. Maitre, "Flexibility and Efficiency in Blackboard Systems: Studies and Achievements in ATOME," in *Blackboard Architectures and Applications*, eds. V. Jagannathan, R. Dodhiawala, and L. Baum (New York: Academic Press, 1989), 366.

[30] V. Lesser, R. D. Fennell, L. Erman, and D. Raj Reddy, "Organization of the Hearsay-II Speech Understanding System," *IEEE Trans. on Acoustics, Speech, and Signal Processing*, ASSP-23 (1975), 11–24.

[31] V. Lesser and L. Erman, "A Retrospective View the HEARSAY-II Architecture," *Proceedings of IJCAI-77* (1977), 790–800. Also in *Blackboard Systems*, eds. R. Engelmore and T. Morgan (Reading, Mass.: Addison-Wesley, 1988).

[32] V. Lesser and D. Corkill, "The Distributed Vehicle Monitoring Testbed: A Tool for Investigating Distributed Problem Solving Networks," *AI Magazine*, 4, no. 3 (1983), 15–33. Also in *Blackboard Systems*, eds. R. Engelmore and T. Morgan (Reading, Mass.: Addison-Wesley, 1988).

[33] V. Lesser, D. Corkill, R. Whitehair, and J. Hernandez, "Focus of Control Through Goal Relationships," *Proceedings of IJCAI-89* (1989), 497–503.

[34] V. Lesser, "A Retrospective View of FA/C Distributed Problem Solving," to appear in *IEEE Transactions on Systems, Man and Cybernetics*, special issue on distributed artificial intelligence (1992).

[35] J. Maksym, A. Bonner, C. Ann Dent, and G. Hemphill, "Machine Analysis of Acoustical Signals," in *Issues in Acoustic Signal/Image Processing and Recognition*, ed. C. H. Chen (New York: Springer-Verlag, 1983).

[36] A. Newell, "A Tutorial on Speech Understanding Systems," in *Speech Recognition,* ed. D. Raj Reddy (New York: Academic Press, 1975).

[37] H. Penny Nii and E. Feigenbaum, "Rule-Based Understanding of Signals," in *Pattern-*

Directed Inference Systems, eds. D. A. Waterman and F. Hayes-Roth (New York: Academic Press, 1978).

[38] H. Penny Nii and N. Aiello, "AGE (Attempt to Generalize): A Knowledge-Based Program for Building Knowledge-Based Programs," *Proceedings of IJCAI-79* (1979), 645–55. Also in *Blackboard Systems*, eds. R. Engelmore and T. Morgan (Reading, Mass.: Addison-Wesley, 1988).

[39] H. Penny Nii, E. Feigenbaum, J. Anton, and A. J. Rockmore, "Signal-to-Symbol Transformation: HASP/SIAP Case Study," *The AI Magazine*, 3, no. 1 (1982), 23–35. Also in *Blackboard Systems*, eds. R. Engelmore and T. Morgan (Reading, Mass.: Addison-Wesley, 1988).

[40] H. Penny Nii, "Blackboard Systems," in *The Handbook of Artificial Intelligence*, *Volume 4*, eds. A. Barr, P. Cohen, and E. Feigenbaum (Reading, Mass.: Addison Wesley, 1989).

[41] J. Pearl, *Probabilistic Reasoning in Intelligent Systems: Networks of Plausible Inference* (San Mateo, Calif.: Morgan Kaufman, 1988).

[42] Y. Peng and J. Reggia, "Plausibility of Diagnostic Hypotheses: The Nature of Simplicity," *Proceedings of AAAI-86* (1986), 140–45.

[43] D. Raj Reddy, "Speech Recognition by Machine: A Review," *Proceedings of the IEEE*, 64 (1976), 501–31.

[44] M. Williams, "Hierarchical Multi-expert Signal Understanding," in *Blackboard Systems*, eds. R. Engelmore and T. Morgan (Reading, Mass.: Addison-Wesley, 1988).

7

INTEGRATED PROCESSING AND UNDERSTANDING OF SIGNALS

- S. Hamid Nawab
  Boston University
- Victor Lesser
  University of Massachusetts

7.1 INTRODUCTION

As the complexity and variety of the investigated signal understanding domains has grown, the need for sophisticated integration of traditional signal processing technology [22, 23] and traditional signal understanding technology [2, 3] has become apparent. For the purposes of this chapter, we consider traditional signal processing technology to be primarily concerned with the theory-based numerical transformations applied to signal data, and we consider traditional signal understanding to be primarily concerned with the heuristic search for interpretation models in order to obtain the most consistent explanations for the signal processing output data. This chapter discusses a novel architecture which integrates signal processing and signal understanding within a unified problem-solving framework. We call this architecture IPUS for *Integrated Processing and Understanding of Signals*. In this architecture, the power of signal processing theories such as filtering theory, time-frequency signal modeling, and parameter estimation is combined in a sophisticated manner with the power of artificial intelligence theories such as blackboard architectures for constructive problem solving, symbolic representations of uncertainties, incremental hierarchical planning for sophisticated control, approximate processing for real-time problem solving, and diagnostic reasoning. An important feature of the IPUS architecture is the bidirectional interaction between the problem solving for processing the input signals and the problem solving for interpreting the results of the signal processing. The practical advantage of such an approach is its suitability for signal understanding domains involving complicated interacting signals. The intellectual thrust of this approach is on the effective integration of knowledge from the two diverse fields of signal processing and artificial intelligence. As detailed in this chapter, the issues that are raised by the IPUS architecture often require that the signal processing and artificial intelligence issues be addressed simultaneously and in close cooperation. Problems that superficially appear to concern only one discipline often turn out to have elegant solutions that utilize knowledge from both disciplines.

This chapter focuses on the domain-independent aspects of the IPUS architecture and how signal processing and signal understanding theories play a role in supporting various architectural features. The chapter also reflects the fact that the principles underlying this architecture still require further research. Therefore, the discussions of various aspects of the architecture vary considerably in depth of treatment. Where the issues still need further research, we point out possible directions of future research. Finally, in the interest of keeping the chapter to a reasonable length, the examples used to illustrate the supporting theories have been kept relatively simple. In particular, the examples generally are in the context of a system that uses short-time Fourier transform (STFT) [5] data for characterizing signals that consist of linearly combined frequency-modulated sinusoidal signals. These examples are a simplified form of the type of problem solving that takes place in the sound understanding testbed we have developed at our laboratories [1]. In section 9.3 of Chapter 9 in this book, an example is presented that illustrates how the IPUS architecture is utilized by the sound understanding testbed for performing a complicated sound understanding task where the class of input signals is significantly more complex. It should be noted that many aspects of the IPUS architecture are equally applicable to other application domains, including those that do not utilize the STFT algorithm and where the input signals are of a very different nature than those encountered in sound understanding.

In section 7.2, we provide the motivation for the development of the IPUS architecture. An overview of the IPUS architecture is presented in section 7.3. We then discuss in section 7.4 the signal processing theories that support the IPUS architecture. In section 7.5, we discuss the artificial intelligence/signal understanding theories that support the IPUS architecture. We conclude the chapter in section 7.6 by evaluating applications of IPUS from the perspectives of signal processing and signal understanding.

7.2 MOTIVATION FOR IPUS ARCHITECTURE

The IPUS architecture has evolved from our research on the design of a sound understanding system [1]. The goal of such a system is to identify the types of sound sources (such as telephones, crying infants, household appliances, etc.) that may have generated the signals received by the system. Specific applications of sound understanding include assistive devices for the profoundly deaf, robotic hearing, speech recognition in environments with nonspeech sound sources in the background, and experimentation with theories of human hearing. The complexity of the sound understanding problem largely arises because of two factors:

- The need to process a tremendous variety of signal types due to the situation-dependent nature of the input. For example, a sound understanding system has to deal with inputs including narrowband, harmonic, noiselike, impulsive, and frequency-modulated signals and combinations thereof.
- The need to change processing goals in a context-dependent way. For example, the goal of a signal understanding system might be to respond to either the sounds of an infant or a ringing telephone and to ignore other sound sources. If an infant sound is detected, the system's main goal may then switch to determining whether the infant is crying or choking and to ignore the telephone.

Because of the above factors, it is very difficult or even impossible to design a single mathematically derived signal processing algorithm that can be applied to all possible input signals to produce the desired information for each input. There are too many types of input signals, each type best suited to a different kind of mathematical formulation. Additionally, the form of the desired information has different mathematical formulations in accordance with the assumed nature of the input signal and the current high-level goals of the signal understanding system. To deal with such complexities, the approach taken in the IPUS architecture is for the signal understanding system to have access to a database of mathematically derived algorithms. This database is indexed by the types of assumptions made about the input signal and the type of output information desired in accordance with the current goals of the signal understanding system. It therefore becomes necessary for the signal understanding system to be able to carry out a context-dependent *search* through the algorithm database.

Traditional signal understanding systems [2, 3] have restricted the class of possible input signals and the high-level goals of the system to ensure that a fixed front-end signal processing algorithm can be applied to all input signals. Thus, the interaction between the interpretation problem solving and the signal processing is confined to that of the former accepting the output data from the latter (see Figure 7.1). Some recent systems [17, 18, 24] have utilized architectures in which the signal processing is not immutable and can be affected by the results of higher level interpretation activity. The IPUS architecture is also designed to achieve such flexibility in the signal processing, with an emphasis on utilizing the sophisticated signal processing theories that underlie many signal processing tasks in order to *structure* the cooperation between signal processing and signal understanding. In particular, the IPUS architecture utilizes the fact that signal processing theories often supply a system designer with a signal processing algorithm (SPA) that has *adjustable control parameters*. For example, the short-time Fourier transform (STFT) may be

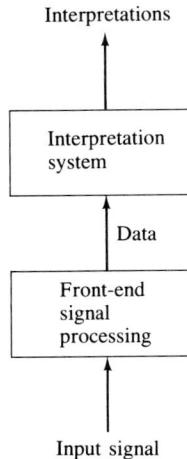

Figure 7.1 Classical Signal Understanding Architecture. Front-end signal processing provides the data to be interpreted by the higher level processes. The higher level processes do not perform any further signal processing.

considered an SPA with adjustable control parameters such as analysis-window length, frequency sampling factor, and temporal decimation factor. Viewed from this perspective, an SPA really denotes a database of SPA "instances," each of which corresponds to a particular set of fixed values for the control parameters. The IPUS architecture is designed to search for appropriate SPA-instances to be utilized in particular situations. Since the search for an appropriate SPA-instance is equivalent to a search for appropriate SPA control parameter values, the theoretical relationships between the control parameters and the SPA performance characteristics (such as frequency resolution and time resolution of an STFT) can be exploited by the search process.

Let us now consider the characteristics that an application domain should have so that it cannot be adequately served by a single SPA-instance.[1] In our experience with the development of a sound understanding testbed [1], we have found that the most important candidate applications are those that involve interacting signals from different sources. Consider the sound understanding problem of recognizing that an input signal was generated by a particular vacuum cleaner. The vacuum cleaner sound may be divided into various sound-segment categories (such as turn-on transient, turn-off transient, forward motion with surface contact, backward motion with surface contact, steady state without surface contact, along with motor speed variations). It is conceivable to use a single SPA-instance to distinguish among the features of all of the sound-segment categories. However, if we were to permit the possibility of two simultaneously operating vacuum cleaners, a combinatorial explosion in the possible categories of sound segments arises. Consider, for example, an observed signal that is composed of the turn-on transient of one vacuum cleaner and the turn-off transient of the other vacuum cleaner. The features of the observed

---

[1] Or even a small number of different SPA-instances. It is for domains that require a large number of different SPA-instances that the IPUS architecture is most suitable.

Sec. 7.2 Motivation for IPUS Architecture

signal not only depend on the simultaneous presence of the two sound-segment categories but also on their relative loudnesses, start times, and end times. If we now widen the scope of the sound understanding system to include not only vacuum cleaners but other sound sources, the combinatorial explosion due to interacting sound-segment categories becomes even more unmanageable. As the combinatorial explosion grows, the viability of using a single SPA-instance to discriminate between various classification categories becomes more and more questionable. To make this idea more concrete, let us consider what happens when signals from two relatively simple categories are combined with each other.

The consequence of combining signals from two relatively simple categories is illustrated in Figure 7.2. Parts (a) and (b) of this figure show representative waveforms for two different categories of signals. The waveform in (a) is from a category of signals, each member of which is a decaying sinusoid whose frequency and amplitude decay rate are within 10% of certain fixed values. The waveform in (b) is from another category if signals, each member of which is a growing sinusoid with a frequency in the same range as the first category of signals and a growth rate that is within 10% of a given value. Parts (c) through (f) of the figure show the results of combining the representative waveforms with possibly different amplitudes, phase shifts, frequencies, and decay/growth rates for each waveform. The goal is to detect the simultaneous presence of both categories from such combined waveforms. Part (c) results when the two waveforms have the same frequencies, amplitudes, and phase shifts. In this case, the only clue to the presence of the two categories is the exponential decay followed by an exponential growth in the waveform envelope. Part (d) results when the two waveforms have the same characteristics as for part (c) but have different phase shifts. Note that this time the envelope is also the only clue *except* that the envelope shows approximately linear growth and decay rates instead of the exponential rates for the growth and decay of the constituent signals. Part (e) results when the growing sinnusoid has a significantly smaller amplitude and a slightly different frequency. In this case, the envelope of the composite waveform seems to indicate the presence of only one category. The only way to discriminate between the two categories is to perform high-resolution frequency separation of the two constituent components. Part (f) results when combining the two categories with the same amplitudes, the same relative phases, but significantly different frequencies. In this case, the two components may be separated using low-resolution frequency separation. The important point is that *different* features are used to differentiate between the same two categories of sound. Furthermore, the extraction of some features precludes the extraction of other features. For example, if the STFT [5] is used with a large value for the STFT window-length control parameter in order to obtain the frequency resolution to separate two frequency components, the resulting decrease in time resolution of the STFT makes it impossible to accurately estimate the waveform envelope from the STFT output data. To obtain an accurate estimate of the envelope, a short window length is needed, resulting in a frequency-resolution distortion which makes it impossible to accurately extract the two separate frequency components. Thus, the STFT algorithm with a fixed set of control parame-

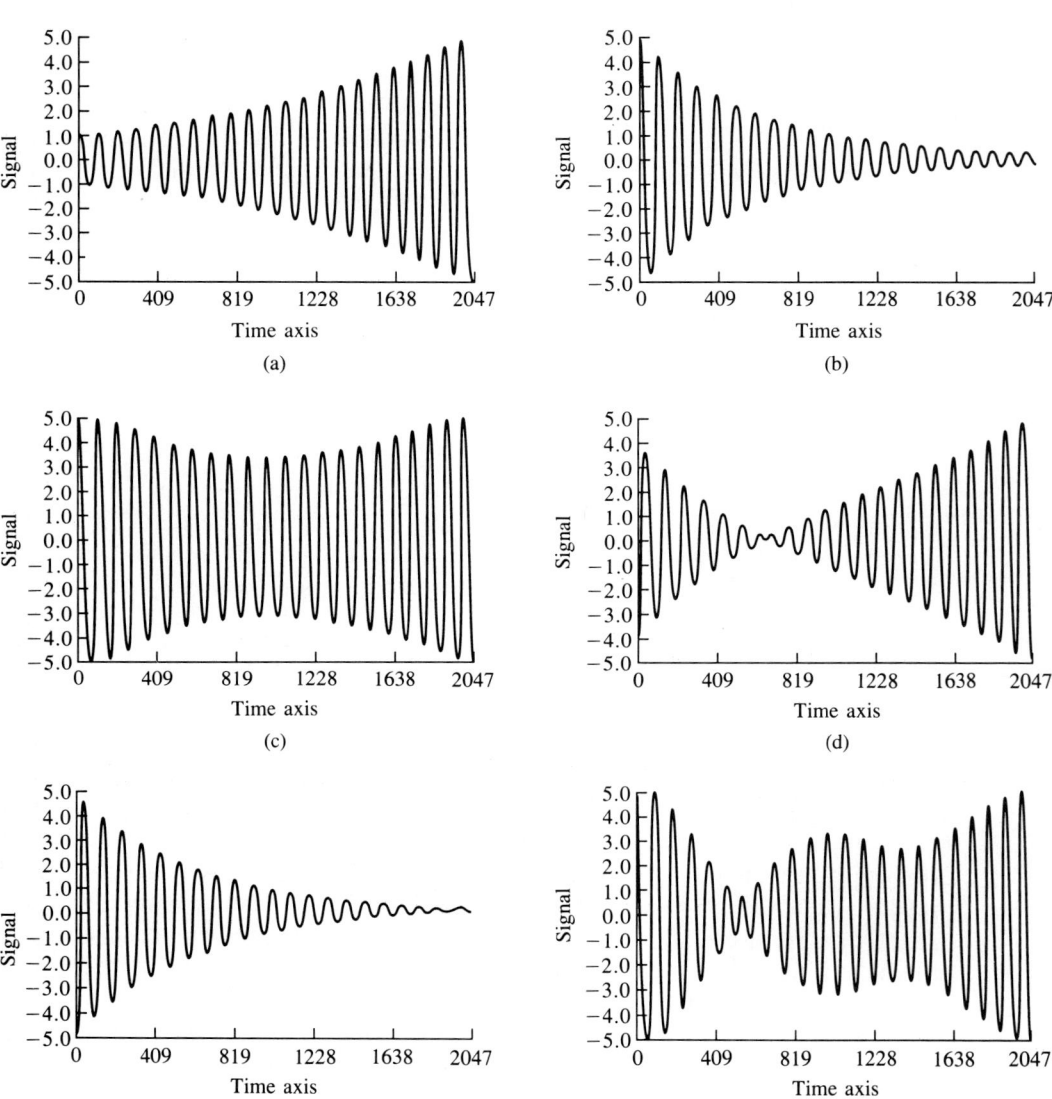

Figure 7.2 Interactions between the characteristics of simultaneously present signals. Parts (a) and (b) are the original two signals, each representing a class of signals with a range of possible frequency characteristics. The remaining four parts represent the sum of the original signals with different relative amplitudes and phases.

ter values would not be able to detect the presence of the two signal categories in all four cases of linear combinations presented in Figure 7.2.

The classical signal understanding approach, which typically uses a single signal processing algorithm with fixed control parameters, also has a weakness in dealing with applications with variable signal-to-noise ratio (SNR) conditions. For example, an algorithm whose output data is guaranteed to provide enough information to discriminate between different sound-segment categories under high SNR conditions will generally fail to provide the necessary discrimination ability under low SNR conditions. When the SNR is low, it is usually necessary to utilize signal processing algorithms that need a priori information about the statistical characteristics of the noise. If the SNR varies and the statistical nature of the noise also varies (often the case in realistic applications), it is necessary to first estimate the SNR and the statistical characteristics of the noise and then to select the most appropriate SPA-instances for further processing.

Given that in sufficiently complicated signal understanding domains it is necessary to use different SPA-instances for different subsets of possible input signals, two basic approaches for carrying out the signal processing suggest themselves:

1. Process the incoming signal with all the SPA-instances that are potentially relevant to the entire class of possible input signals in the application domain and then choose the output data that has the most consistent interpretation. For realistically complex applications (when the combinatorial explosion in possibly applicable SPA-instances is severe), this approach requires vast amounts of signal processing computation and produces vast amounts of signal processing output data to be examined by the higher level interpretation processes.

2. Process the incoming signal with one or a small number of the possibly relevant SPA-instances, then use some mechanism to *recognize* if incorrect processing has taken place. This is followed by determining the nature of the incorrect processing through a diagnostic reasoning process, and finally *changing* the parameter settings of the SPA with the aim of obtaining an SPA-instance that is appropriate for the processing of the input signal. The SPA-instance with adjusted control parameter settings is then used to reprocess the input signal.

The second approach, which involves a *search* for appropriate SPA-instances, is the one we have utilized in an IPUS sound understanding testbed and found it to be quite practical. A description of the testbed is provided in Chapter 9. In the next section, we present an overview of the IPUS architecture with emphasis on its domain-independent aspects.

7.3 OVERVIEW OF IPUS ARCHITECTURE

The IPUS architecture is designed for signal understanding systems where the set of possible input signals and the system goals are so varied that it is unrealistic to use the same SPA-instance for the processing of every input signal. Instead, the signal understanding system has one or more SPAs with adjustable control parameters at its disposal. In order to select appropriate values for the SPA control parameters, the system must consider the current system goals as well as knowledge about certain characteristics of the particular input signal. This leads to the dilemma that choosing the appropriate control parameter values requires knowledge about the signal, but this knowledge can only be obtained by first processing the signal with an algorithm with appropriate control parameter settings. The IPUS architecture uses an iterative technique for converging to the appropriate control parameter values. The technique begins by using the best available guess for the SPA control parameter values (in the worst case, the system may resort to an arbitrary assignment for the control parameter values). The input signal is then processed using the SPA-instance corresponding to the chosen control parameter values. The SPA-instance output is then analyzed through a *discrepancy detection* mechanism for indicating the presence of distorted SPA output data. A *diagnosis* is then performed for mapping the detected discrepancies to distortion hypotheses. A *signal reprocessing planning* phase then proposes a new set of values for the control parameters of the SPA with the aim of eliminating the hypothesized distortions. The SPA-instance corresponding to the new control parameter values is then used to reprocess the input signal. The output from the reprocessing once again undergoes discrepancy detection, and if necessary followed by diagnosis, signal reprocessing planning, and further reprocessing of the input signal. This iterative process continues under the direction of a control framework that is an integral part of the IPUS framework.

The high-level architectural details of the IPUS architecture for accomplishing discrepancy detection, diagnosis, and signal reprocessing in conjunction with the overall signal understanding task is illustrated in Figure 7.3. It should be noted that the signal data and the interpretation hypotheses derived from that data are stored on a blackboard with hierarchically organized information levels. The hypotheses on the blackboard fall into two basic categories: hypotheses posted to explain the signal data and hypotheses posted to specify expectations about the nature of the signal data. The inferencing on the blackboard is performed by various knowledge sources (KSs) for tasks such as discrepancy detection, diagnosis, signal processing, data interpretation, and the formation of expectation hypotheses. The IPUS architecture uses the RESUN [25] framework to control the KSs that perform the inferences on the blackboard.

7.3.1 Discrepancy Detection

The *key to the IPUS approach* lies in a system's ability to carry out discrepancy detection. The purpose of discrepancy detection is to indicate the presence of distorted SPA-instance output data. To illustrate the notion of distorted output from a signal processing algorithm, consider an input signal that consists of two frequency-modulated sinusoidal components to be processed by a STFT algorithm [5]. The desired output is an estimate of the frequency content of the input signal as a function of time. The STFT represents a variety of signal processing strategies. The selection of a particular strategy is accomplished by selecting particular vales for the control parameters (window length, FFT length, temporal-decimation, number of peaks in frequency domain, energy threshold for peak detection). Figure 7.4 shows the output obtained from the STFT algorithm for a particular set of values for the STFT control parameters. The solid lines indicate the actual frequency tracks of the two sinusoids

Sec. 7.3   Overview of IPUS Architecture

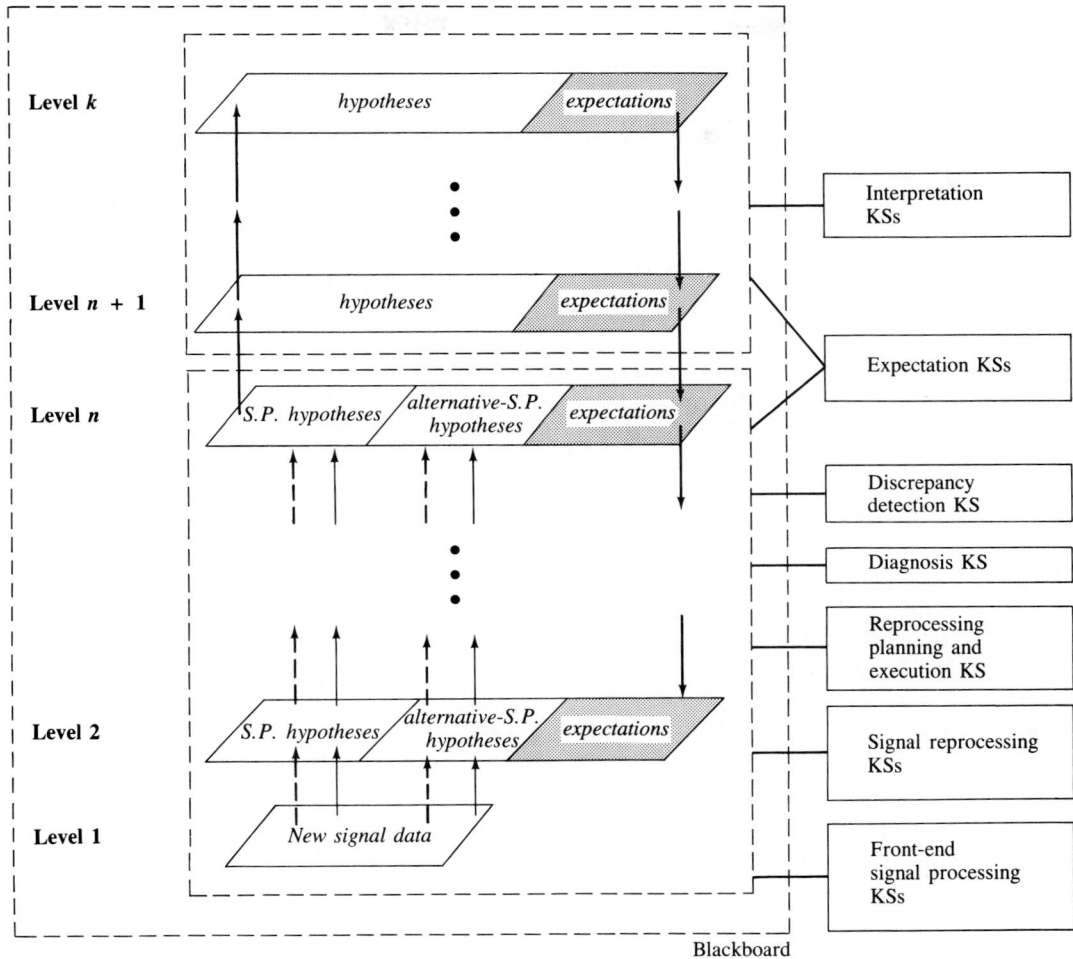

Figure 7.3 IPUS Architecture. Front-End signal processing performs bottom-up hypothesis formation for the first few information levels of the blackboard. Signal Reprocessing makes use of constraints from expectations posted by higher-level problem solving and data hypotheses posted by alternative signal processing algorithms to detect distorted front-end signal processing output data and attempts to eliminate these distortions. The IPUS architecture uses the RESUN [25] control framework to control the knowledge sources that perform the inferences on the blackboard.

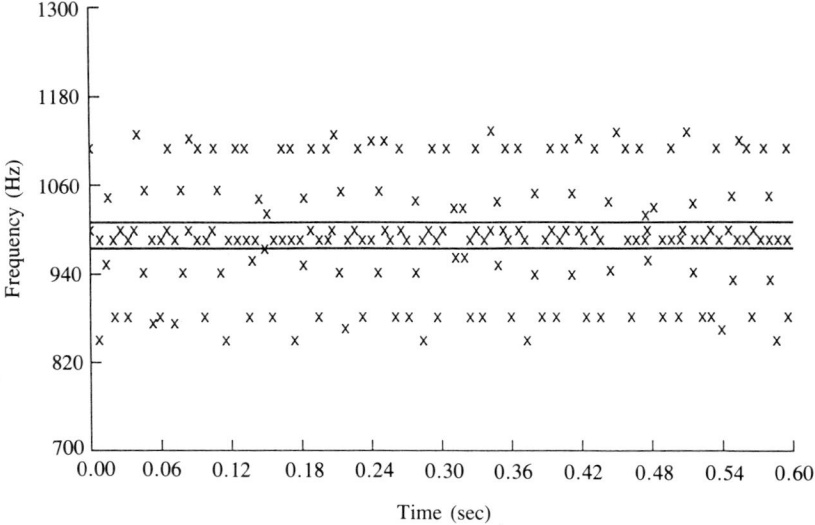

Figure 7.4 Illustration of the frequency-resolution distortion: Actual peak tracks of the input signal are shown as solid lines. The × symbol track shows the peak tracks obtained from signal processing output. STFT Window-length = 128.

in the input signal, while the *x*s indicate estimates of frequencies as a function of time. The STFT output is distorted in that it does not match the actual tracks of the sinusoidal components in the signal data. On the other hand, if the window-length control parameter of the STFT is increased sufficiently, the undistorted output shown in Figure 7.5 is obtained.

The idea behind the discrepancy detection mechanism is that the signal understanding system should be able to detect discrepancies between distorted SPA output data and one or more of the following:

1. The *entire allowable* class of input signals for the application domain. We refer to such discrepancies as *violations*. It should be noted that a violation occurs when the output data has characteristics that are *a priori* known to be absent in the entire class of possible signals in the application domain. For example, if the application domain is known to consist only of signals with frequencies below 500 Hz, output data showing a signal at 700 Hz would give rise to a violation.

2. The *expected* form of the output data, as predicted from interpretations of past data or partially supported interpretations of current data. We refer to such discrepancies as *conflicts*. There are two types of conflicts: conflicts with *unverified expectations* and conflicts with *partially verified expectations*. The former are conflicts with expectations that have no support from current data, and the latter are conflicts with expectations that are partially supported by current

Sec. 7.3 Overview of IPUS Architecture

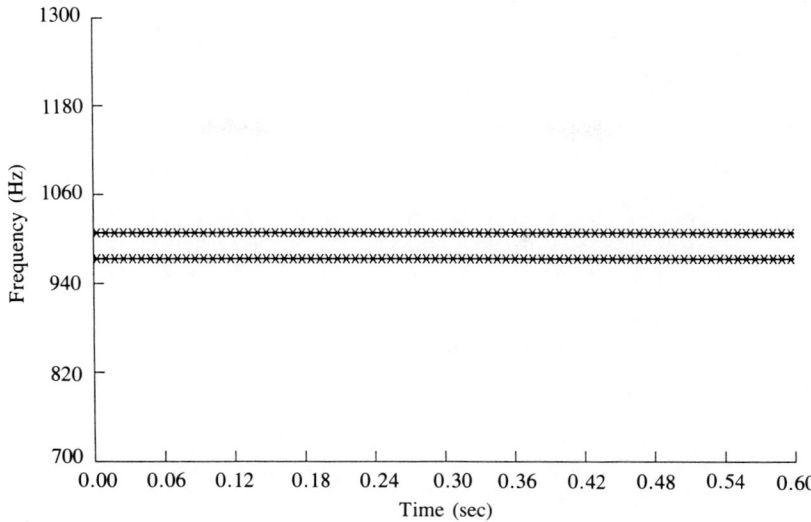

Figure 7.5 Reprocessing to produce undistorted output data. Actual peak tracks of the input signal are shown as solid lines. The × symbol track shows the peak track obtained from signal processing output. STFT Window-length = 1024. Compared to Figure 7.4, distortions have been removed.

data. As an example of a conflict with unverified expectations, consider a situation where the interpretations of past data show two sinusoids at 200 Hz and at 250 Hz respectively with no decline in their amplitudes and the current SPA output data contains neither of the sinusoids. A conflict with partially verified expectations would occur if current signal processing output data contains the 200 Hz sinusoid but not the 250 Hz sinusoid.

3. The *output data* from other SPA-instances applied to the same underlying signal data. We refer to such discrepancies as *faults*. For example, suppose that the signal data is being processed with a zero-crossing analyzer and an STFT. If the zero-crossing analyzer indicates the presence of a sinusoidal signal but the STFT does not, a fault would be declared. Another example of a fault would be the case where the zero-crossing analyzer indicates signficant change in frequency content over a time interval while the STFT produces output data showing no change in frequency content over the same interval.

7.3.2 Diagnosis and Reprocessing

Once discrepancies have been detected, it is necessary to determine the *distortions* in the signal processing that may have led to those discrepancies. To illustrate this concept, let us consider a specific example involving a small electric motor sound with a speed-change transition. Let us assume that previously there had been no speed changes in the electric motor sound and therefore the front-end signal process-

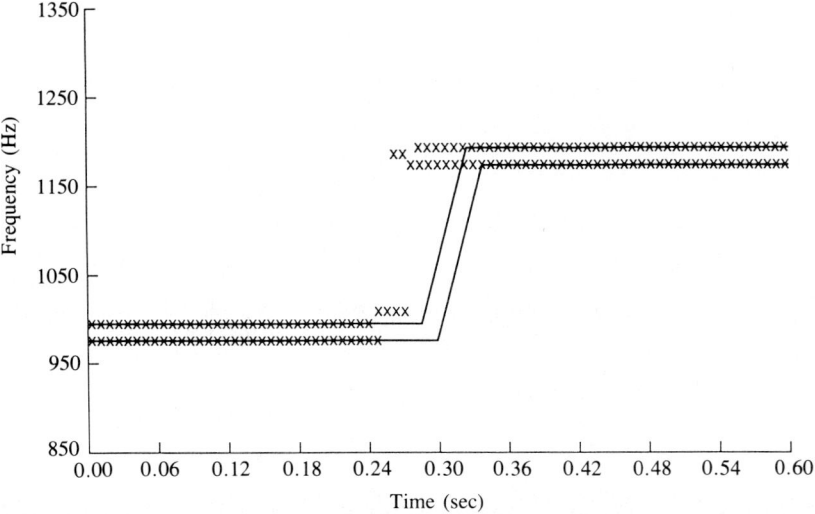

Figure 7.6 A signal processing output for a motor sound. The solid lines represent the actual frequency tracks, while the ×'s represent the STFT output data. Time-resolution distortion causes the motor speed-change interval not to be tracked.

ing has its control parameters set to values that ensure detection of two constant-frequency components. The signal understanding system may not expect speed changes in the motor because the interpretation model for the motor indicates such changes to be infrequent and therefore generally unlikely. However, when the most current signal is processed, the result shown in Figure 7.6 is produced. Note that the signal processing output is in *conflict* with the expectation of two constant-frequency components. This discrepancy is then analyzed by a diagnosis process, which maps symptoms (discrepancies) to hypothesized underlying causes (distortions). In our example, the diagnosis process hypothesizes that the upper two tracks in Figure 7.6 are connected with the lower two tracks and that these connections are missing in the data due to a time-resolution distortion. This distortion arises when the STFT-instance has its window-length parameter set to a relatively large value. A *signal reprocessing planner* then concludes that the window length of the STFT should be decreased, but it also notes that this will result in decreased frequency resolution. Consequently, it forms the expectation that when the input signal is reprocessed with a longer analysis window, the two parallel frequency tracks will be merged into a single frequency track. The consequent execution of the reprocessing plan results in the data shown in Figure 7.7. As expected, evidence is obtained for the speed-change, although the poorer frequency resolution does not discriminate between the two constituent frequency components. The data of Figures 7.6 and 7.7 are then considered to jointly represent evidence for the interpretation model represented by the solid lines in those figures. We refer to this process of using outputs from separate SPA-instances as evidence for an interpretation model as *multiple-outputs-synthesis*.

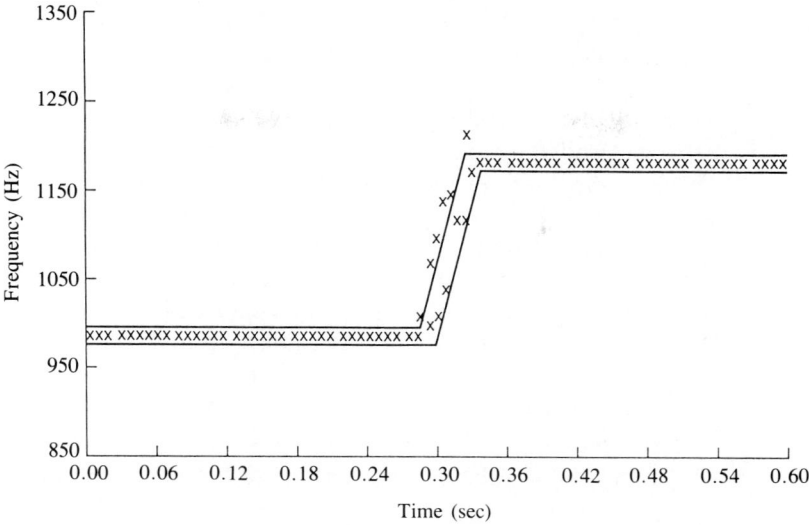

Figure 7.7 Signal processing output for a motor sound. The solid lines represent the actual frequency tracks, while the ×'s represent the STFT output data. Frequency-resolution distortion causes the two frequency tracks to be merged into a single ghost track.

7.3.3 Interpretation Process

The search for an appropriate SPA-instance cannot be carried out independently of the interpretation process for explaining the signal processing output data. The interpretation process may be conceptualized as a search through a space of *sets* of interpretation models. In simpler domains, where the combinatorial explosion in interpretation models is not severe, interpretation may be approached as just a classification process [8, 8, 19]. In complicated domains, the interpretation models search must be a *constructive* problem-solving process [7]. The IPUS architecture employs this more general constructive approach.

Constructive problem solving techniques must be used when the set of possible solutions is too large to be pre-enumerated. In complex domains such as sound understanding, the number of potential explanations for data is very large (or even infinite). This is because there are large numbers of potential interpretation-model hypotheses and explaining the data typically involves multiple hypotheses. To illustrate this for the sound understanding application, consider the following factors:

- There is an enormous variety of individual sound source types. For example, in the household environment, sound sources range from different types of human sounds (shouting, crying, laughing, whispering, walking, running, etc.) to various appliance sounds (refrigerators, blenders, mixers, vacuum cleaners, etc.).

- There are many possible hypotheses of interest even for a single source. For example, "vacuum cleaner being used on a hard surface" has a different model than "vacuum cleaner being used on a rug."
- There are often multiple sources present simultaneously. These sources cannot be considered independently because of interactions; the resulting sound does not necessarily retain the individual characteristics of the sound from each source. As an example, consider a situation involving a telephone ring and a vacuum cleaner sound. The vacuum cleaner sound may "mask" one or more of the frequencies due to the telephone ring. Thus, the model of the telephone sound in the presence of a vacuum cleaner is different from the model of a telephone sound alone. It is impossible to precompute models for all of the possible ways that sources might interact. Instead, the combination models must be constructed from knowledge of how the individual models might interact as these combinations are suggested during interpretation.
- The interpretation data covers a range of time and the time of occurrence of sources and the relative time of sounds may be critical. For example, "loud motor" followed by "crying" versus the opposite.
- There may be multiple instances of each source type (e.g., multiple phones, multiple humans) and each source may be responsible for different instances of sounds. For example, there may be two telephones in the environment so when we identify "telephone ringing" we may be sure which phone. In addition, when we have two successive instances of "telephone ringing" we may not be sure whether it was the same phone or not (correlation ambiguity).

To illustrate the complexity of interpretation models search in complex domains, consider a typical household environment. This sort of environment will contain a wide variety of sound sources (e.g., humans, animals, telephones, appliances, street noises) that may be active from time to time and may be active simultaneously. Among the common sources of sound in the household environment are small electric motors with variable speed. Many household sound source types contain such motors, including blenders, mixers, vacuum cleaners, hair dryers, and toys. With a single, isolated electric motor, STFT processing would produce evidence for frequency tracks shown as solid lines in Figure 7.7. However, in the typical household, the sound from a source containing a small electric motor will be combined with the sound from other motor containing sources as well as other types of sound sources. Having multiple, simultaneously active sound sources complicates the interpretation process because the signature of an electric motor may be difficult to separate from frequency tracks of other sources or it may be obscured by overlapping frequencies from the other sources. In addition, once a "small electric motor" source is identified, a signal understanding system may still have to determine the specific sound source type (e.g., hair dryer vs. blender) and may even have to consider which of multiple instances of the source type is active (e.g., whose hair dryer) in order for the overall system to know how to react.

Because the acoustic signatures of simultaneously active sound sources may

Sec. 7.3 Overview of IPUS Architecture

interact, a signal understanding system cannot consider each possible sound source independent of the other possible sources. In the classification approach to signal understanding, a system compares incoming data with prestored interpretation models to find the model that yields the "best" match with the data. Thus, a system using this approach has to have access to models for each possible combination of sounds. In complex domains, this is highly impractical. This approach first requires the system designer to pre-enumerate all of the potential combinations of sound sources along with all the ways that they could overlap—a very large set. The system designer would then have to determine interpretation models for each of the those possible combinations, taking into account the complex interactions that take place in multiple-source scenarios. In contrast, for a signal understanding system utilizing the constructive problem solving approach, the system designer provides interpretation models for single source scenarios and knowledge about how these models may be combined to construct models for multiple-source scenarios. This information is used by the system to *incrementally* consider potential interpretations; it first focuses on likely interpretation-model hypotheses and then carries out various evidence-gathering strategies to identify additional explanations that should be considered and to prune unsupported interpretations. With this approach, a system has to reason about how different source models may interact only when it is necessary to find explanations for the actual data. Thus, a signal understanding utilizing constructive problem solving must use sophisticated control strategies to focus the attention of the system on the most plausible hypotheses and then gather evidence to refine and validate these hypotheses.

7.3.4 Evidence Gathering in IPUS

The problem-solving supported by the IPUS architecture may be viewed as a process of hypothesis formation coupled with evidence-gathering strategies. At the lower levels, evidence gathering seeks undistorted signal processing output data to increase or decrease the certainty of various higher-level hypotheses. There are two basic modes of evidence gathering provided by the IPUS architecture—evidence aggregation and differential diagnosis. *Evidence aggregation* seeks data for increasing or decreasing the certainty of one *particular* hypothesis. *Differential diagnosis* seeks data with the aim of resolving the ambiguities that may have led to *competing* hypotheses.

The search for appropriate SPA-instances may take place as part of an evidence aggregation process or a differential diagnosis process. To illustrate this, let us first consider a situation where a sound source with three frequency tracks has been hypothesized. However, the front-end signal processing output data may support only two of the frequency tracks. If a high degree of certainty in the hypothesized sound source is important to the system's goals, evidence aggregation would be employed to seek data for supporting the missing frequency track. This may for example lead to a change in the data model in the form of a lowering of the assumed minimum energy level of frequency tracks. An example of a situation where the differential diagnosis mode of evidence gathering may be employed is one where the front-end signal processing data consists of two frequency tracks that support two competing source hypotheses. In this case, the differential diagnosis approach would focus on obtaining data that might potentially help to disambiguate between the competing hypotheses. For example, an important distinction between the interpretation models for the two competing source hypotheses might be that each of the models requires a third frequency track (not currently supported by the data) but each has a *different* frequency for the third track. The signal data might then be reprocessed (using different values for the SPA control parameters) with the specific aim of ascertaining whether a third track is present and which source hypothesis may be supported by it. On the other hand, if the two competing source hypotheses are distinguished by say, the presence of broadband noise in one of the sources, the signal data would be reprocessed with an algorithm specifically aimed at detecting the presence of any broadband noise in the signal data.

In our sound understanding testbed (see Chapter 9) we have employed a *Resolving Sources of Uncertainty* (RESUN) framework [25] for the control of integrated search. This permits both evidence aggregation and differential diagnosis modes of problem solving to invoke signal reprocessing. A key objective of our research is to uncover the principles for the control of problem-solving tasks which utilize data-models search as part of the evidence-gathering process.

7.3.5 Comparison with Adaptive Control

It is important to distinguish between the problem addressed by the IPUS architecture and the problems addressed by classical adaptive control theories [6]. The classical theories use stochastic process concepts to characterize signals. Discrepancies between stochastic characterizations and signal processing output data are used to adapt the further processing of signals. The stochastic characterizations are limited to specifications of the probabilistic moments (usually not higher than second order). In contrast, IPUS uses high-level symbolic descriptions (i.e., interpretation models) as well as discrepancies between the outputs of different SPA-instances to characterize signal data and uses differences or discrepancies between these characterizations and signal processing output data to adjust the further processing of signals. Classical adaptive control may thus be viewed as a special case of an IPUS architecture, where the interpretation models are restricted to characterizations in terms of probabilistic measures and low-level descriptions of signal parameters.

7.4 SP THEORIES TO SUPPORT IPUS

In this section, we focus on the two major roles that classical signal processing theories play with respect to the IPUS architecture. These roles are:

1. To determine whether or not an IPUS architecture is needed in a given application domain.

2. To provide the signal processing knowledge to be encoded in a system using the IPUS architecture.

The IPUS architecture is suitable when a single SPA-instance cannot correctly process all the input signals that can potentially arise in a signal understanding application. How should one go about determining whether or not a single SPA-instance is sufficient in a given signal understanding application? To answer this question, we have formulated an approach that we refer to as *model variety analysis*. To describe this approach, it is first necessary to develop the concept of a *data-model* corresponding to an SPA-instance with respect to a class of signals. The data-model concept is discussed in section 7.4.1. Model variety analysis is described and illustrated in section 7.4.2. A useful byproduct of the framework for model variety analysis is that it conveniently leads to a procedure, discussed in section 7.4.3, for classifying the causes for distorted outputs from SPA-instances. We interpret each category of a cause as representing a signal-processing *distortion*. This categorization of distortions plays an important role in the diagnosis process for hypothesizing distortions on the basis of detected discrepancies. The nature of the mapping from detected discrepancies to corresponding distortions is discussed in section 7.4.4.

7.4.1 Data-Model Concept

In this section, we develop the concept of a data-model for an SPA-instance with respect to a class of input signals. This concept will be useful in section 7.4.2 for the discussion of model variety analysis. We define the data-model of an SPA-instance with respect to a class **S** of input signals as the set of conditions which the various features of the members of **S** must satisfy in order to ensure that the SPA-instance provides output data from which those signal features can be estimated adequately. In this context, we define the features of a signal as the parameters of a signal generation process (SGP) for the input signal. The data-model therefore expresses the relationships that must hold between the SGP parameters of an input signal and the control parameter values of the SPA-instance in order to ensure that the SGP parameters can be estimated to within some specified bounds. When these conditions are satisfied by an input signal, the output from the SPA-instance is said to be *undistorted*. If the data-model conditions are not satisfied by an input signal, the corresponding output is said to be *distorted*. Before illustrating this concept with a specific example, we make a few general observations regarding the utility of the data-model concept. The data-model *partitions* the set **S** into two subsets, one of signals for which the SPA-instance produces undistorted outputs and the other of signals for which the SPA-instance produces distorted outputs. If the subset of **S** that would be incorrectly processed by the SPA-instance is empty, the SPA-instance could serve as the front-end signal processing module of a traditional signal understanding system whose input signals are restricted to the set **S**. To deal with cases where no such SPA-instance can be found, we extend the data-model concept to an SPA instead of just an SPA-instance by noting that in such cases there is a *space* of data-models, each model corresponding to a different setting of the control parameters of the SPA. Changing the values of the control parameters of the SPA changes the partitioning of **S**. It is possible (although not necessary) that the union of the subsets of **S** that can be correctly processed with particular values for the control parameters of the SPA is equal to the set **S**. In other words, the SPA can process all the input signals correctly, provided it is possible to change the values of the control parameters for different input signals. The IPUS architecture provides precisely this capability to a signal understanding system. Furthermore, it can even deal with situations where a signal is not correctly processed by any of the SPA-instances. In such cases, the IPUS architecture provides a *multiple outputs synthesis* capability for obtaining evidence for a hypothesis from the output data of different SPA-instances applied to the same signal. The strategy exploits the fact that although each of the outputs may be partially distorted, their undistorted portions together may provide enough evidence for the correct interpretation.

To make the concept of a data-model more concrete, we now present an example of how a data model is derived in a particular application. In this example, the SPA is the STFT with two adjustable control parameters, the analysis-window length, N_w, and the number of FFT points to calculate, N_{fft}. Each input signal for this application is known to consist of the sum of two frequency-modulated sinusoids. The objective of this application is to determine the time-dependent frequencies of the sinusoids in each input signal. The frequency estimates are required to be within 10% of their true values. Additionally, frequency changes of 1 Hz or higher must be detected. Let us derive the data-model for any STFT-instance that might be used in this application.

The input signals in this application are assumed to have been produced by a signal generation process (SGP) as follows:

$$x(t) = \sin \phi_1(t) + \sin \phi_2(t)$$

The SGP parameters of interest in this application are specified as

$$\omega_i(n) = \left. \frac{d\phi_i(t)}{dt} \right|_{t = nT_s}$$

where T_s is the sampling interval. The output from an SPA-instance is used to form estimates $\hat{\omega}_i(n)$ for the SGP parameters. The estimates are the frequency location values of the two largest maxima in the STFT-instance output data. The criteria for compatibility between the SGP parameter values and their estimates can now be stated as:

$$\left| \frac{\hat{\omega}_1(n) - \omega_1(n)}{\omega_1(n)} \right| \leq 0.1 \qquad (7.1)$$

$$\left| \frac{\hat{\omega}_2(n) - \omega_2(n)}{\omega_2(n)} \right| \leq 0.1 \qquad (7.2)$$

$$\text{if } \omega_1(n) \leq \omega_2(n), \text{ then } \hat{\omega}_1(n) \leq \hat{\omega}_2(n) \qquad (7.3)$$

$$|\omega_i(n) - \omega_i(m)| - 2\pi < |\hat{\omega}_i(n) - \hat{\omega}_i(m)| < |\omega_i(n) - \omega_i(m)| + 2\pi \qquad (7.4)$$

Criteria 7.1 and 7.2 stipulate the requirement that the estimate of each sinusoid's frequency be within 10% of its true value. Criterion 7.3 stipulates the requirement that the estimate tolerances be chosen to ensure that the estimated frequencies of the two sinusoids maintain the order relationship of their actual frequencies. Criterion 7.4 stipulates the requirement that chirp frequency changes of 1 Hz or higher must be detected.

The conditions that must be satisfied by the input signals in order to give "undistorted output" can be derived from the STFT algorithm properties and the compatibility criteria listed above. These conditions and their sources are presented in Table 7.1.

Let us now examine in detail how the conditions in Table 7.1 were derived.

1. **Sufficient-data condition**: This condition states that the window-length N_w imposes a lower bound on the smallest frequency present in the input signal. It is derived on the basis of ensuring that the two spectral peaks due to a single sinusoid may be unambiguously detected from the Fourier transform of a short-time section. A sinusoid of frequency ω_s/N has one frequency component at ω_s/N and the other at $-\omega_s/N$. If the window length has the value N_w, each of these components can be assumed to have significant contributions up to ω_s/N_w away from the actual frequency of the sinusoid. To ensure that the positive frequency component does not significantly interact with the negative frequency component, it is required that

$$\frac{\omega_s}{N_w} < \frac{\omega_s}{N} \Rightarrow N_w > N$$

Clearly, the greatest lower bound for N_w is the value of N corresponding to the lower frequency sinusoid.

2. **Time-resolution condition**: This condition states that the window length N_w imposes an upper bound on the time interval within which any of the sinusoids

TABLE 7.1 DATA-MODEL FOR THE STFT IN A PARTICULAR APPLICATION

The parameters ω_1 and ω_2 respectively represent the time-dependent frequencies of the frequency-modulated sinusoids in each input signal, and ω_s is the sampling frequency.

1.	$\dfrac{\omega_s}{\min(\omega_1, \omega_2)} < N_w$	Sufficient-data condtion
2.	$\lvert \omega_1(n) - \omega_1(m) \rvert \leq 2\pi$ for any n and m such that $\lvert n - m \rvert \leq N_w$	Time-resolution condtion
3.	$\lvert \omega_1 - \omega_2 \rvert > \dfrac{\omega_s}{N_w}$	Frequency-resolution condition
4.	$(0.1) \min(\omega_1, \omega_2) > \dfrac{\omega_s}{N_{fft}}$	Frequency-precision condition
5.	$\lvert \omega_1 - \omega_2 \rvert > \dfrac{\omega_s}{N_{fft}}$	Binwidth-resolution condition

may change its frequency by more than 1 Hz. Note that the STFT assigns only one frequency to the chirp over any short-time section of the signal. Compatibility criterion 7.4 requires that if the chirp frequency changes by more than 1 Hz (or equivalently 2π radians/sec), then the estimate of the chirp frequency must also change. Therefore, the frequency of either sinusoid should not change by more than 2π radians/sec over any short-time section of length N_w or less.

3. Frequency-resolution condition: This condition states that the inverse of the window length N_w places a lower bound on the possible frequency separation between the two sinusoids in order to ensure that the two sinusoids in a short-time section can be unambiguously detected. If the window length has value N_w, each sinusoid has spectral peaks whose width may be approximated by ω_s/N_w. In order to ensure that aliasing does not occur due to interaction between the main lobes of the two sinusoids, the frequencies of the two sinusoids are required to be greater than ω_s/N_w apart from each other.

4. Frequency-precision condition: This condition states that the FFT-length N_{fft} imposes a lower bound on the smallest possible frequency of a sinusoid in order to ensure that the frequency of a sinusoid can be estimated to within the precision required by compatibility criteria 7.1 and 7.2. The binwidth of the FFT is given by, ω_s/N_{fft}. This means that the FFT estimate cannot be more precise than ω_s/N_{fft}. Compatibility criteria 7.1 and 7.2 require that frequency estimates be within 10% of the actual values. This can be accomplished by requiring the FFT binwidth to be, at most, 10% of the smallest frequency sinusoid.

5. Binwidth-resolution condition: This condition states that the inverse of the FFT length N_{fft} places a lower bound on the possible frequency separation of the two sinusoids in order to satisfy the requirements of compatibility criterion 7.4, which stipulates that if the frequency spacing between the two sinusoids is less than the 10% frequency precision requirement, then their frequencies are to be estimated with greater precision. Clearly, this more precise estimation can be accomplished by requiring the FFT binwidth to be less than the frequency distance between the two sinusoids.

7.4.2 Model Variety Analysis

A signal understanding system needs to utilize multiple instances of an SPA if there does not exist a single SPA-instance that can correctly process all the possible input signals in the given application domain. To correctly process an input signal, an SPA-instance must produce undistorted output data. In other words, the input signal must satisfy the conditions in the data model for the SPA-instance. The need for more than one SPA-instance thus translates to the need for a variety of data models. For such applications, we say that the given SPA has a *model variety problem* with respect to the class of possible input signals. It is interesting to note that if an SPA

Sec. 7.4 SP Theories to Support IPUS

has a model variety problem with respect to a class of signals **S**, then it also has a model variety problem with respect to any class of signals that contains the set **S**. This suggests a useful way of analytically showing that an SPA has a model variety problem in a practical application with a possibly open-ended class of possible input signals. We simply choose a mathematically tractable subset of possible input signals and show that the SPA has a model variety problem with respect to that subset. For the sound understanding application, we selected the subset of input signals that can be modeled as the sum of two frequency-modulated sinusoids. As illustrated below, the STFT algorithm has a model variety problem with respect to this class of signals. Consequently, we conclude that the STFT has a model variety problem with respect to any signal understanding application which includes sums of two frequency-modulated sinusoids as possible input signals.

In section 7.4.1, we derived the data model (see Table 7.1) for the STFT with respect to a class of signals, each member of which consists of the sum of two frequency-modulated sinusoids. The model variety problem arises in this application because for some input signals, the time resolution condition and the frequency resolution condition place *conflicting requirements* on the window length N_w. To illustrate this, consider an $x - y$ space where $x \geq 0$ and $y \geq 0$. Furthermore, let any point (x_0, y_0) in this space correspond to those signals produced by the SGP that have the property that the minimum frequency distance between the two sinusoids is y_0 radians and that the frequency of each sinusoid does not change by more than 2π radians over any time interval whose duration is less than or equal to x_0. Setting the window length to the value N_w will set the conditions 2 and 3 shown in Table 7.1 and partitions the set **S** of possible input signals into two subsets, U and D, which respectively contain the signals for which the STFT produces undistorted and distorted outputs. Subsequently, any signal produced by the SGP with $x \geq N_w$ and $y \geq \omega_s/N_w$ would then satisfy these two conditions. These signals are shown in region U of Figure 7.8. Clearly, resetting the window length to $N'_w > N_w$ will repartition the set of all possible input signals, altering the subset U which contains the signals that could satisfy conditions 2 and 3 in Table 7.1. If the set **S** of all possible input signals includes all the signals in the region $\{(x,y)(x > 0, y > 1/x)\}$, it is evident that there is no single value for the window length that could reduce the set D, containing the input signals for which the STFT will produce distorted output, to the null set. On the other hand, there is always a window-length value that would produce an undistorted output for any allowable input signal. We conclude that the STFT has a model variety problem with respect to the set of signals, **S**.

A special case of the model variety problem arises when there is at least one signal in the application domain for which none of the data models in the data-model space produce undistorted output. This occurs in the STFT example given above if the set **S** is extended to include some of the signals in the region $\{(x,y)(x > 0, y < 1/x)\}$. The IPUS architecture provides a *multiple outputs synthesis* capability for dealing with such signals. For example, one STFT-instance may have inadequate frequency resolution, while another STFT-instance may have inadequate time resolution for the same signal. However, each provides partial evidence for

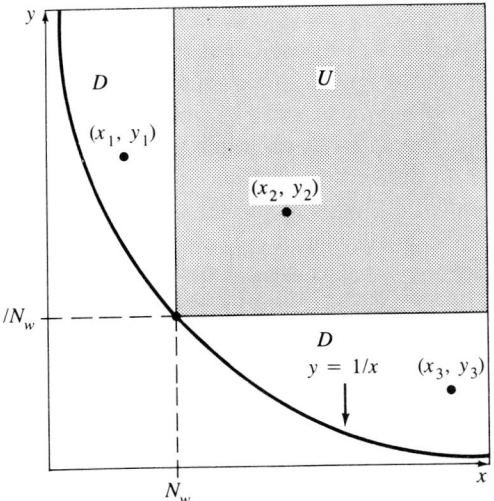

Figure 7.8 Illustration of the partitioning of the set **S** of possible input signals into subsets U and D, which respectively contain the signals for which the STFT produces undistorted and distorted output. Region U: set of input signals that satisfy model-space conditions 2 and 3 (from Table 7.1) for a window length of N_w, for which the STFT will produce undistorted output. Region D: signals that do not satisfy model-space conditions 2 and 3 for a window length of N_w, for which the STFT will produce distorted output.

different aspects of the interpretation model corresponding to the input signal. The combined evidence from the two SPA-instances may then provide sufficient evidence for the interpretation model even though the input signal cannot be correctly processed by any single instance of the STFT.

In addition to ensuring "undistorted output" from an SPA, there is sometimes the concern of minimizing the computational cost in obtaining each undistorted output. For example, for the same situation, a 512 pt. FFT algorithm might give undistorted output, while a computationally more expensive 1024 pt. FFT algorithm would also give undistorted output. In the interest of computational efficiency (for real-time purposes, for example), the 512 pt. FFT algorithm would be more appropriate. Thus, a model variety problem may arise even if there is a single SPA-instance that gives undistorted output for all signals. We refer to this as an efficiency-based model variety problem.

7.4.3 Signal Processing Distortions

A signal processing distortion is said to occur when the input signal to a signal processing algorithm does not satisfy one of the conditions in the data model. We again consider the STFT processing of frequency-modulated sinusoids. The data-model conditions for the STFT in this application are shown in Table 7.1. Violation of the first condition is referred to as an *insufficient-data* distortion, violation of the second condition is referred to as a *time-resolution* distortion, and so on. These distortion categories are explicitly related to specific control parameter values through the data-model conditions such as those in Table 7.1.

Each type of distortion can give rise to various observable discrepancies. We have identified six basic discrepancies that the STFT may produce in the detection of chirps and sinusoids. These discrepancies can occur individually or in various

Sec. 7.4 SP Theories to Support IPUS

combinations. In the following list of the six basic discrepancies, we refer to the signal processing output data as the signal processing view (SPV) and the actual description of the input signal as the signal generation view (SGV).

- track shift: one or more frequency peak tracks in the SPV shift in frequency from their locations in the SGV.
- track extension: one or more frequency peak tracks contained in the SPV are time-extended versions of their counterparts in the SGV.
- track reduction: one or more frequency peak tracks contained in the SPV are shortened versions of their counterparts in the SGV.
- track ghosting: one or more frequency peak tracks contained in the SPV that do not exist in the SGV.
- track break: one or more frequency peak tracks contained in the SPV are not contiguous versions of the tracks in the SGV.
- track removal: the SPV does not contain one or more tracks found in the SGV.

These discrepancies are *caused* by distortions in the signal processing. For example, a frequency-resolution distortion in the STFT processing may cause a combination of track removal and track ghosting discrepancies. In a signal understanding system with the IPUS architecture, an important problem is to try to determine what signal processing distortions may have caused the detected discrepancies.

7.4.4 Discrepancies-to-Distortions Mapping

In the diagnosis phase of the IPUS architecture, it is necessary to perform the inverse mapping from discrepancies to the distortions that may have caused them. The mapping is complicated because of the following factors:

1. Several distortions may be present simultaneously, giving rise to a complicated structure of discrepancies that is difficult to decipher. For example, frequency-resolution distortions in the STFT output data give rise to track removal and track ghosting. On the other hand, time-resolution distortions give rise to track breaks, track extensions, and track removals. Given a combination of discrepancies that indicate the existence of track breaks, track removals, track extensions, and track ghosting, it is difficult to hypothesize the mixture of frequency-resolution and time-resolution distortions that may have given rise to the detected discrepancies.

2. Some distortions may be "ambiguous" in that they may not have given rise to enough discrepancies to uniquely specify them. For example, consider a situation where a new signal source is not detected by the STFT because the peak energy threshold was too high to detect any frequency tracks due to the new source. On the other hand, alternative front-end signal processing (such as zero-crossing analysis) may have detected a change in frequency content of the

input signal. This would indicate that the STFT output is distorted, although the precise nature of the distortion would not be clear (was the discrepancy between STFT processing and zero-crossing processing due to a low peak-energy threshold or was it due to lack of sufficient frequency resolution?).

3. Some discrepancies are "illusionary" in that they may be the result of faulty expectations. For example, consider a situation where a previously detected source was expected to continue its frequency tracks in the current data. If the current signal processing output did not contain those tracks, a conflict type of discrepancy would be declared. One explanation for this would be that the expectation turned out to be incorrect in that the source actually may have been turned off and thus there were no signal processing distortions.

4. Limited computational resources may impose limitations on the process performing the inverse mapping. For example, real-time constraints may not allow the detailed examination of the inverse mapping; instead we might have to opt for a reasonable "guess" as to the underlying distortion based on a very abstract view of the discrepancies.

The above factors lead to the conclusion that in most realistic applications, a certain amount of heuristic search is inevitable in the process that carries out the mapping from observed discrepancies to hypothesized distortions.

7.5 AI THEORIES TO SUPPORT INTEGRATED SEARCH

In this section, we discuss how various artificial intelligence theories support the IPUS architecture. We begin with a discussion of how the RESUN blackboard framework discussed in Chapter 6 of this book is utilized in IPUS for controlling the problem-solving activity of the system. We then discuss how artificial intelligence results on real-time processing are incorporated within the IPUS framework. The section concludes with discussions on the design of knowledge sources (KSs) that carry out the diagnostic reasoning and the interleaved planning and execution of signal reprocessing within a system with an IPUS architecture.

7.5.1 Control Issues

The control structure for any system that performs the integrated search of data and interpretation models is an important issue in the IPUS architecture. In the existing sound understanding testbed, we have utilized the Resolving Sources of Uncertainty (RESUN) control structure [25] because it is particularly suited to differential-diagnosis type of problem solving for eliminating ambiguities due to competing interpretation hypotheses. See Chapter 6 for a description of the RESUN control structure. In this section we point out an important relation between the discrepancy detection mechanism in IPUS and the concept of *sources of uncertainty* utilized by the RESUN control structure.

The RESUN framework is based on a view of interpretation as a process of gathering evidence to resolve *particular sources of uncertainty* in the interpretation hypotheses. The key components of the approach are an *evidential representation*, which includes explicit, symbolic statements of the sources of uncertainty in the evidence for the hypotheses and a script-based, *incremental control planner*. The control plan schemas that define the interpretation methods contain explicit *information gathering actions* which examine the symbolic sources of uncertainty associated with particular hypotheses. This information is used to post goals for resolving specific uncertainties in the current interpretations. These goals then direct the system to expand methods that are appropriate for resolving the uncertainties represented in the goals. Strategy knowledge is defined in focusing heuristics that are applied during the planning process to select the best methods and method instances to be pursued. Sometimes focusing decisions may not be able to be made definitively at the appropriate point during planning because there is insufficient information to select the most appropriate alternative. In these cases, the decisions may be postponed and multiple alternative methods partially expanded in order to accumulate sufficient information to select between them. This is accomplished with a *refocusing* mechanism that allow focusing decisions to be reconsidered once the alternatives have been expanded appropriately. Thus, the control process can be viewed as a search for the best methods to pursue as well as a search for the correct interpretations. The refocusing mechanism also provides the goal-directed planner with a data-directed, opportunistic control capability. This framework is an alternative to a conventional blackboard system. It extends the blackboard representation of interpretation hypotheses and replaces the agenda-based control with a planner.

The crucial link between IPUS and the RESUN control framework lies in the relationship between discrepancy detection and the characterization of sources of uncertainty (SOUs). The basic classes of SOUs in the control framework are described as follows:

- partial evidence—denotes the fact that there is incomplete evidence for the hypothesis.
- possible alternative support—denotes the possibility that there may be alternative evidence that could play the same role as a current piece of support evidence. This reflects the fact that though the uncertainty in a hypothesis may be low, there can still be uncertainty over the correctness of individual pieces of evidence.
- possible alternative explanation—denotes the possibility that there may be alternative explanations for the hypothesis. These SOUs explicitly identify the valid explanation types for the hypothesis based on the characteristics.
- alternative extension—denotes the existence of a competing, alternative version of the same hypothesis. In other words, an alternative version of the hypothesis has been created using one or more pieces of evidence that are inconsistent with the existing versions of the hypothesis, i.e., using alternative

support and/or an alternative explanation. This is one of the key representations of the relationships between hypotheses.

- negative evidence—denotes the failure to be able to produce some particular support evidence or to find any valid explanations. Negative evidence is not conclusive because it also has sources of uncertainty associated with it, e.g., that sensors may have missed some data.
- uncertain constraint—denotes that a constraint associated with the inference could not be validated because of incomplete evidence or uncertain parameter values. This SOU represents uncertainty over the *validity* of an evidential inference whereas the other SOUs are concerned with the *correctness* of inferences. Constraint SOUs are actually associated directly with inferences rather than with hypotheses.
- uncertain evidence—technically, this is not another source of uncertainty *class*. Uncertain evidence SOUs merely serve as placeholders for the uncertainty in the evidence for a hypothesis because the sources of uncertainty are not automatically propagated as evidential inferences are made. They denote the fact that an evidential inference is uncertain because the inference contains uncertain constraint SOUs and/or the hypothesis extension which is the basis for the inference contains SOUs.

In developing the sound understanding testbed, we were able to conceptualize the relationship between discrepancies and SOUs. It should be noted that this conceptualization is largely empirical and needs to be refined and formalized. Our current view of this relationship between discrepancies and sources of uncertainty is described as follows:

1. Violation-type discrepancies and SOUs. A violation-type discrepancy occurs when signal processing output data violates the a priori known characteristics of the entire class of possible input signals in the application domain. When such an output data hypothesis is posted on the interpretation blackboard, a second output data hypothesis is also created, which contains a description of the condition that has been violated. These two hypotheses are then connected by a negative evidence link, labeled VIOLATION-NEGATIVE-EVIDENCE-SOU. Further problem solving can then attempt to remove this SOU by creating extensions of the first hypothesis based on reprocessing the underlying signal with different data-models.

2. Conflict-type discrepancies and SOUs. An unverified expectation hypothesis posted on the interpretation blackboard is labeled with a partial evidence type of SOU. In particular, a NO-SUPPORT-SOU is attached to the hypothesis. Further problem solving can then attempt to find data-models that will lead to the reprocessing of the signal to produce output data than can support the expectation.

A partially verified expectation hypothesis has some of its components at lower information levels supported by the signal processing output data. The other components have NO-SUPPORT-SOUs attached to them. The hypothesis itself is labeled with a partial evidence type of SOU, indicating that it has only partial support. The partial evidence SOU can trigger further problem solving to find support for the lower level hypotheses with no support. In such situations, it is also possible for output-data hypotheses to exist that are not consistent with the expectations. In such cases, the output-data hypotheses are linked with the corresponding expectation hypotheses through negative-evidence SOU links. Furthermore, the unexplained data hypotheses are labeled with partial evidence SOUs indicating the lack of explanation evidence. They would also be labeled with uncertain constraint SOUs, indicating that the data-model used to obtain the data hypothesis is possibly incorrect. Further problem solving would then attempt to find alternate data-models which replace these data hypotheses with data hypotheses compatible with the expectations. If no such data-models are found, negative evidence SOUs (in particular, MISSING-SUPPORT-SOUs) would be attached to the expectation hypotheses.

3. Fault-type discrepancies and SOUs. Fault-type discrepancies arise when two different signal processing algorithms produce conflicting hypotheses about the same underlying signal data. In such a case, a composite hypothesis is created which is a copy of the more reliable of the two data hypotheses and is considered to be an extension of that hypothesis. A link labeled with a negative evidence SOU (in particular, a SUPPORT-LIMITATION-SOU, which indicates that support for a hypothesis is limited until results of further processing are obtained) connects the less reliable hypothesis to the composite hypothesis. Further problem solving attempts to remove the negative-evidence SOU by reprocessing the signal using a different data-model for the less reliable signal processing algorithm. Two outcomes are then possible. Either the negative evidence SOU is eliminated or it is replaced by another negative evidence SOU, namely the NO-SUPPORT-SOU. In the latter case, further problem solving can attempt to reprocess the signal with the more reliable signal processing algorithm but a different data-model.

7.5.2 Real-Time Issues

Another important issue in the IPUS architecture is how the integration of signal processing and signal understanding is affected by real-time constraints. This problem arises because practical systems often do not have sufficient processing power to carry out the "optimal" problem solving for a particular situation in a timely fashion. It therefore becomes necessary to reorganize higher level objectives to get acceptable solutions in such overloaded situations. The resultant problem solving can be viewed as pursuing a "satisficing" approach with respect to the original objectives that did not take the computational resource limitations into account.

An approximate processing framework [21, 12] has been developed for organizing the satisficing problem solving dictated by real-time considerations. This framework consists of well-formed strategies for utilizing approximations in the search processes, approximations in the data for whose explanation the search processes are carried out, and approximations in the knowledge utilized by the search processes. *Approximate search* limits the areas of the search-space that are to be explored by either (1) eliminating processing that would otherwise provide additional support for an already existing solution or (2) characterizing the relationships between potential solution paths and then pruning paths that lead to potential solutions that are somehow inferior to a mutually exclusive alternative. *Data approximation* limits the characteristics of the data to be inspected by the search process and consequently results in solutions that are less precise and more uncertain. *Knowledge approximation* eliminates or simplifies the constraints utilized by the search process. In this case, certainty in the answer obtained is reduced because it may be different from the answer obtained with the full set of constraints. The Approximate Processing framework uses these approximations to decrease the variability in processing time both by changing the criteria for acceptable solutions and by using algorithms that abstract data to minimize the characteristics of the data that can cause significant processing time variability. By representing both the average processing time and the variance, the system can decide that if the deadline for a task is closer than the average time for a task then approximations must be used, but if the deadline is farther away than the average time (but still within the variance), then the current processing strategy can be continued and monitored closely so that approximate strategies can be exploited if necessary.

The approximate processing approach is suitable for signal understanding domains where the situation-dependent objectives can be used in conjunction with the surface characteristics of the input data and results of intermediate processing to set deadlines for meeting the objectives to within certain degrees of precision, certainty, and completeness. In contrast to anytime algorithms [20], approximate processing algorithms are not guaranteed to produce meaningful answers anytime *before* a specified deadline. However, after the deadline has passed, approximate processing algorithms act like anytime algorithms, incrementally increasing the quality of the solutions obtained before the deadline. The advantage of this approach is that the answers obtained by the deadline are [12] in many cases better in quality than those obtained from an anytime strategy. Thus, approximate processing may be viewed as utilizing anytime algorithms *after* a prespecified deadline. Of course, if the deadline time is zero, anytime algorithms may be considered to be a special case of approximate processing. This approach is useful as long as the temporal rate of change in situations being interpreted and system objectives is much smaller than the average processing time for reasonable satisficing of system objectives. For example, in the sound understanding application, it is reasonable to assume that situations and system objectives remain constant over intervals of several seconds, while the processing time (with appropriate hardware) for interpreting the events during that interval may be much smaller than that. Thus, the objective of identifying the sound sources present during an interval of, say, 3 sec, may be given a deadline of 2 sec from the moment that the signal begins to arrive. The remaining second would then be used to improve on the interpretation of the events that occur during the entire interval. Let us now discuss in greater detail how this approximate processing framework interacts with the IPUS architecture.

The two fundamental searches being carried out in the IPUS architecture are: the search for data-models and the search for interpretation models. As we have seen previously, these two searches are highly interdependent. To control these search

Sec. 7.5 AI Theories to Support Integrated Search

processes in the face of real-time constraints requires the formation of criteria for predicting the complexity of any particular search that the system needs to carry out at a given time as well as criteria for predicting the effects on the resulting answer if any approximations are invoked during the search process. As we discuss next, the integration of the data-models search and the interpretation models search in the IPUS architecture precludes us from incorporating approximate processing *separately* into each of these searches. Instead, it is necessary to view the IPUS problem solving as consisting of one basic search process, which includes *both* signal processing and signal understanding components.

In the traditional front-end signal processing architecture for signal understanding, the complexity of the search for interpretation models that can explain signal processing output data in accordance with the system objectives is generally a monotonically increasing function of the amount/complexity of signal processing output data to be interpreted at any given time. This is based on the assumption that the amount/complexity of signal processing output data generated depends on the complexity of the phenomena that produced the input signal. For example, the STFT processing of a signal may be assumed to produce a number of frequency tracks proportional to the actual number of frequency tracks in the input signal. However, in domains suffering from data-model variety, the amount of output data generated by the front-end signal processing depends not only on the complexity of the phenomena represented in the input signal but also on whether or not the data-model governing the front-end signal processing is appropriate. For example, the number of frequency tracks in the STFT output depends not only on the number of actual frequency tracks in the input signal but also on whether or not the frequency resolution of the STFT was suitable for the accurate extraction of each of those tracks. It therefore follows that in such domains the complexity of the search for interpretation models cannot be gauged very accurately from simply the amount of data produced by the front-end signal processing. The estimation of a deadline must also take into account the possibility that the amount/complexity of signal processing output data may be artificially high or low due to the use of an inappropriate data model. The approximate processing framework [12] does include strategies for dealing with distorted data produced by a faulty sensor. However, these strategies need to be tailored to systems where the distorted data can be replaced by possibly undistorted data by utilizing signal reprocessing. What makes this even more interesting is the fact that the reprocessing has a supporting underlying theory that can be utilized to predict relevant deadlines. The exploration of this concept is a major focus of our research on real-time issues for the IPUS architecture.

Given a deadline dictated by high-level considerations and given the degree of desirable precision/certainty/completeness of the signal processing output data as dictated by higher level system objectives, the underlying signal processing theory can be used to design a signal processing strategy with the appropriate data model. To illustrate this, let us consider a situation where a sound segment has already been interpreted with a high degree of certainty to contain two simultaneous sources, say A and B. Suppose that both the sources are typically expected to last at least 5 sec each. Furthermore, suppose source B is much louder than source A, while the signal processing output data due to source A is much more complex and generally requires significant amounts of signal reprocessing to explain accurately. If source B (which may be a smoke alarm) is more important to the objectives of the system than source A (which might be a radio broadcast from a baseball game), the system might decide to focus more of its resources on source B instead of on source A. If new sources can be ignored for up to, say, 1 sec, a deadline of 1 sec may be imposed on the simultaneous tracking of sources A and B without consideration of new sources. The important question is how to structure the system's tasks within the specified deadlines to ensure that source B is tracked with more precision/certainty than source A. A possible strategy might be to adjust the signal processing so that only a very abstract view of source A is represented in the signal processing output data, while a detailed view of source B is obtained. This can, for example, be accomplished (1) by using a long STFT window length (which causes time-resolution distortion in source A data (because its frequency components vary rapidly) but not in source B data, whose frequency components are essentially stationary; and (2) by keeping the peak-detection energy threshold high in most STFT window segments (which eliminates the low-energy peaks of source A). Consequently, the STFT output data represents an abstract (in frequency and time) view of source A data while maintaining a detailed view of source B. The savings in signal processing time in such a strategy can be precisely estimated using the underlying theory of the STFT. At the end of the 1-sec deadline, the system may decide to inspect any discrepancies that may have arisen during the 1-sec period and decide whether or not there is evidence for the arrival of a new source or changes in the important characteristics of the existing sources.

7.5.3 Diagnostic Reasoning

The diagnostic reasoning process is needed to carry out the discrepancies to distortions inverse mapping. The diagnostic process reported in [10] is based on modeling the reasoning of a signal processing expert once various discrepancies are detected. We found that a major part of the reasoning makes use of knowledge regarding the underlying Fourier *theory* for the signal processing algorithms. This diagnostic reasoning is implemented through a means–ends analysis type of search [16], utilizing multipler levels of abstraction. Furthermore, the reasoning is carried out with a qualitative description of the various quantities involved in order to deal with uncertain and approximate information. The task is formalized as that of generating a sequence of distortion "operators" that can explain the discrepancies between an *initial signal state,* which represents the believed information (properties of the input signal class, expectations, or outputs from alternative SPA's whose outputs are less precise but more reliable) and a *goal signal state,* which represents the SPA output data. The knowledge base of the diagnosis process includes *operators* that model various kinds of distortions that can result from improperly tuned signal processing control parameters. The diagnosis process searches through this KB to hypothesize a sequence of operators, which when applied (in order) to the initial signal state will yield the distorted goal state. The search is carried out using progressively more complex representations (lower abstraction levels) of the initial and final states, until finally an abstraction level is reached where an operator sequence can be generated using no more information than is available at that level. That is, the diagnosis process mimics the diagnostic reasoning of signal processing experts in that they offer explanations (operator sequences) that are as uncomplicated as possible.

Once a candidate sequence of operators has been proposed, the "verify" phase of the diagnostic strategy takes place. The abstraction level of the verification is the lowest one at which a description of the initial state is known. Verification proceeds as a degenerate case of the GPS algorithm at the lowest abstraction level, except that no "real" operator search is carried out: the algorithm simply selects the operators in accordance with the plan to be verified. If verification succeeds, the diagnosis system returns the operator sequence as its final answer. If verification fails, however, the diagnosis system attempts to "patch" the explanation depending on the nature of the failure. This sequence of operators is then used by the signal reprocessing planning process to plan the reprocessing of signal data in a way that seeks to avoid the distortions uncovered by the diagnosis process.

7.5.4 Interleaved Planning and Execution

Once the distortions have been hypothesized by the diagnostic reasoning process, the next task is to search for data-models that would remove those distortions. This part of the architecture requires interleaved Planning and Execution. It consists of the following major components: *situation assessment, reprocessing-plan creation,* and *reprocessing-plan execution*. These are described below.

The input to the planning and execution stage consists of descriptions of the input and output signal states (see diagnostic reasoning section above), the sequence of distortion operators hypothesized by the diagnosis stage, and a description of the discrepancies present between the input and output signal states.

The situation assessment phase uses case-based reasoning to generate multiple reprocessing plans, each of which has the potential of eliminating the hypothesized distortions present in the current situation. Plans for eliminating various categories of distortions are stored in a knowledge base. The situation assessment phase constructs a plan schema pertaining to the current overall situation. It should be noted that various components of the plan schema may contain several alternative plans since the knowledge base may contain several alternative strategies for the removal of a particular distortion.

The plan-creation phase selects one overall plan by selecting among the alternative strategies contained within the plan schema generated by the situation assessment phase. The selection is based on criteria such as compatibility between strategies for eliminating different distortions and computational cost.

The execution of a reprocessing plan consists of incrementally adjusting the SPA control parameters, applying the SPA to the portion of the signal data that is hypothesized to contain distortions, and testing for the removal of discrepancies. The incremental process is necessary because the situation description is at least partially qualitative, and therefore it is generally impossible to predict exact values for the control parameters to be used in the reprocessing. The reprocessing continues until the goal of distortion removal is achieved or it is concluded that the reprocessing plan has failed. In the latter case, the diagnosis process is reinvoked to produce an alternative hypothesis for the distortions present in the reprocessed signal data.

7.6 CONCLUSIONS

In this chapter, we have discussed features and supporting theories of a novel architecture for integrating signal processing and signal understanding technologies. This architecture supports sophisticated interaction between signal processing and signal interpretation and places them under the umbrella of a single problem-solving paradigm. The architecture is based on combining the powers of signal processing theories such as filtering theory, time-frequency modeling, and parameter estimation with artificial intelligence theories for constructive problem solving, problem-solving control, approximate processing, diagnostic reasoning, and interleaved planning and execution. The practical application of this architecture to sound understanding problems is discussed and illustrated in section 9.3 of Chapter 9 in this book.

The development of the IPUS architecture has so far focused on providing a signal understanding system the means of utilizing different instances of the same SPA. However, we believe that with appropriate extensions this architecture may also be utilized for applications where SPA-instances corresponding to different SPAs may be needed. For example, in the sound understanding application we have found that processing based on the Pseudo-Wigner distribution [30] can aid the basic STFT processing. For example, in certain situations a Pseudo-Wigner SPA-instance provides simultaneous time and frequency resolution superior to that provided by any STFT instance. On the other hand, the Pseudo-Wigner distribution can produce large cross-terms [31] that result in "ghost" data in the time-frequency plane. Consequently, the Pseudo-Wigner SPA-instances are not generally appropriate for bottom-up processing, but they can be useful for obtaining evidence to support interpretation models that have already been hypothesized by the signal understanding system. These issues are being investigated in our ongoing research related to the IPUS architecture. In fact, the IPUS architecture provides many interesting issues for further research from both signal processing and signal understanding viewpoints.

From a signal processing perspective, the IPUS architecture has led to the formulation and illustration of a new paradigm for designing signal processing techniques. The new paradigm focuses on the control parameterization of signal processing algorithms and developing supporting theories that relate the control parameters to formalized data-models for various application domains. In our research, we have formalized the data-models that relate the control parameters of the well-known STFT algorithm with a class of input signals that correspond to a large number of common sounds. We hope that the results from this research and our ongoing efforts should motivate further research of this type for other signal processing algorithms and other classes of signals. Furthermore, we hope our results will lead to consideration of designing new signal processing algorithms with the aim of having their control parameters well suited to the class of input signals for which they are to be used. In other words, a new consideration in the design of signal processing algorithms will be the ease with which the data-model governing the processing of input signals can be varied.

From a signal understanding perspective, the research in this area aims at producing a generic architecture for operationalizing the knowledge associated with signal understanding tasks that require integrated search for data and interpretation models. The most appropriate application domains are those that utilize signal processing techniques (including the possibility of controllable sensors) which have control parameters and underlying theories. The generic architecture incorporates control mechanisms that utilize symbolic evidential representations, incremental planning, and approximate processing control for real-time applications. This generic architecture will permit developers of signal understanding systems for realistic applications to focus their efforts on domain-specific knowledge gathering and be guided in that process by the formal structural specification of the generic architecture.

From an applications perspective, the IPUS architecture is relevant to the design of assistive devices for the hearing impaired, speech recognition in noisy everyday environments, and industrial sound understanding applications. Assistive devices could utilize sound understanding technology to notify the hearing-impaired user of important sound sources such as fire alarms, police and fire sirens, and the cries of an infant. Current speech recognition technology does not take into account the presence of non-speech sounds in practical environments; hence, sound understanding technology will play an important role in the development of next-generation speech understanding systems. Industrial applications of the IPUS architecture include diagnostic analysis of engine sounds.

ACKNOWLEDGMENTS

We would like to acknowledge the contributions of our students at Boston University and the University of Massachusetts to the research upon which the material in this chapter is based. In particular, we would like to mention Frank Klassner, Erkan Dorken, Izaskun Gallastegi, Avi Weiss, Malini Bhandaru, and Zarko Cvetanovic. Finally, we would like to acknowledge the many fruitful discussions we had with Dr. Norman Carver of the University of Massachusetts in regard to the applicability of the RESUN blackboard framework in the IPUS context. The work reported in this chapter was sponsored in part by the Rome Air Development Center (RADC) of the Air Force Systems Command under contract number F30602-91-C-0038, in part by the Office of Naval Research under a University Research Initiative grant, ONR N00014-86-K-0764, and in part by NSF under contract number CDA 8922572.

REFERENCES

[1] H. Nawab and V. Lesser, "High-Level Adaptive Signal Processing," *NAIC Final Report,* 17 (October 1989).

[2] L. Erman, R. Hayes-Roth, V. Lesser, and D. Reddy, "The Hearsay II Speech Understanding System: Integrating Knowledge to Resolve Uncertainty," *Computing Surveys,* 12 (June 1980), 213–53.

[3] H. Nii, E. Feigenbaum, J. Anton, and A. Rockmore, "Signal-to-Symbol Transformation: HASP/SIAP Case Study," *AI Magazine,* 3 (Spring 1982), 23–35.

[4] R. Reddy and R. Watkins, "Use of Segmentation and Labeling in Analysis-Synthesis of Speech," *ICASSP* (May 1977), 28–32.

[5] H. Nawab and T. Quatieri. "Short-Time Fourier Transform," in *Advanced Topics in Signal Processing.* Edited by Jae S-Lim and A.V. Oppenheim. (Englewood Cliffs, N.J.: Prentice Hall, 1988).

[6] D. E. Seborg et al. "Adaptive Control Strategies for Process Control: A Survey," *AIChE Journal,* 32, no. 6 (June 1986), 881–913.

[7] W. Clancey, "Heuristic Classification," *Artificial Intelligence,* 27 (1985), 289–350.

[8] B. Buchanan and E. Shortliffe, eds., *Rule-Based Expert Systems* (Reading, Mass.: Addison-Wesley, 1984.)

[9] Y. Peng and J. Reggia, "Plausibility of Diagnostic Hypotheses: The Nature of Simplicity," *Proceedings of AAAI-86* (1986), 140–45.

[10] H. Nawab, V. Lesser, and E. Miliois, "Diagnosis Using the Underlying Theory of a Signal Processing System," *IEEE Trans. on Systems, Man and Cybernetics. Special Issue on Diagnostic Reasoning* May/June 1987, 369–79.

[11] E. Hudlicka and V. Lesser, "Meta-Level Control Through Fault Detection and Diagnosis," *Proceedings of AAAI-84* (1984), 153–61.

[12] K. S. Decker, V. R. Lesser, and R. C. Whitehair, "Extending a Blackboard Architecture for Approximate Processing," *Journal of Real-Time Systems,* 2 (1990), 47–79.

[13] J. Tenenbaum, and H. Barrow, "Experiments in Interpretation Guided Segmentation," *Artificial Intelligence* 8(3) (1977), 241–74.

[14] A. Weiss, "Model Variety Problem in Signal Processing" (M.S. thesis, Boston University, September 1990).

[15] V. Lesser and L. Erman, "A Retrospective View of the HEARSAY-II Architecture," *Proceedings of IJCAI-77* (1977), 790–800.

[16] A. Newell and H. Simon, "GPS: A Program that Simulates Human Thought," in *Computers and Thought,* eds. E. A. Feigenbaum and J. Feldman (New York: McGraw-Hill, 1963), 279–93.

[17] W. Dove, "Knowledge-Based Pitch Detection" RLE Technical Report 518, Cambridge Mass.: MIT (June 1986).

[18] B. Hayes-Roth et al., "Intelligent Real-Time Monitoring and Control," Technical Report KSL-89-05, Computer Science Department, Stanford University (1989).

[19] J. Pearl, *Probabilistic Reasoning in Intelligent Systems: Networks of Plausible Inference* (Morgan Kauman, 1988).

[20] E. J. Horvitz, "Reasoning Under Varying and Uncertain Resource Constraints," *Proceedings of the Seventh National Conference on Artificial Intelligence* (August 1988).

[21] V. R. Lesser, J. Pavlin, and E. Durfee, "Approximate Processing in Real-Time Problem Solving," *AI Magazine,* 9, no. 1 (Spring 1988), 49–61.

[22] A. V. Oppenheim and R. W. Schafer, *Discrete-time Signal Processing* (Englewood Cliffs, N.J.: Prentice Hall, 1989).

[23] J. S. Lim and A. V. Oppenheim, eds., *Advanced Topics in Signal Processing* (Englewood Cliffs, N.J.: Prentice Hall, 1988).

[24] C. A. Kohl, R. Hanson, and E. M. Reisman, "A Goal-Directed Intermediate Level Executive for Image Interpretation," *Proceedings of IJCAI-87* (811–14).

[25] N. Carver, "Sophisticated Control for Interpretation: Planning to Resolve Uncertainty" (Ph.D. thesis, University of Massachusetts, 1990).

[26] J. Hendler, A. Tate, and M. Drummond, "AI Planning: Systems and Techniques," *AI Magazine,* 11, no. 2 (1990), 61–77.

[27] D. Wilkins, *Practical Planning: Extending the Classical AI Planning Paradigm* (Morgan Kaufman, 1988).

[28] R. Davis, "Meta-Rules: Reasoning about Control," *Artificial Intelligence,* 15 (1980), 179–222.

[29] A. Bregman, *Auditory Scene Analysis* (Cambridge: MIT Press, 1990).

[30] T.A.C.M. Classen and W.F.G. Mecklenbrauker, "The Wigner Distribution–A Tool for Time-Frequency Signal Analysis. Part II: Discrete-time Signals," *Phillips J. Research,* 35 (1980), 276–350.

[31] L. Cohen, "Time-Frequency Distributions—A Review," *Proceedings of the IEEE,* 77, no. 7 (July 1989), 941–81.

8

SIGNAL ABSTRACTION CONCEPT FOR SIGNAL INTERPRETATION

- Evangelos E. Milios
  York University
- S. Hamid Nawab
  Boston University

8.1 INTRODUCTION

Since digital signal processing was first introduced, the nature of the applications that it addresses has gradually changed. In the late 1970s, typical applications [6] were in telecommunications (speech and image encoding, echo canceling, etc.), computing and transduction hardware, image enhancement, image reconstruction from projections, target detection and parameter estimation using radar or sonar, and seismic signal analysis. More recently, attention has shifted to applications of digital signal processing to signal interpretation. In these applications, we wish to perform the task of extracting information about a physical phenomenon by processing a signal generated by that phenomenon. It is typically very difficult to perform this inverse task.

In active signal interpretation (e.g., sonar and radar applications), a controlled signal is emitted, transformed during its propagation and reflection from objects in the environment, and recorded thereafter. The objective of interpretation is to extract information about the environment in terms of the transformations (e.g., time delay) to which the received signal may have been subjected. In passive signal interpretation, signals are emitted by the objects, and, since these signals are not controlled, they are potentially more difficult to interpret.

A close analogy exists with the fields of computer graphics and computer vision: computer graphics is now in a position to generate rather sophisticated images and animation sequences, but much of the technology is not "invertible." The primary reason is that the generation process involves several many-to-one mappings, the most prominent of them being the projection operation, which make the inverse problem of going from generated images to objects in three-dimensional space underdetermined.

In this chapter, we take a closer look at issues involved in signal interpretation. We start by recognizing the fact that signal interpretation is inherently underdetermined and therefore additional constraints must be imposed for finding a solution. Such constraints are obtained if we think of the input signal not as a numerical sequence, but as consisting of distinct parts, which are more immediately related to the physical phenomenon that generated the signal. For example, the image of a human face is a two-dimensional array of integers representing brightness. For face recognition, however, we have to focus on distinct parts of the face, such as eyes, nose, lips, and forehead, because they are the ones that have meaning in relation to the person's face. This is in agreement with Gestalt psychology [1,13], which claims that a homogeneous perceptual input (such as a raw image or acoustic signal) contains no units. Only when the input is partitioned by some type of discontinuity is it possible to organize it into units. Thus, the perceptual unit itself is formed by a process of perceptual organization. After being formed, units can be grouped by similarity and other factors to form higher-order organizations.

Based on the above intuitions, in this chapter we suggest a view of a signal as consisting of multiple levels, with higher levels of abstraction obtained by suppressing information (i.e., detail) at lower levels. A high level of abstraction supports a smaller set of inferences than a lower level of abstraction, but more elegantly, i.e., with simpler programs, and with less computation. With such signal abstractions, our goal is to provide a framework that enables us to represent signals and to organize the computation for signal interpretation in a manner that is consistent with the need for perceptual organization. To substantiate our framework, we implemented a working system for the interpretation of passive acoustic signals containing sound from harmonic acoustic sources.

In section 8.2 of this chapter, we provide a precise definition of signal abstractions. Section 8.3 presents the extended spectrum as an example of the application of these concepts to harmonic spectra. In section 8.4, we illustrate the power of signal abstractions in the problem of adjustment of spectral estimation parameters to balance the trade-off between noise reduction and signal nonstationarity, as applied to the case of real acoustic helicopter data. In section 8.5, we discuss the relation of the signal abstractions framework to other work. Section 8.6 provides a summary of the chapter.

8.2 SIGNAL ABSTRACTIONS

Signal abstractions [14] provide a framework for signal representation designed to explicitly capture the grouping operations that are fundamental to a signal interpretation task. The specific grouping operations used are problem-dependent, and therefore are not part of the framework. A basic assumption we make in the framework is that each unit at a given abstraction level can be described by a number of features in a manner that is adequate for performing the grouping that generates the units at the next higher level of abstraction. Another assumption is that the features associated with each unit are sufficient to determine whether the unit should participate in the grouping operation. Units at the lowest level of abstraction are simply the (indexed) numerical values of the signal, while units at higher abstraction levels correspond to conceptually meaningful entities.

To put the signal abstraction definitions in a more concrete context, we consider an application of the framework to the description of spectra with harmonically related sets of peaks (harmonic spectra). Such spectra arise often in applications involving periodic or oscillatory phenomena. The numeric signal in this case is a power spectrum (power as a function of discrete frequency). For harmonic spectra, we have defined an extended spectrum representation (discussed fully in section 8.3) with three levels of abstraction, the numeric spectrum level, the peak level, and the harmonic set level. The numeric spectrum level consists of all frequency/power pairs that are associated with the power spectrum as a function of frequency. The peak level consists of selected peaks of the power spectrum, where each peak corresponds to frequency/power pairs and features associated with a local maximum. Finally, the harmonic set level consists of harmonic sets, where each harmonic set is composed of a number of peaks, the frequencies of which are multiples of the fundamental frequency.

An abstraction level A_i consists of a set of conceptual units, a grouping operation, a feature computation operation, and a qualitative characterization operation. Each of these components will now be discussed in greater detail. A formal specification of these components can be found in [14].

- *A set of conceptual units*, where the jth unit at level i is $u_i(j)$, with $0 \leq j < N_i$, where N_i is the total number of conceptual units at abstraction level i. The *internal view* of a conceptual unit $u_i(j)$ is the subset of conceptual units at abstraction level $i - 1$ which have been grouped together to define $u_i(j)$. The *external view* of a conceptual unit $u_i(j)$ is a collection of features and qualitative labels computed from its internal view. A qualitative label is assigned to a unit if its features satisfy the classification rule(s) associated with the label.

The internal view of $u_i(j)$ allows us to trace the composition of that conceptual unit in terms of the lowest level of abstraction, the numeric signal. The external view is a concise description of $u_i(j)$ that lacks detail but is sufficient for many tasks involving $u_i(j)$, most important the grouping operation performed at level j. In the extended spectrum, conceptual units at the peak level are frequency/power pairs associated with local maxima of the spectrum.

- *A grouping operation* G_i, which completely specifies how the conceptual units at level i are being computed from the conceptual units at level $i - 1$, where $i = 0$ corresponds to the numeric signal. G_i relies only on the external view of the units at level $i - 1$, and, in general, it depends on a set of parameters, which are part of the specification of G_i. In the extended spectrum, the grouping operation G_1

specifies how local maxima and minima of the power spectrum are detected and how spectral values are grouped together to form conceptual units corresponding to individual peaks.

- A *feature computation operation* F_i, which specifies how the features of a unit at level i, $u_i(j)$, are computed. In the extended spectrum, examples of peak features are the power of the local maximum of the peak, the power of the two local minima of the peak, and the frequency difference between the two local minima. Features of peaks rely solely on features of indexed spectral values (i.e., frequency and power pairs).
- A *qualitative labeling operation* Q_i, which specifies how the qualitative labels of the elements $u_i(j)$ are computed. Examples of qualitative labels of peaks are: "prominent" (if the ratio of the peak power to the average of the power of the two local minima is large, while the ratio of the powers of the two local minima is close to unity), and "locally strong" (if the peak is a local maximum over a sufficiently large neighborhood around the peak).

It is possible for a total order and a distance metric to exist for the conceptual units $u_i(j)$ at abstraction level i. In the case of harmonic spectra, peaks are ordered by their peak frequency and harmonic sets are ordered by their fundamental frequency.

8.3 THE EXTENDED SPECTRUM

Spectra of periodic signals are characterized by harmonically related sets of peaks. In interpreting such spectra, two levels of grouping take place, corresponding to a representation that involves three levels of abstraction, the numeric signal, the peak level, and the harmonic peak set level. In this section, we describe the extended spectrum, which results from the application of the signal abstraction formalism to the description of harmonic spectra. In section 8.4, we will present an example of the use of the extended spectrum for adaptive spectral estimation based on the averaged periodogram method. In Chapter 9 of this volume, the grouping of harmonic sets at successive time intervals will be presented as another application of the extended spectrum.

The Numeric Spectrum Level

This is the lowest level of abstraction. It is a set of conceptual units $u_0(j)$, where each conceptual unit is a pair $(f(j), p(j))$, where $p(j)$ is the spectral power at frequency $f(j)$. We use a separate index j, with a one-to-one correspondence with frequency $f(j)$, to emphasize the concept of an individual spectral value as a single unit with two features, frequency and power. We assume that $j_1 < j_2$ if and only if $f(j_1) < f(j_2)$. Therefore, the $u_0(j)$'s are ordered by increasing frequency and the distance between $u_0(j_1)$ and $u_0(j_2)$ is equal to the absolute value of the difference of the corresponding frequencies, i.e., $|f(j_1) - f(j_2)|$.

The Peak Abstraction Level

This is the next higher level of abstraction after the numeric spectrum level in the extended spectrum. Peaks correspond to local maxima of the spectrum, therefore they are detected by finding the zero crossings of the first difference of the spectrum. Local maxima alternate with local minima, so for each local maximum $u_0(j_0)$ the nearest local minima before and after it are well defined (except perhaps for the first and last local maximum). Let them be $u_0(j_l)$ and $u_0(j_r)$ respectively. In terms of the framework presented in section 8.2, the internal view of a peak $u_1(k)$ is the set of conceptual units or spectral points $\{u_0(j), j_l \leq j \leq j_r\}$, as shown in Figure 8.1.

The grouping operation G_1 consists of the following steps:

1. **Extrema computation:** Find local maxima and minima in the numeric spectrum.
2. **Grouping:** Group all conceptual units $u_0(j)$ associated with each local maximum as described in the previous paragraph.
3. **Peak selection:** Construct A_1 by selecting peaks from the output of step 2.

Peak selection is based on the set of qualitative labels $q_1(k)$ for each peak $u_1(k)$. A peak is selected if and only if it has at least one qualitative label associated with it.

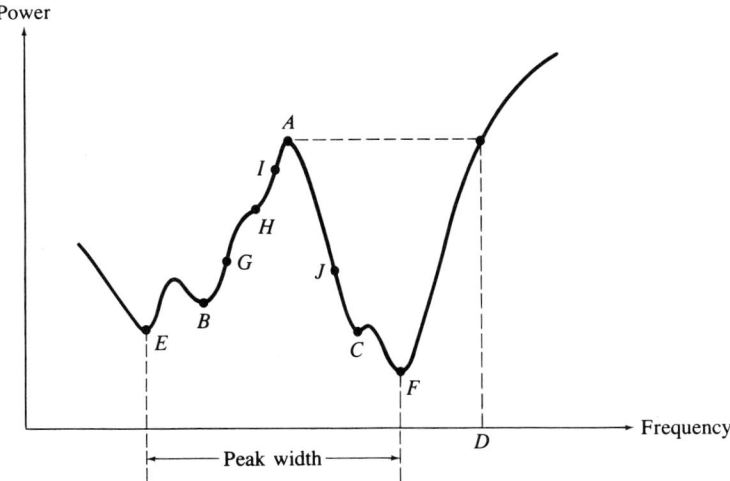

Figure 8.1 Characteristic points associated with a local maximum, which are used in defining the peak features in the extended spectrum. These include local minima B and C, the point D closest to the peak at which power matches the peak power, and local minima E and F closest to the peak with power below a certain fraction of the peak power, in this case 0.4. These local minima help characterize rippled peaks.

Sec. 8.3 The Extended Spectrum

Equivalently, a peak is ignored if and only if there is no qualitative label attached to it. The intuitive rationale for this selection criterion is that qualitative labels are associated with peaks that have some "special" character. Peaks with no special character are assumed to be random variations of the signal and, as such, not worthy of further consideration.

The features defined for a peak $u_1(k)$ (Figure 8.1) are the following:

- Frequency and power of the local maximum and the adjacent local minima $(f(j_0), p(j_0))$, $(f(j_1), p(j_1))$, and $(f(j_2), p(j_2))$ (points A, B, and C, respectively, in the figure).
- The radius of the maximal symmetric neighborhood around the local maximum, over which it is a local maximum. It is defined as $\min_j |f(j_0) - f(j)|$ such that $p(j) = p(j_0)$. It is the difference between the frequency at point D and point A in the figure. If the point D is not defined, for example when A is a global maximum, the radius of the neighborhood is infinite.
- The width of the peak, defined as the frequency difference between the local minima surrounding the peak with power below a certain fraction of the peak power (local minima E and F in the figure). More formally, the width of the peak is equal to $\max |f(j_a) - f(j_b)|$, where $u_0(j_r)$ and $u_0(j_s)$ is the local minimum before and after $u(j_0)$ respectively and closest to it such that $p(j_a)/p(j_0) < K$ and $p(j_b)/p(j_0) < K$. With this definition, shallow local minima that occur in peaks with rippled sides are ignored. It is the difference between the frequencies of points E and F in the figure.

Associated with the feature computation at level 1, F_1, there is a single parameter K, a number between 0 and 1, which is used to compute the two local minima that surround the peak, are nearest to it, and their power is less than a fraction K of the peak power. A value for K that leads to results plausible for the human user can be determined by a training process.

The qualitative labels for a peak are:

- Large average prominence: the criterion is that the quantity $p(j_0)/(p(j_1) + p(j_2)/2)$ is large. The intuition is that peaks that satisfy this criterion visually stand out in a neighborhood around them (Figure 8.2).
- Prominent unbalanced: the criterion is that the quantity $p(j_0)/(p(j_1) + p(j_2)/2)$ is not large, but the quantity $p(j_0)/\min(p(j_1), p(j_2))$ is large. The intuition is that such peaks visually stand out in a neighborhood around them, but only on one side (Figure 8.2).
- Strong: the criterion is that $p(j_0)/\max p(j)$ is high, where the maximum is over all possible values of j. The intuition is that the power of the peak is high in absolute terms.
- Locally strong: the radius of the neighborhood over which the peak is a local maximum is large.
- Prominent but maybe rippled: the width of the peak is defined and it is small.

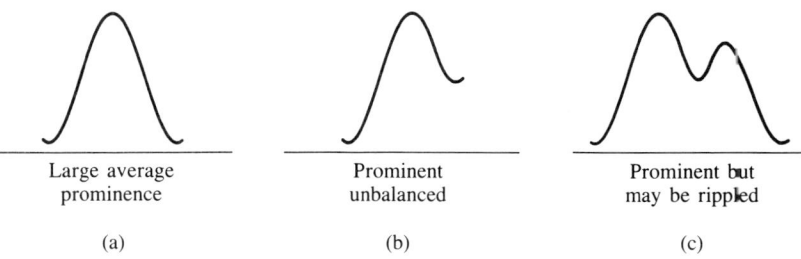

Figure 8.2 Examples of peak classes: (a) large average prominence (b) prominent unbalanced (c) prominent but maybe rippled.

The qualitative labeling operation depends on thresholds, one per each subjective notion of "large" or "small" in each of the above qualitative labels. Plausible values for these thresholds are determined via a training procedure [14].

The Harmonic Set Abstraction Level

This is the highest level of abstraction in the extended spectrum. A set of harmonically related peaks consists of peaks with the property that the frequency at the local maximum of each peak is a multiple of the fundamental frequency characteristic of the set. The grouping operation G_2 for the harmonic set abstraction level is based on the frequencies of the local maxima associated with peaks. We assume that $f(k)$ is the frequency of the local maximum associated with peak $u_1(k)$.

The internal view of a harmonic set $u_2(m)$ is a set of peaks $\{u_1(k)\}$ with the property that there exists f_m such that for each $u_1(k)$ of the set, there exists $i_{mk} \in Z$ satisfying the relation $|f(k)/f_m - i_{mk}| \leq \epsilon$. The frequency f_m is the fundamental frequency of the harmonic set $u_2(m)$, while the integer i_{mk} is the order of the peak $u_1(k)$ as a harmonic of f_m. The parameter ϵ is the maximum deviation from exact multiplicity allowed for characterizing $f(k)$ as an (approximate) multiple of f_m. Note that the same peak $u_1(k)$ may be present in two different harmonic sets under different orders. The rationale for the above criterion, as opposed to a criterion of the form $f(k) - f_m i_{mk} \leq \epsilon$ is that, in the latter case, a small error in f_m gets amplified by i_{mk}, and, therefore, it is inappropriate to compare it with a threshold ϵ that is independent of the amplification factor.

A plausible grouping operation G_2 consists of the following steps:

1. Find candidate fundamental frequencies f_m.
2. For each f_m, group together all the peaks $u_1(k)$ for which $|f(k)/f_m - i_{mk}| \leq \epsilon$ has a solution for i_{mk}.
3. Construct A_2 by pruning or merging harmonic sets obtained from the previous step.

The similarity between the grouping operation G_2 for harmonic sets and the grouping operation G_1 for peaks is worth observing. Both consist of first finding "seed" information, which consists of local maxima for peaks or candidate fundamental frequencies for harmonic sets, then forming a conceptual unit at the next abstraction level around each seed, and finally pruning and/or modifying the resulting conceptual units.

Within the above framework, several problem-dependent heuristics have been added to perform the three steps of the grouping operation G_2. In the first step of the grouping operation, the problem dictates that we are interested in fundamentals in a well-known frequency range. We also know from experience that for harmonic sets of interest, at least one of the first three harmonics is present and is highly prominent. These observations lead to the following rule for finding candidate fundamental frequencies, where $[f_{min}, f_{max}]$ is the range of fundamental frequencies of interest.

If
$u_1(k)$ is a peak **and**
$u_1(k)$ has LARGE-AVG-PROMINENCE as one of its qualitative labels
and
for either $i = 1$ or $i = 2$ or $i = 3$, $f(k) \in [i*f_{min}, i*f_{max}]$
then
$f(k)/i$ is a candidate fundamental frequency

The second step of the grouping operation G_2 amounts to testing each peak against the approximate multiple criterion. The third step is based on a number of rules shown in Figure 8.3. These rules act like knowledge sources in a single-level blackboard: they keep modifying the list of harmonic sets obtained from the second step in an opportunistic manner, as described in Chapter 7 of this volume and in [7] until the harmonic sets obtained cause no rule to trigger.

8.4 APPLICATION OF THE EXTENDED SPECTRUM TO ADAPTIVE SPECTRAL ESTIMATION

Practical power spectrum estimation has a strong empirical flavor, because typically less information is available than required by optimum estimation theoretic techniques [16]. Different techniques involve various trade-offs, and there is no general agreement on the best method. The magnitude-squared of the Fourier transform of a finite-length discrete signal is the spectrum estimate known as the periodogram. The periodogram is not a consistent estimate of the spectrum, i.e., its variance does not approach zero as the length of the signal increases, but instead it approaches the square of the spectrum, leading to large fluctuations about the true spectrum values. It can be shown that lower variance can result if the power spectrum is estimated not as a single periodogram of the whole signal, but as an average of periodograms of shorter, possibly overlapping and windowed sections of the signal. Variance is then reduced by a factor equal to the number of sections used.

If we fix section size, it appears that the optimal spectrum estimate is obtained

1. If no peak in a harmonic set has large average prominence, the set is pruned out.
2. If a harmonic set consists of a single peak, it is pruned out.
3. If both the second and third harmonic of a harmonic set are missing, and its fundamental is not a prominent peak, the set is pruned out.
4. If two harmonic sets have very similar fundamentals, they are tentatively combined into a single harmonic set with fundamental frequency equal to the average of the two fundamentals. If the resulting set has at least as many peaks as each of the original sets or it has at least 3 peaks, it replaces both of the original sets. If not, the original set with the fewest peaks is pruned out, if it has at least 2 fewer peaks than the other set and at most 2 peaks in total.
5. If a harmonic set is a subset of another harmonic set and if its average prominence is much smaller than that of the latter set, the former harmonic set is pruned out.
6. If a harmonic set shares all its prominent peaks with another harmonic set, the former harmonic set is pruned out.
7. If the fundamental frequency of a harmonic set is equal to half the fundamental frequency of another set, and the former set shares all its prominent peaks except at most one with the latter, and the fundamental of the latter set is a prominent peak, whereas the fundamental of the former set is not, the former set, i.e., the set with the lowest fundamental frequency, is pruned out.
8. If the fundamental frequency of a harmonic set is equal to half the fundamental frequency of another set, and the latter set shares all its prominent peaks except at most one with the former set, and the fundamental of the former set is a prominent peak, whereas the fundamental of the latter set is not, the latter set, i.e., the set with the highest fundamental frequency, is pruned out.

Figure 8.3 Harmonic set pruning and merging heuristics.

by using as much of the signal as possible. However, if the spectrum changes over time, then a trade-off needs to be achieved between noise reduction by averaging as many periodograms as possible and spectrum distortion due to the shifting of spectral peaks within the time between the first and last periodogram. In the case of noisy harmonic spectra, reducing the variance of the spectral estimate implies fewer and weaker noise peaks relative to the signal peaks. Distortion of the spectral estimate due to shifting of signal peak frequencies with time is associated with peak broadening, weakening of their power, and peak splitting.

These perceptual characteristics of an averaged periodogram are captured by the extended spectrum, which can be used to achieve a solution to the trade-off of the previous paragraph. The idea behind the algorithm is to adapt the number of periodograms being averaged, or, equivalently, the length of signal used, so as to achieve the sharpest spectral peaks possible. We consider only the spectral peaks associated with signal information, not noise, and to determine such peaks we need to compute the harmonic set abstraction level of the extended spectrum.

Sec. 8.4 Application of the Extended Spectrum

Figure 8.4 shows the overall structure of the algorithm for adapting the number of periodograms in spectral estimation. The spectral estimate used is given by:

$$S(L) = \frac{1}{L} \sum_{k=0}^{L} |\mathcal{F}[x(n)w(n - n_0 - kT)]|^2 \quad (8.1)$$

where the only variable quantity is L, the total number of periodograms averaged. T is the offset between two successive blocks of data or, equivalently, between two successive positions of the analysis window $w(n)$. We assume that $w(n) = 0$ for $n < 0$. The start of the data under consideration is represented by n_0. $\mathcal{F}$ is the discrete Fourier transform operation.

The algorithm seeks to determine a value L^* for L, for which the resulting spectral estimate achieves an optimal trade-off. At a given iteration of the algorithm, three spectral estimates are considered: $S(L - N_0)$, $S(L)$, and $S(L + N_0)$, where L is the current estimate of L^* and N_0 is a small integer determining how fine the sampling of L, the number of periodograms averaged, is going to be.

At each iteration, the algorithm computes the three spectral estimates for the current value L and orders the pairs $(S(L - N_0), S(L))$ and $(S(L), S(L + N_0))$ according to the overall "peakiness" of a specific harmonic set selected in advance by the user of the algorithm (a human or another program). Thus, the algorithm focuses on a specific harmonic set in the spectrum and it is possible that different optimal values L^* will be found for different harmonic sets. The outcome of ordering pairs of spectra with respect to peakiness of a specific harmonic set is that the first spectrum of the pair is MORE, LESS or EQUALLY peaky to the second spectrum of the pair.

What happens at the next iteration depends on the outcome of the orderings of the two pairs $(S(L - N_0), S(L))$ and $(S(L), S(L + N_0))$. If $S(L)$ is a "maximum" of peakiness, then the algorithm terminates and L is returned as the optimal number L^* of periodograms. Otherwise, the current value of L is changed to either $L - N_0$

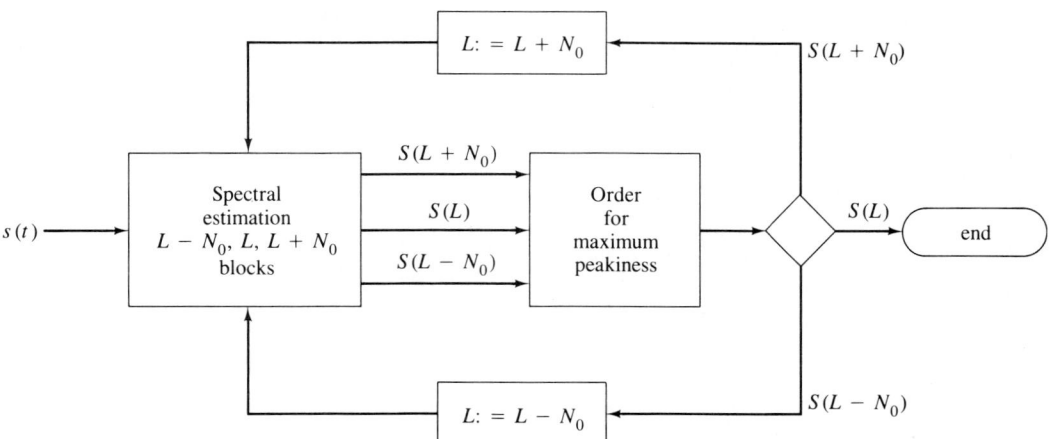

Figure 8.4 The algorithm for adapting the number of periodograms in spectral estimation. $S(L)$ is the average of L periodograms.

A is ___ peaky than B	B is ___ peaky than C	Peakiest of A, B, C
LESS	MORE	B
EQUALLY	MORE	B
MORE	MORE	A
LESS	EQUALLY	B
EQUALLY	EQUALLY	B
MORE	EQUALLY	A
LESS	LESS	C
EQUALLY	LESS	C
MORE	LESS	ERROR

A: Harmonic set in $S(L - N_0)$
B: Harmonic set in $S(L)$
C: Harmonic set in $S(L + N_0)$

Figure 8.5 Table of outcome of the comparisons between three versions of a harmonic set in three spectral estimates obtained by averaging three different numbers of periodograms.

or $L + N_0$, and the process of ordering is repeated. Figure 8.5 shows all possible outcomes of the two orderings and the resulting maximum in terms of peakiness. Note that the expectation built into the table is that there is only one (global) maximum of peakiness as L varies. Therefore, $S(L)$ cannot have a local maximum.

The interesting aspect of the above algorithm is the ordering of two spectral estimates according to the relative peakiness of the two versions of a harmonic set present in both of them. Ordering takes place at multiple levels of abstraction, making use of the extended spectrum representation of the two spectral estimates. Ordering is first attempted at the highest level of abstraction (harmonic set level) by relying on the features of the two harmonic sets. If comparison at this level is not conclusive, then ordering takes place at the next lower level of abstraction (peak level) on a peak-by-peak basis and voting takes place among peaks to determine the outcome of ordering, as shown in Figure 8.6.

The program outlined in Figure 8.6 includes three sets of heuristics:

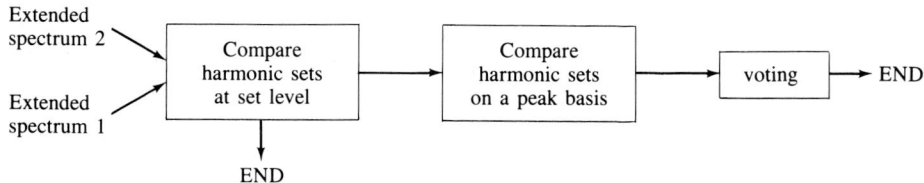

Figure 8.6 Block diagram of the determination of relative peakiness of two harmonic sets by matching at multiple levels of abstraction.

Sec. 8.4 Application of the Extended Spectrum

The first set of heuristics implements comparison of harmonic sets directly based on their features and is shown in Figure 8.7. Each rule implements a condition on features of the two harmonic sets that indicates that one harmonic set is more peaky than the other. The overall strategy is that in order to make a clear decision, two such conditions need to be satisfied, showing that set A is more peaky than set B, and no condition indicating that set B is more peaky than set A is satisfied. Figure 8.7 shows the list of heuristic conditions in the present status of the program. This list is not intended as the "best" collection of heuristics for the purpose, but as indicative of the nature of heuristics that are possible within the framework of the extended spectrum.

The second set of heuristics implements comparison of the features of two peaks for deciding which one is more peaky. We make that decision based on the prominence of the peaks (defined as the ratio of the peak power to the average power of the two enclosing valleys). The associated rule declares that one peak is peakier than the other if the ratio of the prominence of the two peaks is sufficiently high. The possible outcomes are that one peak is MORE, LESS, EQUALLY peaky as the other peak, or that a peak has NO-MATCH in the other harmonic set.

The third set of heuristics implements the voting mechanism for deciding about the relative peakiness of two harmonic sets based on a peak-by-peak comparison. Figure 8.8 shows the rules currently in the program. Again, this list is intended to be indicative of the nature of the heuristics that can be conveniently expressed within the extended spectrum framework. Although voting was found to be adequate for our application, it should be noted that more elaborate schemes can be used instead, which employ confidence factors and means for combining them [5]. A careful consideration of the trade-offs between simple and elaborate schemes must precede system design in signal processing and interpretation applications.

Results of the above algorithm applied to real helicopter data, involving two acoustic sources (helicopters), is presented in [14]. This type of real data is very appropriate for the algorithm, because it exhibits several phenomena. First, it contains two sets of harmonically related peaks, one per acoustic source, whose power and fundamental frequency varies with time. Second, the power depends on

HARMONIC SET A IS MORE PEAKY THAN SET B IF:

- A has a fundamental peak and B has none.
- A has a prominent fundamental peak and B has not.
- A has a lot more prominent harmonics than B.
- A has much higher average prominence than B.
- A has much higher total prominence than B.

Figure 8.7 Heuristic rules indicating that harmonic set A is more peaky than set B. At least two of them must be satisfied, and none must be satisfied in the opposite direction, in order to make a clear decision concerning whether set A is more peaky than set B.

HARMONIC SET A IS MORE PEAKY THAN SET B IF:

- The number of peaks that are peakier in A is higher by at least 2 than the number of peaks that are peakier in B, and there is no more than one peak in B with no match in A.
- At least two peaks in A have no match in B, and no more than one peak in B has no match in A.

Figure 8.8: Heuristic rules implementing the voting mechanism for deciding about the relative peakiness of two harmonic sets based on a peak-by-peak comparison.

the distance between the source and the sensor, while the fundamental frequency at the sensor is equal to the frequency emitted by the source, modified by a factor that depends on the speed of the source according to the Doppler effect. Finally, the signal-to-noise ratio depends on the distance between the source and the sensor.

8.5 RELATION TO OTHER WORK

In this section, we will relate signal abstractions with other work, both computational and psychological. We will review psychological work based on Gestalt theories which lends support to our approach, and we will relate signal abstractions to other approaches based on abstraction, as described in Chapters 1 and 6 of this volume. We will also relate signal abstractions to object recognition using decomposition.

8.5.1 Auditory Signal Perception

The Gestalt theory of perception [1] has as its central point that perception at all levels is organized, and that the organization at each level is established by the brain in a way that is consistent with that at more global levels as well as at more microscopic levels. More specifically, in auditory signal perception, two kinds of grouping have been proposed [1]. The first is grouping together events that occur at the same time but belong to different parts of the spectrum (simultaneous grouping). The second involves putting together events that follow one another in time (sequential grouping). Computing the extended spectrum representation is essentially equivalent to simultaneous grouping, by considering spectral peaks as events that occur at the same time but at different parts of the spectrum. Furthermore, the extended spectrum supports sequential grouping, by facilitating the connection of events in time, where events are harmonic sets, i.e., the result of simultaneous grouping. An example of the use of the extended spectrum to perform sequential grouping is shown in Chapter 9 of this volume. It is interesting to note that people, who routinely analyze acoustic spectra by visual inspection, do so by identifying high-level features consisting of harmonically related sets of peaks, essentially performing visually simultaneous and sequential grouping.

Sec. 8.5 Relation to Other Work

In regard to the application of simultaneous grouping to the case of a harmonic signal, psychology research [1] indicates that our auditory system seems to have evolved a method of finding out what the fundamental is, even if only some of the higher harmonics are present. As part of the extended spectrum computation, we have proposed techniques for accomplishing this task, described in [14].

8.5.2 Peak-Based Methods in Various Signal Interpretation Tasks

Peak-based methods have been used in the correlation of well logs [20], in the description of ECG signals [12], in the pitch detection of speech signals [18], and in the speaker separation of voiced speech [17]. [20] attempts to match well logs by matching curve elements, which are peaks, troughs, spikes, steps, and levels. Each element is assigned a pattern vector characterizing the shape of the element. Correspondence of similar elements in the two well logs is established based on the correlation of their pattern vectors, and is then tested against additional geological criteria. Parsons [17] represents spectral peaks in terms of peak model parameters, and uses the harmonic constraint to separate the spectral peaks of the two speakers, which form two distinct harmonic sets. The whole approach has a very heuristic flavor to it. Signal abstractions can be viewed as a framework for formalizing and organizing approaches like that of [20] or [17].

8.5.3 Abstraction in Blackboard Architectures

Blackboard architectures (Chapter 6 of this volume), as applied to the interpretation of ship harmonics from underwater acoustic sensors in the HASP/SIAP system [15] and to speech understanding in the HEARSAY system [7], involve hypothetical interpretations at multiple levels of abstraction. There are important differences between blackboard-based systems and signal abstractions. The first difference is control. Blackboard-based systems have opportunistic control, in the sense that hypotheses at a specific abstraction level are used to opportunistically trigger knowledge sources, which generate hypotheses at higher abstraction levels. Signal abstractions, on the other hand, have control that is much more structured, and is opportunistic only during the grouping operation. Once grouping is done, we obtain a stable representation of the signal, which can change only in a systematic manner by adjusting the parameters governing the grouping and qualitative labeling operations, and the computation of the numeric signal (e.g., the spectrum). The second difference is that both HASP/SIAP and HEARSAY had a straightforward signal processing front-end, which was activated once and computed multiple competing symbolic hypotheses at the lowest abstraction level. Symbolic constraints were then applied at higher levels to select the correct hypothesis. In contrast, signal abstractions resolve conflicting hypotheses one level at a time, and are intended as part of a closed loop operation which includes numeric signal processing as well.

Extensions to blackboard architectures that impose different types of structure to the totally opportunistic control of earlier schemes are described in Chapter 6. The chapter also points out the need for different abstraction levels of describing the signals, and considers the process of abstraction as filtering out noise as we move to higher abstraction levels. Generation of more abstract hypotheses is performed by opportunistic knowledge sources, as in earlier schemes.

8.5.4 Object Recognition Using Decomposition

Motivated by the Gestalt principle, a general approach to object recognition has been explored, which relies on the decomposition of objects into constituent parts, followed by the classification of the parts [9]. Recognition is then based on using the identified parts. The decomposition process can take place at multiple levels, by identifying certain part subgroups as higher order parts, thus building a part hierarchy. A simple example of such a simple part hierarchy is to detect straight line segments in an image as the most primitive parts, and to then define higher order parts such as corners or polygons, using the already detected line segments. Grouping of primitive parts can be accomplished by using the known relationships between the parts. Syntactic pattern recognition [10] can be used to perform the grouping operation, if part relationships can be captured by rules of an appropriate formal language. The representation of data used in systems performing object recognition using parts decomposition is thus similar in nature to signal abstractions. The emphasis in such systems is on the matching of the features derived from the data with the available models. The presence of occlusion (only part of the model is present in the image), deformation of the scene object with respect to the model, and a scene representation different from the model (as is the case when the scene is an intensity image associated with a full three-dimensional precise geometric model such as those used in computer-aided design), are some of the reasons why object recognition is an extremely difficult problem, for which no general solution exists to date [19].

8.5.5 Data Abstraction in Signal Processing

Data abstraction is a concept in programming languages and it has been applied to the representation of discrete-time signals in the form of abstract signal objects (Chapter 1). Abstract signal objects can be manipulated symbolically, much like the manipulations of algebraic expressions by systems like MACSYMA (Chapter 3), and have led to sophisticated search techniques for the discovery of novel signal processing structures (Chapter 2). There is a conceptual difference between abstract signal objects and signal abstractions. Abstract signal objects are ideally suited to the representation of signals described by a small number of parameters. The term "abstraction" here refers to the separation of the functional form of the signal (equivalent to a parameterized mathematical expression) from the signal values themselves. Such abstraction does not involve loss of detail, since the functional form of the signal together with the parameter values is equivalent to the set of signal values. In fact, the functional form is more powerful than the set of signal values, since it is much more compact and efficient to store and manipulate, and allows the complete representation of signals, the cardinality of the set of values of which is infinite.

8.6 CONCLUSIONS

In this chapter, we advanced an extended notion of a signal, as consisting of multiple levels of abstraction. At each level of abstraction, conceptual units are explicitly represented as subsets of the set of conceptual units at the immediately lower level of abstraction. We defined multilevel signal abstractions and we demonstrated how the concept can be used in organizing signal processing software including both algorithmic and perceptual techniques for adaptive spectral estimation of periodic signals with time-varying characteristics. Signal abstractions, as applied to the description of the spectrum of periodic signals, are in close agreement with the laws of Gestalt psychology, and are useful in computationally implementing experimental findings in auditory perception.

ACKNOWLEDGMENTS

This research was sponsored at the MIT Lincoln Laboratory by the Defense Advanced Research Projects Agency under contract F19628-85-C-0002, at the MIT Research Laboratory of Electronics by the Advanced Research Projects Agency monitored by ONR under contract N00014-81-K-0742 NR-049-506, and at the University of Toronto by strategic and operating grants from the National Sciences and Engineering Research Council of Canada. The views expressed are those of the authors and do not reflect the official policy or position of the U.S. government.

REFERENCES

[1] A. Bregman, *Auditory Scene Analysis: The Perceptual Organization of Sound* (Cambridge: MIT Press, 1990).

[2] N. Carver and V. Lesser, "Blackboard Systems for Knowledge-Based Signal Understanding," Chapter 6 in this volume.

[3] M. Covell, C. Myers, and A. Oppenheim, "Computer-Aided Algorithm Design and Rearrangement," Chapter 2 in this volume.

[4] E. Dorken, E. Milios, and H. Nawab, "Knowledge-Based Signal Processing Applications," Chapter 9 in this volume.

[5] W. Dove. "Knowledge-based Pitch Detection", RLE Technical Report 518, Cambridge Mass.: MIT (1986).

[6] A. V. Oppenheim, ed., *Applications of Digital Signal Processing* (Englewood Cliffs, N.J.: Prentice-Hall, 1978).

[7] L. Erman, F. Hayes-Roth, V. Lesser, and R. Reddy, "The Hearsay-II Speech Understanding System: Integrating Knowledge to Resolve Uncertainty," *Computing Surveys*, 12, no. 2 (June 1980), 213–53.

[8] B. Evans and J. McClellan, "Symbolic Analysis of Signals and Systems," Chapter 3 in this volume.

[9] M. Fischler and O. Firschein, *Readings in Computer Vision: Issues, Problems, Principles and Paradigms* (Los Altos, Calif.: Morgan Kaufmann, 1987).

[10] K. S. Fu, *Syntactic Methods in Pattern Recognition* (New York: Academic Press, 1974).

[11] G. Kopec, "Signal Representations for Numerical Processing," Chapter 1 in this volume.

[12] H. Lee and N. Thaker, "Frame-Based Understanding of ECG Signals," *1st IEEE Conf. Artif. Intell. Applic.* (1984), 624–29.

[13] D. Marr, "Visual Information Processing: The Structure and Creation of Visual Representations," *Phil. Trans. R. Soc. London B*, 290 (1980), 199–218.

[14] E. Milios and H. Nawab, "Signal Abstractions in Signal Processing Software," *IEEE Trans. on Acoustics, Speech and Signal Processing*, 37, no. 6 (June 1989), 913–28.

[15] P. Nii, E. Feigenbaum, J. Anton, and A. Rockmore, "Signal-to-Symbol Transformation: HASP/SIAP Case Study," *The AI Magazine*, (Spring 1982), 23–35.

[16] A. V. Oppenheim and R. Schafer, *Digital Signal Processing* (Englewood Cliffs, N.J.: Prentice Hall, 1975).

[17] T. Parsons, "Separation of Speech from Interfering Speech by Means of Harmonic Selection," *Journal of the Acoustical Society of America*, 60 (October 1976), 911–18.

[18] L. Rabiner and R. Schafer, *Digital Speech Processing* (Englewood Cliffs, N.J.: Prentice Hall, 1978).

[19] S. Ullman, "Aligning Pictorial Descriptions," in *Artificial Intelligence at MIT: Expanding Frontiers*, eds. P. Winston and S. Shellard (1990), pp. 344–404.

[20] P. Vincent, J. E. Gartner, and G. Attali, "An Approach to Detailed Dip Determination Using Correlation by Pattern Recognition," *Journal of Petroleum Technology* (February 1979), 232–40.

9

KNOWLEDGE-BASED SIGNAL PROCESSING APPLICATIONS

- Erkan Dorken
  Boston University
- Evangelos Milios
  York University
- S. Hamid Nawab
  Boston University

9.1 INTRODUCTION

In this chapter we present an overview of application-specific knowledge-based signal processing (KBSP) systems that have been developed in recent years. Our intention is to give the reader a flavor of the kinds of applications that have motivated researchers in the KBSP area to consider the design of systems in which there is sophisticated interaction between numeric and symbolic signal processing. The emphasis in this chapter is on characterizing the application areas and the reasons why these applications are particularly suitable for KBSP investigations. It is important to note that these are not the only possible KBSP applications nor are they necessarily the most significant ones. Rather, their distinction is a historic one in that they were among the first application areas considered by researchers who were explicitly interested in KBSP issues.

The *Pitch Detector's Assistant*, discussed in section 9.2, was designed to provide an interactive facility for pitch detection which permits the user to experiment with changes in the input data, the derived data, and the problem knowledge used for the tracking of pitch in speech signals. The *Sound Understanding Testbed*, discussed in section 9.3, was designed to experiment with techniques for extracting information about sound sources from signals recorded on a microphone placed in a household environment. This system makes use of the RESUN blackboard framework and the IPUS architecture presented respectively in Chapters 6 and 7 of this book. The *Helicopter Signal Tracking System*, discussed in section 9.4, was designed to track the harmonic components in a moving-helicopter acoustic signal recorded at a fixed microphone. This system makes use of the Extended Spectrum concept discussed in Chapter 8 of this book. In the fifth section, we discuss some of the common themes that emerge from consideration of the KBSP applications discussed in this chapter.

9.2 PITCH DETECTOR'S ASSISTANT (PDA)

The Pitch Detector's Assistant [1] is a KBSP system that was developed for a pitch detection problem whose solution involves both symbolic and numeric information processing. From a traditional signal processing viewpoint, pitch detection is usually formulated as the problem of numerically processing a speech waveform in order to divide the waveform into regions that are estimated to be either voiced or unvoiced, and to determine the pitch (or more precisely, the fundamental frequency F0) in the voiced regions. The developers of the PDA system consider a "modified pitch detection" problem in which the data to be processed includes not only a speech waveform but also a phonetic transcript of the speech utterance (including word and syllable markings) as well as the sex and age of the speaker. The consequence of this modification is that the design of the PDA system involves a combined use of symbolic and numerical information. In particular, W. Dove [1], the main developer of the PDA system, cites three major ways in which the modified pitch detection problem can serve as a vehicle to address issues related to the goal of combining symbolic and numerical information processing in balanced proportions:

- The amount of symbolic input information balances the amount of numeric input information (the waveform) already present in the traditional pitch detection problem, so there are opportunities to combine the two.
- Providing symbolic input information can reduce the tendency for early processing to be dominated by a numerical approach, as has been the case with a number of expert systems for processing signals.
- Providing a phonetic transcript as input alleviates what would otherwise be a major task, the analysis of the waveform to generate that information. Such a difficult symbolic analysis problem would probably result in a system architecture in which symbolic processing is dominant.

A system devised to solve the modified pitch detection problem, discussed in more detail in section 9.2.1, has a number of potential uses. These include [1] generation of pitch tracks for use in the enhanced reconstruction of archival speech material, pitch analysis for talking computer databases, and reference pitch tracks for testing other pitch detectors.

Sec. 9.2 Pitch Detector's Assistant (PDA)   **305**

9.2.1 PDA Objectives

The PDA system was designed to address the modified pitch detection problem of extracting pitch information from the waveform as well as from symbolically encoded inputs such as a phonetic transcript of the speech utterance. An example of the type of input that may be provided to the PDA system is illustrated in Figure 9.1 (reproduced from [1]). The input includes the digitized waveform and aligned phonetic transcript for the utterance "which tea party did baker go to." The four types of transcript marks (phrase, words, syllables, and phonemes) are shown in the labeled vertical strata of the picture, and each mark is identified by a string of characters with its extent depicted by the line below it. The widths of the boxes that border these lines signify the uncertainty with which these boundaries are known. Identical boxes that lie one above the other are images of the same box viewed from the different strata. The type of output produced by the PDA system when given the input depicted in Figure 9.1 is shown in Figure 9.2 (reproduced from [1]). All four plots in this figure span time horizontally (in samples at 10 Khz). The first plot shows the "confidence" that the speech is glotally excited. The second is a pseudo-intensity plot[1] of the probability density of F0 as a function of time. The next plot

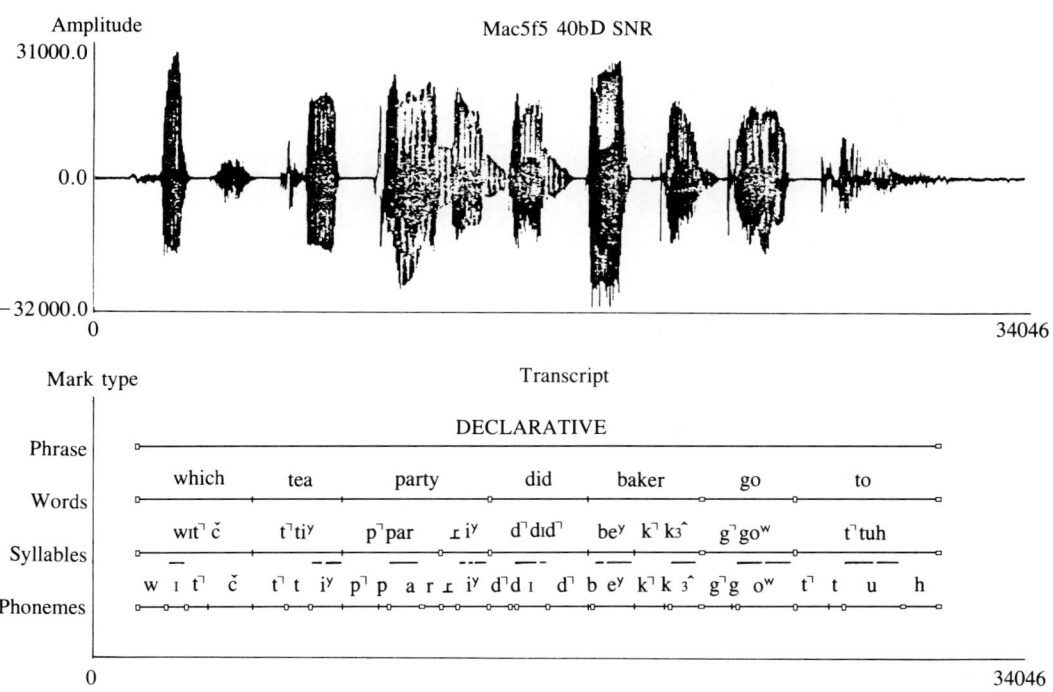

Figure 9.1 PDA Inputs: Symbolic Transcript and Waveform.

---

[1] The density of dots in this plot corresponds to F0 with black being a high probability density and white being low.

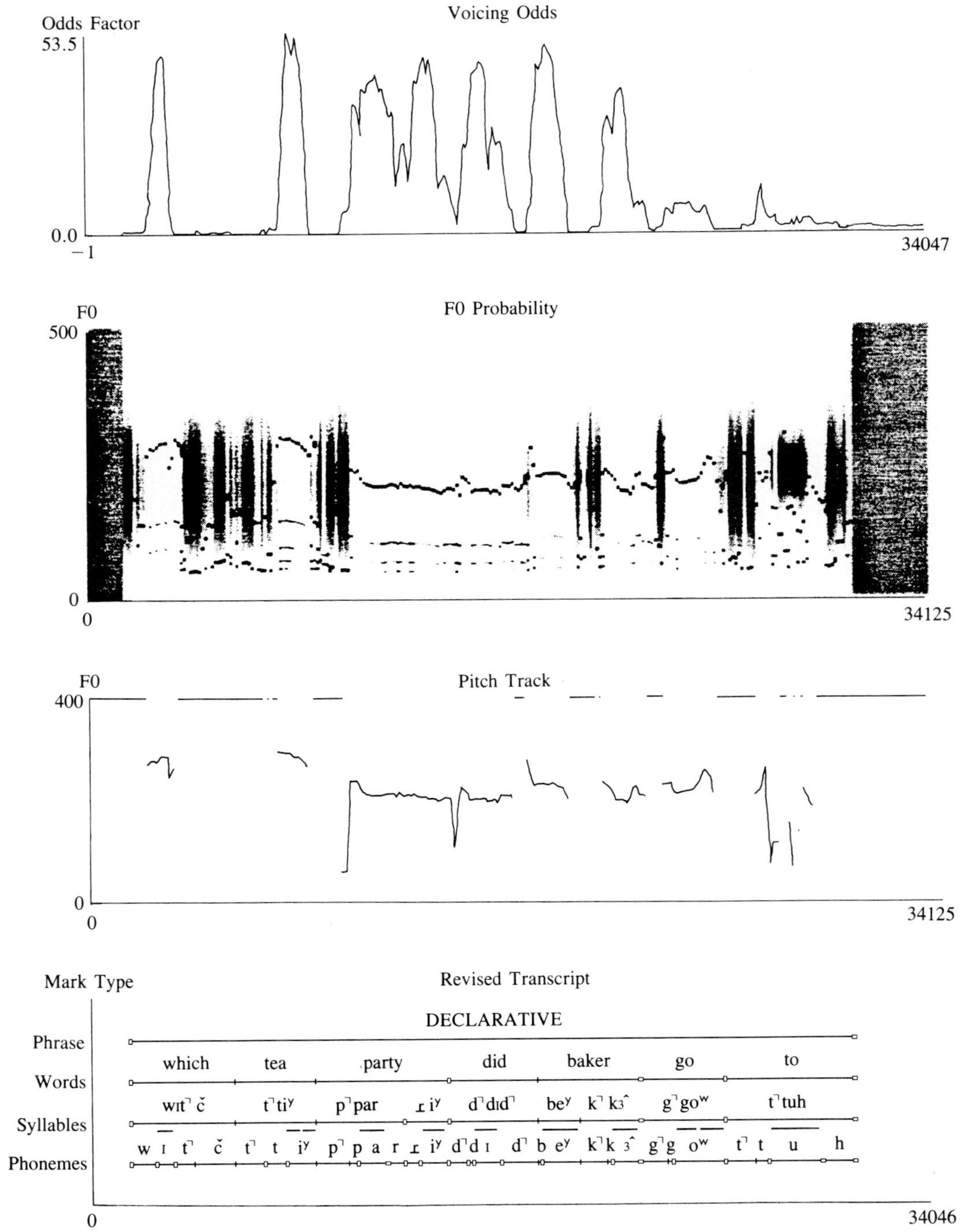

306   Knowledge-Based Signal Processing Applications   Chap. 9

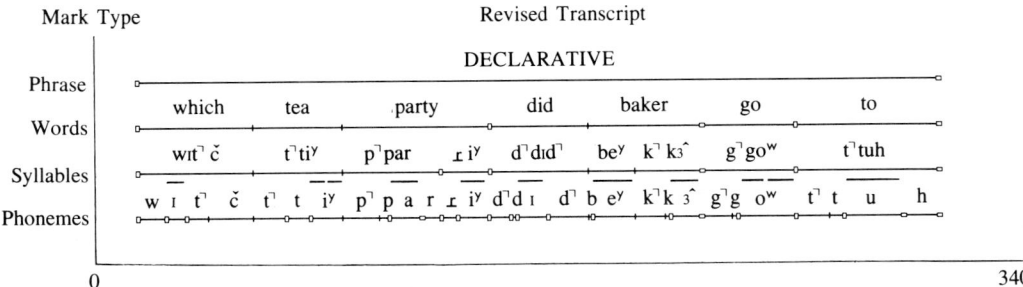

Figure 9.2 The Outputs of the PDA.

shows the revised alignment of the phonetic transcript (where the depicted boundaries include information from numerical measurements), and the last plot is a conventional pitch track derived from the F0 probability density and the voicing confidence at each location.

From an applications perspective, the goal of the PDA system is to provide an interactive computing environment in which pitch detection is carried out through mechanisms for conveniently combining the information extracted through signal processing with information extracted from symbolically encoded inputs to the system. In particular, the system must be able to effectively "merge" the results of the symbolic and the numeric processing, and it must possess a control structure that can direct the symbolic and numeric processing in a context-dependent fashion. The developers of the PDA system were also interested in keeping the system computationally efficient and allowing the system user to interrupt the processing, make desired changes in the system's knowledge base, and have the system resume its processing. The PDA system can thus also be viewed as an aid for pitch detector design.

9.2.2 PDA System Details

In the PDA system, the modified pitch detection problem is viewed as being composed of two subproblems: (1) finding the voiced and unvoiced regions and (2) finding the pitch period of the utterance as a function of time. The two subproblems are solved using the same basic mechanism of making "assertions," verifying them, and combining them to form "composite assertions." An assertion is a data structure that represents a statement about some aspect of the solution to the modified pitch detection problem. In addition, an assertion includes some statistical information regarding support for that assertion from other assertions. At the end of this process, the answers to both of the subproblems are given in the form of composite assertions, the assertion called VOICED for the voicing subproblem and the assertion called FINAL-PITCH for the pitch period subproblem. The assertion VOICED represents the final probability of voicing as a function of time for the entire utterance. The assertion FINAL-PITCH represents the probability density function of the pitch period as a function of time for the entire utterance.

The flow of analysis to make the assertions VOICED and FINAL-PITCH is shown in Figures 9.3 and 9.4 respectively. Both of these assertions span the entire utterance. In both figures, ovals represent a group of assertions and rectangles represent a group of rules. Figure 9.3 depicts determination of voicing in the PDA system. The flow of analysis starts with two input assertions, *phonetic transcript* and the *speech waveform* for the entire utterance. Phoneme-mark assertions in the phonetic transcript are converted to *voicing-mark* assertions using the *voicing analysis* rule. For example, phoneme /a/ in the phonetic transcript is converted to a voicing-mark assertion which asserts that the region corresponding to /a/ in the utterance is voiced. The confidence associated with this assertion is calculated by the *phonetic-verify* rule, which applies numerical procedures to the waveform. The

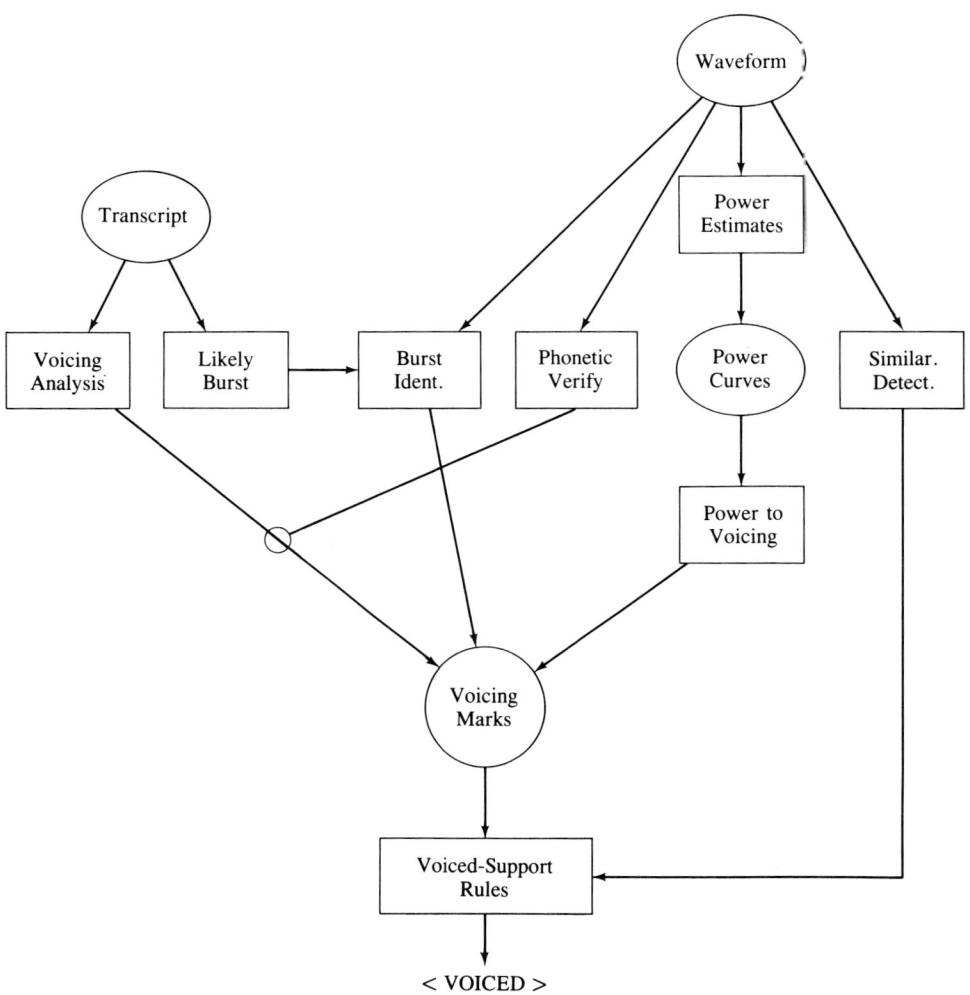

Figure 9.3 Determining Voicing in the PDA.

likely-burst rule finds the stop bursts in the phonetic transcript and executes another rule to locate the bursts in the waveform numerically. Stop bursts are characterized by their brief high intensity, which may lead numerical voicing determination rules to generate inappropriate voicing support. For this reason, stop bursts are identified from the phonetic transcript and their exact location is obtained by means of a numerical process depicted as the burst identification rule in Figure 9.3. The *power estimates* rule calculates the broadband power in low and high frequencies. The *power curves* assertion, based on broadband spectral power estimation, is analyzed to generate numerically derived voicing assertions through the rule depicted as *power to voicing* in the figure. The *similarity detection* rule performs a short-time

Sec. 9.2 Pitch Detector's Assistant (PDA)

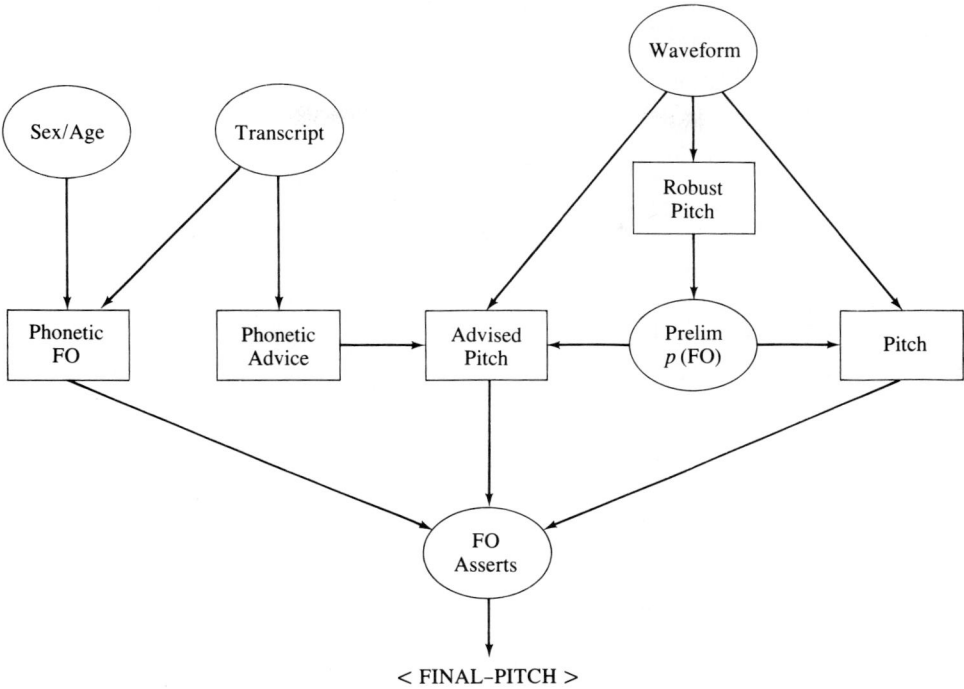

Figure 9.4 Determining Pitch in the PDA.

autocorrelation based algorithm which is used as a measure of waveform similarity (and in turn periodicity). All the assertions (indicated as *voicing-marks* in the figure) derived from the phonetic transcript and the *speech waveform* support a unique assertion VOICED. This assertion spans the entire utterance and represents the probability of the speech waveform being voiced as a function of time.

Figure 9.4 explains the determination of a FINAL-PITCH assertion in the PDA. A FINAL-PITCH assertion is determined through the interaction of numerical estimates of the spacing between similar parts of the waveform, pitch prediction from phoneme identity, and speaker sex/age information. The numerical pitch estimator is based on the short-time autocorrelation function. In Figure 9.4, the sex/age of the speaker, the phonetic transcript, and the speech waveform are input level assertions. *Phonetic F0* rules in the figure assert pitch period from the phoneme identity. The PDA system includes tables specifying the expected pitch period for each phoneme associated with the sex/age group of the speaker. The *phonetic advice* rule provides guidance for the numerical pitch detection procedure (e.g., autocorrelation lag direction is specified, etc.). The *adviced pitch* rule combines a numerical pitch detection procedure with the phonetic advice. The *preliminary pitch* assertion is the estimated pitch period range from the previous pitch estimates. The rule, depicted as *pitch* in the figure, calculates numerical pitch estimation without the phonetic advice. All the pitch-related assertions based on the above defined rules then support the FINAL-PITCH assertion.

The confidence of voicing-related assertions is represented in the form of numerical *odds-factors* [1]. The odds of an assertion is defined as the ratio of the probability of that assertion is true to the probability of the assertion is false. The odds-factor of an assertion is defined as the ratio of current odds (odds of an assertion given what is known) to the a priori odds. The confidence of a composite assertion is computed from the odds-factors of its supporting assertions by using a combination rule derived from *Bayesian probability* theory. The confidence of pitch-related assertions is represented in terms of the mean and variance of a *Gaussian probability density function*. The Gaussian probability density for the FINAL-PITCH assertion is given as a sum of weighted products of Gaussian densities associated with the supporting assertions [1].

The PDA system contains three major subsystems: the Epoch system, the Rule system, and the Knowledge Manager (KM). The Epoch system is used to post estimates regarding the positions of objects such as words, syllables, phonemes, and segments. It is also used for the precise alignment of symbolic and numerical information about the utterance. This alignment is used to guide the operation of some numerical procedures. For example, if the numerical period estimator is to be advised about a likely F0 fall after a uv-consonant/vowel boundary, then there must be a single statement about where it is expected to occur. Thus, the position information from separate numerical and symbolic sources must be combined before it can be used to guide the period estimator. The Rule system in the PDA supports the following features: networked objects, mutable objects, a dynamic rule base, and test-directed [1] rather than pattern-directed rule invocation. A particularly attractive feature of the PDA rule system is that it allows rule changes to be made by interrupting a system run, changing rules or adding new rules to the rule base, and restarting the system from the point at which it was interrupted. This is made possible by an extensive binding mechanism that keeps a "history" of how the various assertions depend on the firing of the various rules. Mechanisms are provided for propagating changes in rules to the assertions that depend on those rules. The Knowledge Manager (KM) is a part of the PDA system that acts as the intermediary between the rules and the asserted information in the PDA assertion database. It runs the rule system, maintains a "dependency network" and a "support network," and computes confidences for assertions. The following sections give more detail on each of the three subsystems.

The Epoch System

The "epoch" is a special data structure in the PDA system used to represent the temporal boundaries of assertions. For instance, a phoneme-mark in the phonetic transcript is an assertion with two epochs associated with the start and end times of the phoneme. The uncertainty in the location of the temporal boundaries is represented by a Gaussian distribution function. The epoch associated with this

Sec. 9.2 Pitch Detector's Assistant (PDA)

boundary includes four basic types of information: the *name* of the epoch, the *mean* of the temporal location, the *variance* in the mean of the temporal location, and pointers to *neighbor* (temporally adjacent) assertions.

Whenever a new assertion is created in the PDA assertion database, the Epoch system is triggered to create a simple epoch associated with the new assertion. The Epoch system then checks this new epoch against other existing ones and decides if an *epoch merging* is possible. The Epoch system calculates the probability of a group of epochs being generated from the same underlying event, and uses this probability as a criterion for deciding whether a new *composite* epoch should be created by merging the constituent epochs.

Once the PDA system performs numeric analysis on the speech waveform, new assertions and new epochs are created. For instance, when a numerically derived voiced assertion is created, the Epoch system creates an epoch associated with it. Merging the epochs derived from numeric processing with the epochs hand-marked by a phonetician, the locations of the boundaries are determined more accurately and the uncertainty in the boundary locations is reduced. It is this merging process that allows the PDA system to refine the input phonetic transcript.

The Epoch system also supports the Rule system (see next section) by answering the rule condition questions and providing backtracking to undo epoch merges. For example, if a rule trigger condition requires a /s/t/ phoneme configuration, then the rule is activated whenever the Epoch system merges the ending epoch of an /s/ phoneme with the starting epoch of the /t/ phoneme. When a change in one of the constituent epochs of a composite epoch is made, the Epoch system "removes" the changed epoch from the composite epoch. This is achieved by backtracking to the constituent epochs by utilizing the information in the "constituents" slot of the composite epoch.

The Rule System

The phonetic and signal processing knowledge used in the PDA system is represented in terms of rules. The PDA system uses many different sources of knowledge to make the VOICED and FINAL-PITCH assertions. The supporting rules for the voicing-related assertions includes rules for extracting information from the speech waveform, rules for extracting information from the symbolic transcript, and rules for verifying symbolic transcript assertions through numerical analysis of the speech wave form. Examples of each type of rule are given in Figures 9.5, 9.6, and 9.7. The rule in Figure 9.5 is for making a voicing decision from sonorant power measurements in the speech waveform. The rule in Figure 9.6 is for marking silence gaps on the basis of the symbolic transcript. The rule in Figure 9.7 is for numerically detecting a stop-burst on the basis of the symbolic transcript. Note that the PDA rules have three major parts: the *conditions*, the *premise-odds*, and the *actions*. The conditions part of a rule is composed of tests that have to be satisfied to trigger the rule. Once the tests are satisfied, the rule is triggered and a *binding* is created which describes the association of the assertions that triggered the rule with rule condition

```
(defrule < numeric-voiced1 >
  CONDITIONS
  (type 'utterance utt)            ;$utt$ is the waveform
  ; $pwr$ is assertion concerning the maximum sonorant power
  (type 'max-sonorant-power pwr)
  ; $pwr -value$ is the numerical value of that assertion.
  (let pwr-value (value pwr))
  PREMISE-ODDS
  ACTIONS
  (scan-for-voicing (sonorant-log-power utt) pwr-value))
```

Figure 9.5 PDA Rule: Voicing from Sonorant Power.

```
(defrule < phrase-end-gap >
  CONDITIONS
  (type 'word-mark word)       ;a transcript word mark
  (type 'utterance utt)        ;the waveform
  (type 'phrase-mark phr)      ;the transcript phrase mark
  (end-of phr word)            ;$word$ ends at phrase end.
  (let start (end word))       ;the ending epoch of $word$
  (let end ($end utt))         ;$utt$ is not bounded by epochs.
                               ;Therefore, this is just a number.
  ( < start end)               ;Smart ' < ' fcn can compare the epoch
                               ;$start$ with the number $end$
  PREMISE-ODDS
  ;binding confidence is given by the confidence in $word$ and $utt$
  (lambda (word utt) (min (odds word) (odds utt)))
  ACTIONS
  ;Since there was no epoch at the end of $utt$, we must make one.
  (assert (gap start (make-simple-epoch :mean end :sd 100 :support utt))
          .9 .4))
```

Figure 9.6 PDA Rule: Finding Silence Gaps.

```
(defrule < numeric-burst0 >
  CONDITIONS
  (type 'p-mark x)                     ;$x$ is a phoneme-mark
  (type 'utterance utt)                ;$utt$ is the waveform
  (isa ':unvoiced-stop-release x)      ;$x$ must be a stop release
  (let start (start x))
  (let end (end x))
  PREMISE-ODDS
  ACTIONS
  (scan-for-burst utt x))
```

Figure 9.7 PDA Rule: Finding Stop Bursts.

variables. The premise-odds part of a rule calculates the confidence of a binding. For example, if a rule is triggered by a phoneme-mark, the confidence of the binding is equal to the confidence in the phoneme-mark that triggered the rule. The actions part of a rule alters the status of the existing assertions in the assertions database or makes new assertions. The Epoch system is triggered if a new assertion is created in order to determine the associated epoch.

The PDA rule system was developed with the idea that the PDA system works interactively with an operator. The operator may enter input information in batches. For instance, the operator may first enter phoneme-marks and observe the results. Then by entering syllable-marks, the operator can observe the impact of this additional information on the performance of the system. In addition, the operator may change assertions during run time to see the resulting effects. The operator may also experiment with rules by adding and/or removing rules to the rule base. The basic idea is that it should be possible for the operator to alter the system during run time without having to restart the system to maintain consistency between the asserted objects. To meet these specifications, the PDA rule system has the following properties:

- When a data configuration satisfying the conditions part of a rule is detected, the rule is always activated.
- When the data structure that triggered a rule ceases to exist, then the changes caused by that rule's firing are always undone.
- If a rule is retracted, then the changes caused by that rule's firing are always removed.
- Anytime a new rule is entered into the system, the rule is tested against the current data in the PDA assertion database and executed on all appropriate data configurations.

The consequence of the above defined properties is that the *rest-state* of the system (the state in which no rules can be fired) is dependent on the rules and the present data but not dependent on the history of the rule and data entry. Another property of the rule system is that there is no need for *conflict resolution*. This is because all the rules whose test conditions are satisfied are all always fired. The computational price one pays for a history-independent rule system is the cost of maintaining a network of dependency links between the objects in the system.

Objects in the PDA system are related to each other through two networks called the dependency and support networks. The dependency network is designed to represent absolute dependence between the objects and the notification of change. The dependency network is composed of servers and users; if the server is retracted, then the user (and the user of the user, etc.) is retracted as well. For example, the speech waveform assertion and the power estimate assertion based on the speech waveform are related to each other by a server/user relation. If the speech waveform assertion is removed, then the power estimate assertion and all of its users are removed from the network.

The support network represents the probabilistic support relations between the assertions. This network is composed of supporters and supportees. If a support is removed from an assertion, the confidence of the assertion is changed. For example, if one of the supporters of the VOICED (final voiced assertion) assertion is removed, the confidence is changed without removing the VOICED assertion.

The KM, described in the next section, maintains both of the networks in order to keep all the data consistent whenever data and/or rules are modified or retracted.

The Knowledge Manager (KM)

The KM is composed of parts of the PDA system that handle interactions between the assertions and the operation of the rule system. The KM performs three main tasks: running the rule system, operating the dependency and support networks, and computing the confidence and other statistics of the assertions. The KM runs in a cycle of four phases.

1. The KM is a user of all assertions in the dependency network. When an assertion is changed or a new one is generated, the KM is notified by the assertion. This is how the KM keeps track of changed and new assertions. In this phase, the KM collects bindings dependent on the changed assertions and marks these bindings as potentially invalid. Newly generated assertions are also collected by the KM in this phase.
2. The KM runs the conditions of the rules (specified by the potentially invalid bindings of the previous phase) against changed assertions. There are three possible outcomes of the rule condition checking. The first possibility is that a potentially invalid binding remains invalid. This happens if the changed assertion does not satisfy the rule conditions anymore. Another possible outcome is that a potentially invalid binding is made valid. This possibility reflects the situation in which the rule conditions are still satisfied by the changed assertion. The last possible outcome is that a new binding may be generated. This outcome arises when the changed assertion satisfies a rule condition that it did not previously satisfy.
3. In the third phase, the KM retracts the invalid bindings and all their results, updates the dependency and support networks to maintain the PDA system in a self consistent configuration.
4. In the last phase of the cycle, the KM executes actions of all the rules associated with the newly generated bindings generated in phase two. As a result of rule firing, new assertions are created. If an epoch associated with a newly generated assertion is needed, the Epoch system is triggered and a new simple epoch is created. Also, the confidences of the newly generated assertions are calculated in this phase. The dependency and support networks are updated to reflect the changes due to the new assertions in the system.

The KM then returns to phase one to check bindings dependent on the changed and newly generated assertions. These bindings are verified in phase two by checking the rule conditions. The bindings that are not verified are removed and their results are retracted. Rules that are associated with the newly generated bindings are fired. This cycle is run until there is no new binding generated in the system. This state of the system is called the rest state, reflecting the fact that there is no rule to be fired in the system.

9.3 SOUND UNDERSTANDING TESTBED (SUT)

The Sound Understanding Testbed (SUT) [3] provides a computational environment which allows the user to experiment with techniques for sound understanding in a household setting (telephone rings, electrical appliance sounds, smoke alarms, etc.). Sound understanding involves the processing of acoustic signals in order to determine the sound sources that generated those signals. Specific applications of sound understanding include assistive devices for the profoundly deaf, robotic hearing, speech recognition in environments with nonspeech sound sources in the background, and experimentation with theories of human hearing.

The complexity of the sound understanding problem largely arises from the fact that at any given time multiple sources of sound may be present in the environment; that is, the signals from each of these sources overlap in time. Furthermore, in most cases there is significant overlap in their frequency content as well. Development of the sound understanding testbed was predicated on the belief that to *separate* signal components from different sound sources and to *identify* those sources is a problem that requires a sophisticated combination of symbolic and numeric processing techniques.

The specific interpretation techniques that have been incorporated in the sound understanding testbed are guided by results from studies of human auditory processing [4]. These studies indicate that sound understanding by humans involves the decomposition of the input signal into time-dependent spectral components and the use of certain "integration" criteria for assigning groups of these spectral components to individual sound sources. The details of this complicated process are the subject of on-going research. However, it has been suggested [4] that human auditory processing involves a sophisticated interaction between a numerical data-driven *perceptual* mode and a goal-directed *cognitive* mode. The perceptual mode involves processing that is best modeled by numerical computational mechanisms, while the cognitive mode appears to be most conveniently modeled as a symbolic information processing mechanism. The perception mode may be viewed as the generator of data whose explanation requires the formation of symbolic source hypotheses. The cognitive mode may be viewed as the generator of symbolic source hypotheses that can be used to direct perceptual processing aimed at searching for data to support the symbolic source hypotheses. The sound understanding testbed was designed to experiment with strategies for integrating the cognitive and perceptual aspects of sound understanding within a single computer-based system.

9.3.1 SUT Objectives

The sound understanding testbed was designed to provide a convenient computational environment for experimentation with the sound understanding problem. In particular, the testbed allows the user to conveniently specify and modify the interpretation techniques, the data and interpretation representations, and the problem-solving control strategies, as well as the specific goals of the interpretation process. The testbed utilizes the IPUS architecture (see Chapter 7) as the framework within which most of the interaction between signal processing and signal interpretation takes place. The testbed also utilizes the RESUN blackboard framework (see Chapter 6) for organizing its problem solving. In addition to these generic features, the testbed also includes specific data and interpretation representations and associated knowledge sources that SUT developers have found useful in attacking the sound understanding problem. In the following discussions of the testbed, we focus our attention on an example run of the sound understanding testbed and how this example illustrates the architecture of the testbed and the domain-specific problem-solving knowledge contained in the testbed.

An example of the type of results that are produced by the sound understanding testbed is shown in Figure 9.8. This figure does not show many of intermediate results computed by the system. In particular, we caution the reader not to interpret the hierarchical display of the results as implying a bottom-up (data-driven) processing strategy. The sound understanding system uses both bottom-up and top-down (goal-directed) processing strategies. In this example, the input to the SUT is a digitized acoustic waveform, corresponding to three rings of a telephone and the hum of a hairdryer being turned on. A portion of the waveform is displayed in the lowest box of Figure 9.8. The box above it displays the spectrum of a short-time section or "frame" of the waveform. The SUT performs STFT analysis on the input signal to obtain spectral information. A peak detection procedure is applied to each short-time spectrum. The peaks are characterized in terms of their times, frequencies, and powers. The box at the "contour" level in Figure 9.8 shows the time evolution of the various spectral peaks over a representative time interval. Peaks are considered to belong to the same contour if their times, frequencies, and powers are determined to be "close" according to the criteria in the system's knowledge base. This process of grouping spectral peaks together at different temporal locations has been referred to as *sequential integration* [4] in the sound understanding literature. It should be noted that the contour hypotheses have to be consistent with the peaks that support them as well as the higher level hypotheses (such as sound "microstreams") that explain the contours. The box at the "stream" level consists of hypotheses that correspond to groups of microstreams that have temporal overlap as well as certain other characteristics in common. This process has been referred to as *spectral integration* [4] in the sound understanding literature. In the case of Figure 9.8, there

Sec. 9.3 Sound Understanding Testbed (SUT) **317**

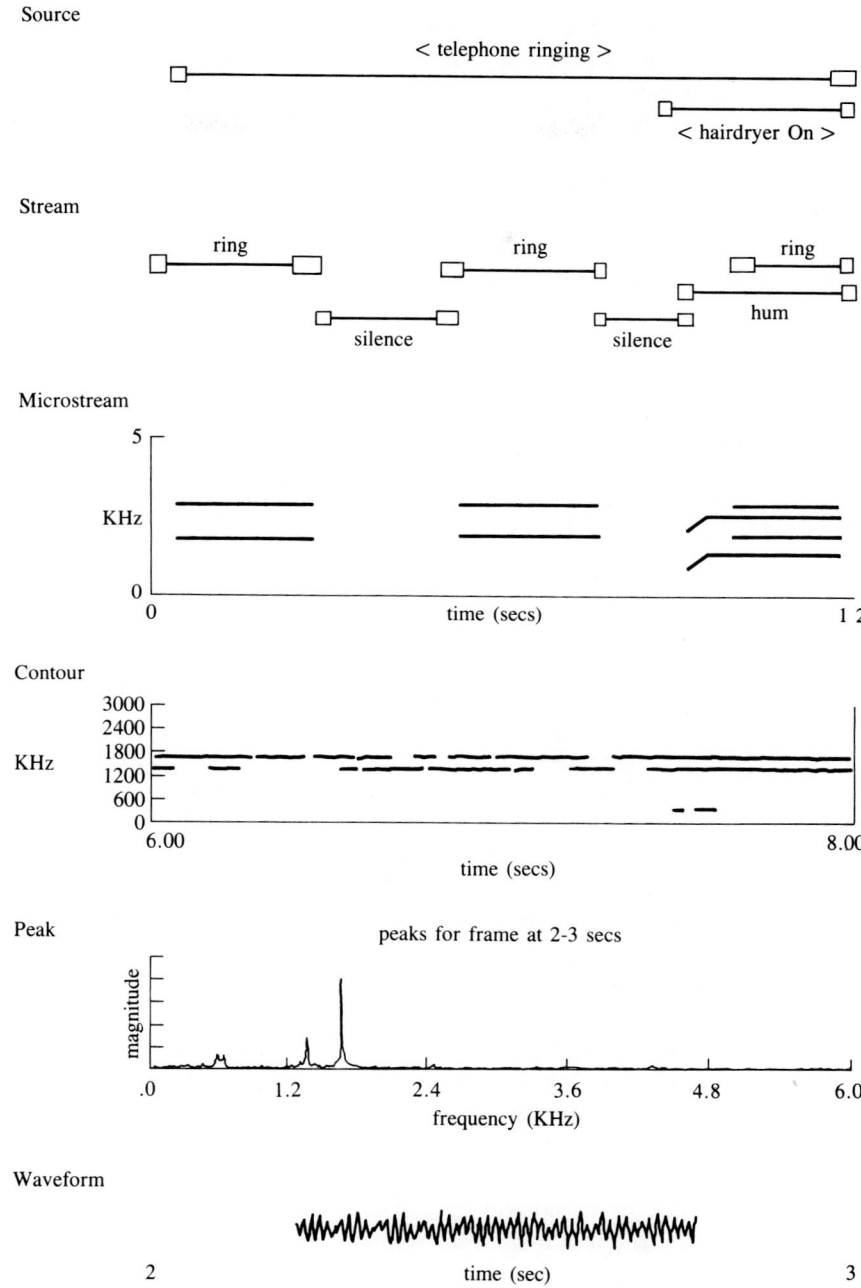

Figure 9.8 Illustration of the input waveform, the intermediate processing results, and the final interpretation results produced by the SUT for the example discussed in the text.

are three types of sound streams hypothesized: "rings," "hums," and silences. Silence is a special type of sound stream that is supported by the *absence* of any contours in a time interval. Finally, streams are grouped together to form source hypotheses. For example, ring and silence streams are grouped together to support a "ringing telephone" hypothesis, while the "hum" stream supports the "hairdryer on" hypothesis. It should be noted that this process is analogous to the speech understanding process of forming phrase hypotheses by combining word hypotheses. In the next section, we discuss some of the features of the sound understanding testbed that enable it to generate the kind of output illustrated in Figure 9.8.

9.3.2 SUT System Details

In this section, we describe some of the system details of the sound understanding testbed. This description is given in the context of the example output illustrated in Figure 9.8. That figure hides many of the complexities involved in the process through which each of the hypotheses illustrated in the figure are actually formed. Our purpose is to elaborate on some of the more important hidden details and relate them to how the sound understanding system is organized.

The sound understanding testbed may be conceptualized as consisting of a blackboard database that is divided into various information levels. The various boxes in Figure 9.8 correspond to the six information levels (waveform, peak, contour, microstream, stream, and source) in the SUT blackboard database. In addition, there are knowledge sources (inference procedures) that can be triggered to use hypotheses at one information level to generate hypotheses at another information level of the blackboard database. Some of the important knowledge sources which will be referred to in the description given below are described in Chapter 7.

Let us now examine some of the more detailed aspects of how the SUT output of Figure 9.8 was actually produced. The input waveform was 12 sec long. The system analyzed the waveform sequentially over 6 consecutive 2-sec blocks. The waveform for each 2-sec block was first processed with an STFT algorithm with fixed values for the analysis-window length, the FFT length, and the temporal decimation rate. A peak-picking algorithm was then applied to each of the spectra in the STFT output. This was followed by a procedure for linking temporally consecutive peaks that are within a specified frequency radius and a specified power radius of each other. Any set of linked peaks constitutes a contour hypothesis at the contour level of the blackboard. The contours formed this way for each of the six blocks are shown in Figure 9.9. Although Figure 9.9 shows the contours produced by the initial processing for all the blocks, it should be kept in mind that the testbed actually performs the higher level interpretations for each block before proceeding to the initial processing for the next block.

For this example, the adequate interpretation of the contour data produced by the initial signal processing in the second through fifth blocks required no further signal processing. However, an account of how the interpretation process proceeds

Sec. 9.3 Sound Understanding Testbed (SUT)

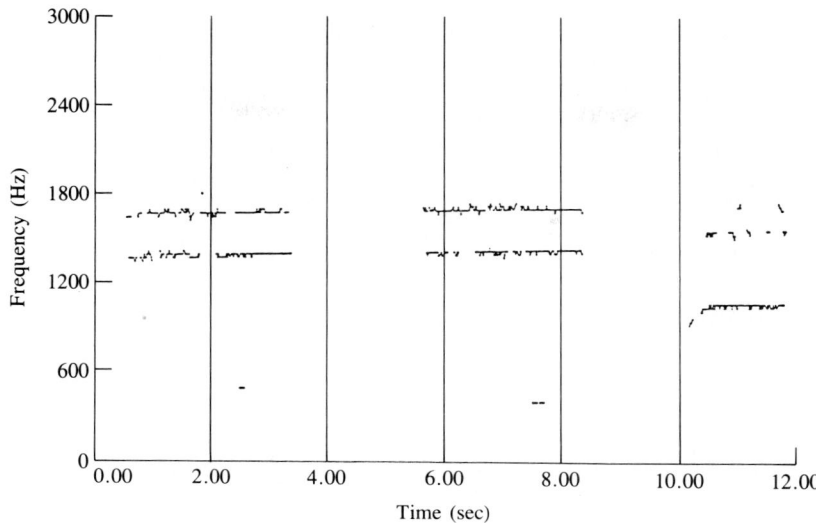

Figure 9.9 Hypotheses at the contour level produced by the initial signal processing of six consecutive 2-sec blocks of the input waveform for the example described in the text. The initial processing for each block is performed only after interpretation of the contour data from the previous block has been completed.

in these blocks helps to illustrate how interpretation models are hypothesized and verified and how expectations are formed for the data to be analyzed in forthcoming blocks. The contours derived from the waveform for the first 2-sec block triggered a search for sources that could potentially have produced such contours. This was accomplished by determining which frequencies contain significant contributions from the contours. The frequencies with most contributions were then compared against a stored database of possible sources and the frequencies at which those sources are likely to contribute significantly. In our current example, the frequencies with dominant contributions correspond to various source hypotheses including a "ringing-telephone" source. For each hypothesized source, stored models for its microstreams were compared against the contour data. Each such comparison resulted in a numeric "confidence measure" reflecting how well the microstream characteristics were matched by the contour data. In this example, the highest confidence measures were obtained for the ringing-telephone hypothesis. Since the contour data started about 0.5 sec into the first block, it was hypothesized that the ring started at that time. Furthermore, the model for the telephone dictates that the ring microstream should last about 3.0 sec, followed by a silence period of about 2.5 sec, and then possibly by other rings. This leads to the expectation that the two ring microstreams in the first block would continue for about 1.5 sec in the next block. This expectation was posted on the blackboard for the processing of the second block of data. Once the initial bottom-up formation of the contours in the second block had been completed, the resulting contours were compared against the posted expectations. In our example, the posted expectations were verified. The system then checked for the existence of any contours that had not been explained and found none. This process continued in a similar fashion through the third, fourth, and fifth blocks. So far, the example has illustrated the basic signal understanding in SUT as it proceeds when the scenario is relatively simple (consisting of a single source). The need for sophisticated interaction between signal understanding and signal processing arises in the sixth block, which corresponds to data from multiple sources.

For the sixth block, the system posted the expectation for a telephone ring to start near the beginning of the block. The contours resulting from the initial bottom-up processing can be seen in Figure 9.9. Clearly, most of these contours do not lie in the regions where they would be expected for the telephone. Thus, the system recorded that a discrepancy had been detected between an unverified expectation and the contour data produced by the initial signal processing. Before proceeding to diagnose this discrepancy, the system's control plans dictated that it must first try to explain the remaining contour data from the initial processing. Following the type of strategy that was utilized in the first block, the system hypothesized that most of the contours from the initial processing correspond to a hairdryer being turned on.

However, to confirm that this data corresponds to a hairdryer, it is necessary to obtain contour data that supports an initial frequency modulation in at least one of the two hairdryer microstreams. Since the contours from the initial signal processing do not provide this support, the system recorded that yet another discrepancy had been detected. This discrepancy is between a verified expectation and the current contour data. The cause for this discrepancy was then hypothesized by the diagnosis knowledge source as being due to inadequate linking of STFT peaks where the hairdryer microstreams move rapidly in frequency as a function of time. Consequently, the signal reprocessing planner decided to reprocess the STFT peaks in those regions by utilizing a greater frequency radius as its criterion for linking the peaks. Figure 9.10 shows how the frequency-modulation region for one of the microstreams is better supported by the resulting contour data. The hairdryer sound was thus confirmed. The model for the hairdryer includes two narrowband microstreams and accompanying broadband noise at a loudness level about 4 times lower than those of the narrowband microstreams. The broadband noise component of the hairdryer sound is not considered important in verifying the presence of a hairdryer sound. However, the broadband noise has to be taken into consideration when analyzing the interaction of the hairdryer sound with other sounds that are not as loud.

Responding to the discrepancy due to the absence of significant supporting contours for the expected telephone ring, the diagnosis knowledge source then analyzed how the components of the hairdryer sound may have interacted with the components of the expected telephone ring. Since the hairdryer was about twice as loud as the expected telephone ring, it was found that there would be significant interaction between the broadband noise from the hairdryer and the telephone microstreams. The diagnosis knowledge source concluded that the discrepancy was

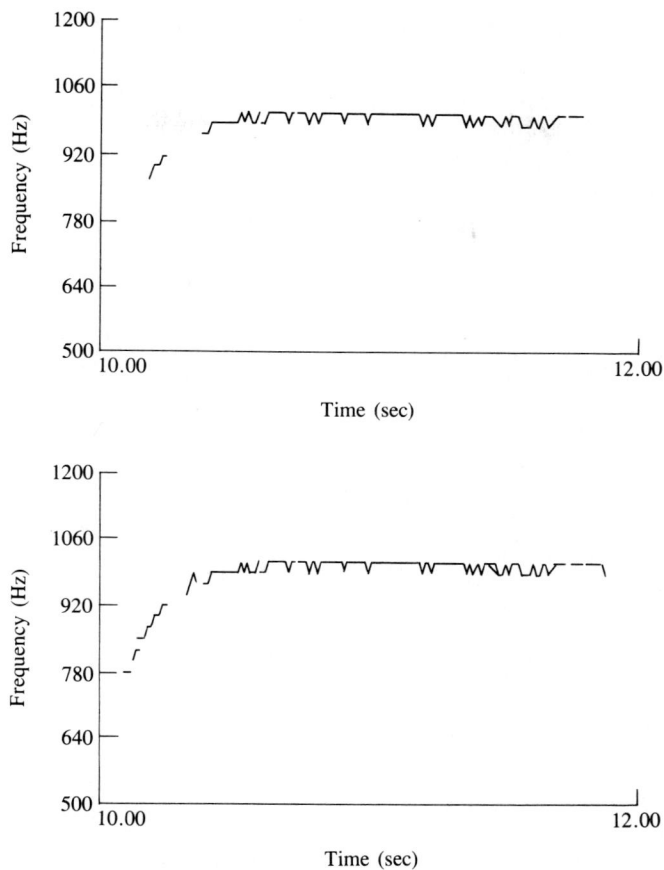

Figure 9.10 Plots showing the contours that lend support to the hairdryer hypothesis during the sixth block. The upper plot shows the contours obtained from the initial signal processing. The lower plot shows the contours obtained after signal reprocessing. The contours obtained from the reprocessing lend support to the initial frequency modulation required by the microstream model for the hairdryer sound.

caused due to insufficient SNR reduction by the STFT. The signal reprocessing planner responds to this diagnosis by hypothesizing that the STFT analysis-window length should be increased to improve the SNR for the telephone's narrowband components. For such situations, the planner sets up an iterative procedure for reprocessing the signal with analysis-window lengths that are progressively increased by factors of 2 until either the discrepancy is removed or the maximum analysis-window length allowed by the STFT program is reached. In this case, increasing the window length by a factor of 4 produces sufficient contour data to support the expectations for the telephone microstreams. The improvement in the contour data for one of the microstreams is illustrated in Figure 9.11.

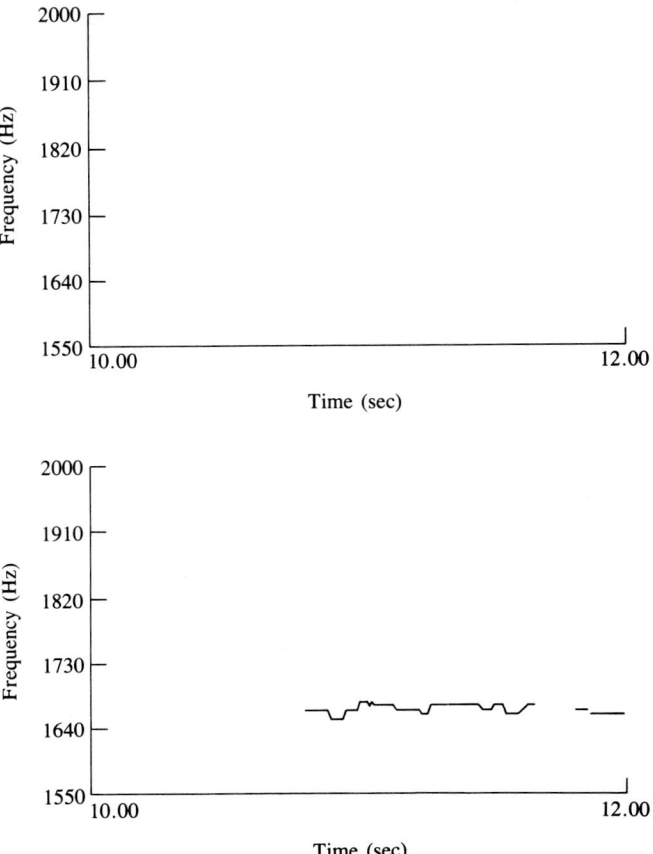

Figure 9.11 Plots showing the contours that lend support to the expected telephone ring during the sixth block. The upper plot shows the contours obtained from the initial signal processing. The lower plot shows the contours obtained after signal reprocessing. The contours obtained from the reprocessing lend greater support for the expected telephone ring.

The preceding example illustrated how various aspects of the SUT can be utilized to experiment with sound understanding applications that require sophisticated integration of signal processing and signal understanding. The SUT continues to be an integral part of ongoing sound understanding research at Boston University and at University of Massachusetts, Amherst.

9.4 HELICOPTER SIGNAL TRACKER (HST)

The Helicopter Signal Tracker is a KBSP application system that was designed to track the acoustic signal (received on a microphone at a fixed ground location) from one or more flying helicopters. The main causes of a helicopter sound are aerodynamic in nature and are the result of fluctuations and motions of the distributed pressures on the rotor blades. These pressures are in turn due to the the rotor blade/wake interactions. Since the wake is periodic, the resulting sound is periodic, and hence its spectrum is characterized by a fundamental frequency and its harmonics. There are two such harmonically related peak sets present in a helicopter sound, due to the main rotor and the tail rotor. The tail rotor fundamental is usually several times higher than the main rotor fundamental and by design not an integer multiple of it. Typically, the main rotor fundamental ranges between 8 and 20 Hz, with harmonics up to 100 Hz. The tail rotor fundamental ranges between 70 and 140 Hz, with harmonics up to about 800 Hz. When a helicopter is moving (in flight), the harmonic components change in power (according to the inverse of the square of the distance of the helicopter from the microphone) as well as in frequency location. The frequency locations change due to the well-known Doppler shift phenomenon. The HST system is designed to track the power and frequency changes in the harmonic sets contained in the helicopter sound signal. It is also designed to separate the components due to different helicopters flying in the vicinity of the same microphone. The problem is particularly difficult when the signal-to-noise ratio is low. In the following material, we discuss the specific objectives of the designers of the HST system as well as details of the system they designed and implemented for the helicopter signal tracking problem.

9.4.1 HST Objectives

The HST system was designed to illustrate a practical application for the use of the Extended Spectrum concept, described in the signal abstractions chapter of this book. The multiple levels of abstraction in the extended spectrum are utilized to apply a tight set of conditions on whether to associate two harmonic sets at successive times with the same acoustic source. This is important for the system to be able to distinguish between the tracks of different helicopters that are in the vicinity of the microphone at the same time. In addition to using the extended spectrum representation, the HST system has other architectural aspects designed specifically for tracking the power and frequency changes in the harmonic sets contained in the sound of each moving helicopter.

Consider the two-helicopter situation depicted in Figure 9.12. Two helicopters, one trailing the other, are flying in a straight path past a fixed microphone. The point where each helicopter is the closest to the microphone is referred to as the Closest Point of Approach (CPA). The signal received at the microphone during this two-helicopter flight path consists of harmonic sets due to each of the helicopters. The basic problems addressed by the HST system are (1) to estimate the harmonic spectra over each short-time section of the received signal and (2) to associate harmonic sets in different short-time sections with each other if they are judged to arise from the same helicopter. The spectral estimation problem is discussed in detail in the signal abstractions chapter in this book. In the following section, we consider details of the HST system that are related to the problem of associating harmonic sets in different short-time sections with each other.

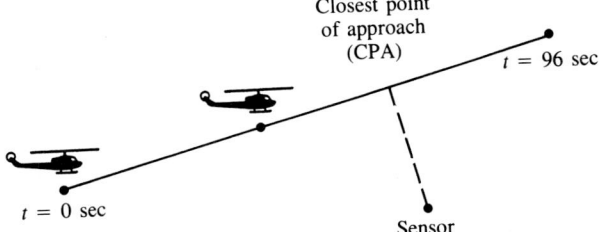

Figure 9.12 A straight source path and two helicopters one following the other. The acoustic signal contains two harmonic sets, one from each helicopter. Power is maximum at the closest point of approach. Fundamental frequency decreases from the asymptotic value $\frac{1}{1 - v/c}$ far before CPA to another asymptotic value $\frac{1}{1 + v/c}$ far after CPA, where v and c are the speed of the helicopter and sound, respectively.

9.4.2 HST System Details

In this section we illustrate the architectural characteristics of the HST system that enable it to efficiently associate harmonic sets at different times with their respective acoustic sources. The discussion is presented in the context of the HST analysis of the acoustic signal corresponding to the two-helicopter scenario depicted in Figure 9.12. The processing results presented in this section were obtained by running the HST system on an actual signal recorded from two real helicopters that flew the path illustrated in Figure 9.12. The experiment involving the two helicopters was performed at MIT Lincoln Laboratory.

The problem of associating harmonic sets at different times with a single source is made difficult because the spectra obtained from the experimental helicopter data are noisy and the two helicopters in the experiments have fundamental frequencies relatively close to each other. A simple "nearest-neighbor" technique, which associates two harmonic sets at successive times on the basis of the difference between their fundamental frequencies, relies on a single threshold value. If the difference between the fundamental frequencies is less than the threshold, then an association is established. The extended spectrum permits us to augment the nearest neighbor criterion with other criteria that take into account more detailed features of the harmonic sets in establishing an association. Examples of such features include the number of harmonics present in the two sets and their qualitative characterizations.

Another advantage of the extended spectrum in harmonic set tracking is that it provides detailed support at lower levels of abstraction for the output tracks. This "record-keeping" is useful in building a "meta-level" program for iteratively improving the solution by applying higher-level contextual information. This is demonstrated via an implemented system for harmonic source pitch and power tracking, shown in Figure 9.13. Furthermore, the record keeping facilitated by the extended spectrum approach is useful in debugging the heuristics used in the program.

Sec. 9.4 Helicopter Signal Tracker (HST)

The input to the system shown in Figure 9.13 is an acoustic signal, the spectrum of which contains multiple harmonic sets. The output of the system is a set of harmonic chains, i.e., linked harmonic sets at successive times in the short-time spectrum of the acoustic signal.

The system operates as follows: The first iteration is an open-loop pass (without chain pruning) that leads to the formation of chains of harmonic sets that are considered as potential parts of harmonic chains sources present in the signal (islands of certainty). These chains are then refined (i.e., completed and extended) by later iterations that perform a focused search for appropriate harmonic sets. The iterations have converged when the focused search does not result in any more changes to the existing chains. At this point, chains that are not sufficiently long are pruned out, and the chains that remain are the final result of the process.

Open-loop Operation

The open-loop operation of the system of Figure 9.13 includes estimation of a "short-time" numeric spectrum, computation of the extended spectrum representation for each of its sections and linking of harmonic sets in neighboring sections based on their fundamental frequencies.

Linking heuristics depend on the expected types of acoustic source motion relative to the sensors. In the implemented system, we have incorporated heuristics related to the fundamental frequency of the sensor signal for acoustic sources moving in straight lines with constant speed. For such acoustic sources, two harmonic sets belonging to spectra at successive times t and $t + T$ are linked together if their fundamental differences $f(t)$ and $f(t + T)$ obey the constraints: $-K_1 T < f(t) - f(t + T) < K_2 T$, where K_1 and K_2 are two positive constants with $K_2 \gg K_1$. Intuitively, K_2 corresponds to the maximum expected frequency drop due to the Doppler phenomenon per time unit. K_1 is the maximum increase in fundamental frequency that may override the Doppler drop due to noise.

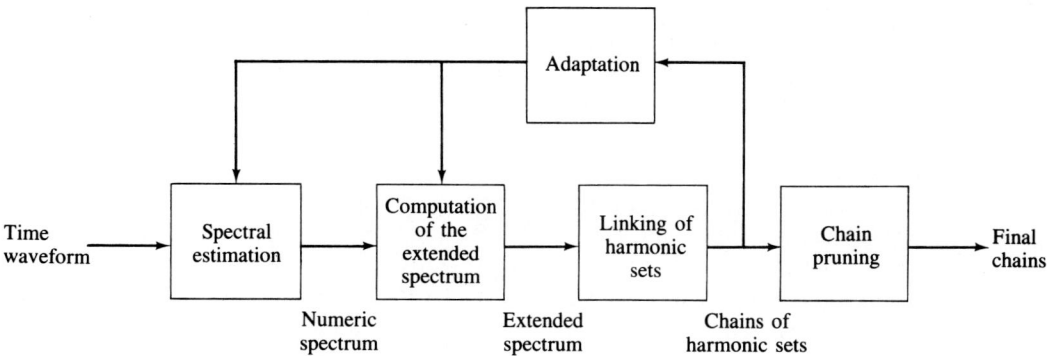

Figure 9.13 Architecture of the helicopter signal pitch and power tracking system.

Closed-loop Operation

In the closed-loop operation of the system shown in Figure 9.13, the search for new harmonic sets that extend or complete harmonic chains becomes more focused, i.e., it takes place at specific times and frequencies. This is equivalent to adaptation of the extended spectrum, guided by the harmonic chains already constructed.

The open-loop operation of the system shown in Figure 9.13 (without chain pruning) has stringent criteria for linking harmonic sets into chains, so the chains it forms can be viewed as "islands of certainty." Adaptation then seeks to fill in the gaps between chains that could potentially be linked together and looks specifically for harmonic sets that might have been missed by the open-loop pass.

The focused search for a harmonic set is conducted among the peaks of the extended spectra along the chain link (Figure 9.14), with a hypothesized fundamental frequency as dictated by the frequencies of the linked chains. All peaks that are approximate harmonics of the hypothesized fundamental frequency are collected and the harmonic set thus formed is tested against heuristic criteria for acceptance, similar to, but weaker than the criteria used in pruning the harmonic sets in the original extended spectrum computation. If the harmonic set passes at least one of those criteria successfully, it contributes to linking the chains on either side of the chain link of Figure 9.14. The heuristic criteria in the implemented system are shown in Figure 9.15. Very similar is the search for harmonic sets to extend existing harmonic chains.

Figure 9.16 shows the result of the above process on a segment of real harmonic source data. The scenario includes two harmonic sources, both traveling the same straight-line path with constant speed, with the second harmonic source following the first. Part (a) of the figure shows the acoustic signal (pressure as a function of time), of a total duration of 95.7 sec, at a 2 KHz sampling frequency. The harmonic source is closest to the sensor at about 60 sec, while the time the second harmonic source is closest to the sensor is outside our data segment (after 95.7 sec). Part (b) shows the harmonic chains formed by a straightforward "nearest-neighbor" tech-

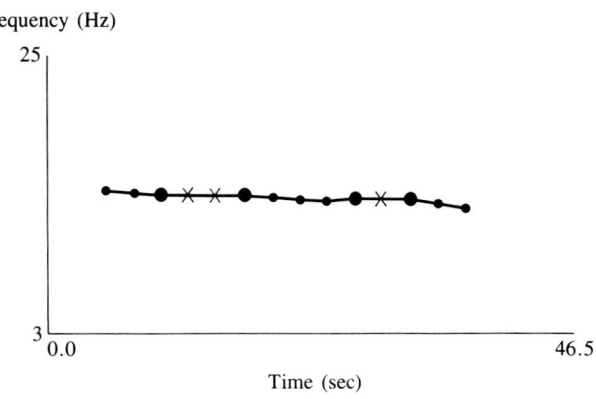

Figure 9.14 Completion of gaps in pitch tracks by focused search for harmonic sets in times along the gap. Newly found harmonic sets are shown as 'x's. Chain ends that are linked are denoted by larger circles than the normal harmonic sets. The chain links themselves are denoted by dotted lines.

Sec. 9.5 Conclusions   327

> **A HARMONIC SET IS ACCEPTED IF AT LEAST ONE OF THE FOLLOWING CONDITIONS IS SATISFIED:**
>
> - It has at least two harmonics of order less than or equal to 3.
> - At least its first or second harmonic has the characterization LARGE-AVG-PROMINENCE.
> - It has at least three harmonics in total.

Figure 9.15 Criteria for acceptance of harmonic set formed after focused search based on a fundamental hypothesized by a chain.

nique. Each dot represents a harmonic set, with the vertical axis being frequency in Hz and the horizontal axis being time in seconds. Links between dots in the graph represent explicit links between harmonic sets established by the method. Part (c) shows the resulting chains from the initial, open-loop pass of the system shown in Figure 9.13. Part (d) shows the result of the first iteration of focused search for harmonic sets to complete and extend chains found so far. Linked thick dots in the graph correspond to chain links while "x" denotes harmonic sets found after focused search. Part (e) follows from part (d) by incorporating the newly found harmonic sets into the harmonic chains. Part (f) shows the result of focused search on the chains of part (e). In this example, the next iteration does not change the harmonic chains at all, so the process has converged after three iterations, including the initial open-loop pass. Parts (g) and (h) show the resulting chains after chain pruning, both as frequency/time and as power/time graphs. The longer chain in (g) corresponds to the first harmonic source, and we can easily see the steep drop in fundamental frequency and the maximum in power corresponding to the time when the harmonic source is closest to the sensor. The fundamental frequency of the second harmonic source is approximately constant while its power increases within our data segment.

9.5 CONCLUSIONS

The most obvious common theme in the development of the Pitch Detector's Assistant (PDA), the Sound Understanding Testbed (SUT), and the Helicopter Signal Tracker (HST) is the rejection of classical signal processing and classical signal understanding approaches to the design of applications systems. In the classical signal processing approach to system design, the starting point is to formulate the overall objectives of the system in mathematical terms. This mathematical formulation is then used to design a numeric-processing algorithm for achieving the system objectives. The input data, the intermediate data, and the output data of the algorithm are all represented numerically. This is done in order to exploit the available battery of efficient procedures for carrying out numeric computations that are common to many signal processing applications. In contrast, the classical signal understanding approach to system design seeks to exploit heuristic search strategies

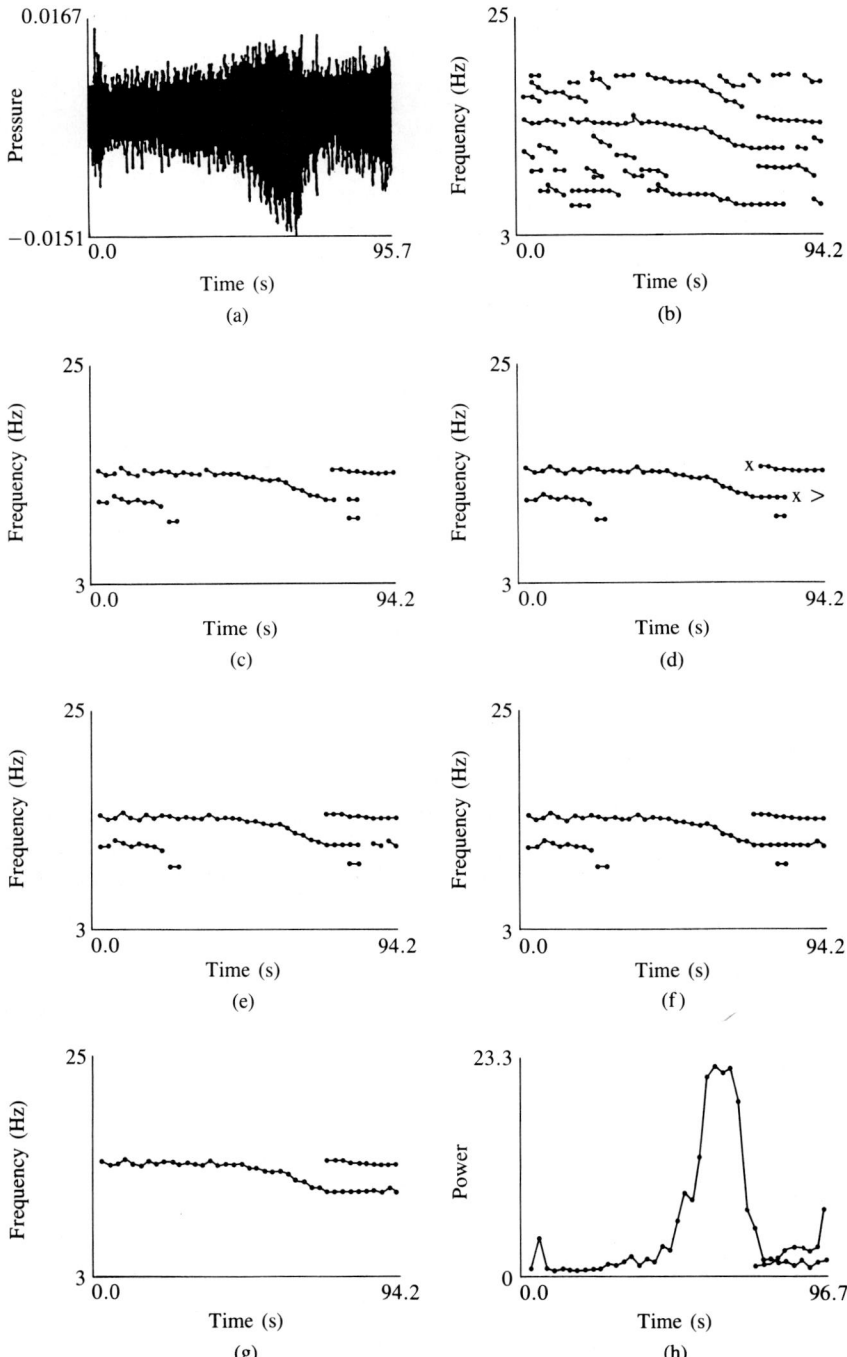

Figure 9.16 Fundamental frequency and power traces in the 3–25 Hz region from 95.7 sec of real data from a two-helicopter scenario.

that have been developed in the field of artificial intelligence. In this approach, the system objectives are formulated in terms of a *search* to be conducted through a space of possible solutions. How to represent the space of possible solutions and how to represent the information that can potentially aid the search process become the major focus in this "knowledge-based" approach. The battery of such representation formalisms available in the artificial intelligence field are generally most suited to symbolic processing environments. The designers of classical signal understanding systems have therefore tended to neglect how judicious selection of the numeric processing of the input signal may aid the system in achieving its overall objectives. The systems described earlier in this chapter serve as an illustration of how a system design strategy different from the classical approaches may be beneficial and even necessary in some applications. Let us take a closer look at this issue for each of those systems.

The Pitch Detector's Assistant tackles a problem in which one of the inputs is a symbolically encoded speech transcript and one of the desired outputs is a numeric pitch track. The classical signal processing system designer would be at a loss to deal with the input transcript unless it was first somehow translated into a numeric representation. Even if that were accomplished, it is very unlikely that the algorithms needed to extract appropriate information from the transcript would have much in common with traditional signal processing algorithms, thereby losing the computational efficiency advantage. The classical signal understanding system designer would similarly have difficulty formulating efficient symbolic representations for the search space of possible pitch tracks. The mix of symbolic and numeric specifications in the problem statement are thus a primary motivation for rejecting the classical system-design approaches for the Pitch Detector's Assistant.

The Sound Understanding Testbed tackles an application where the classical system design approaches are inappropriate due to the enormous variety of input signal types and system goals in that application. It is contended that in the face of such variety, it is necessary for the signal understanding system to have many different signal processing algorithms at its disposal. Consequently, the focus of system design is on providing the capability of utilizing the most appropriate algorithm on a when-needed basis as well as providing mechanisms for deciding which algorithm is needed in particular situations. Neither the classical signal processing approach nor the classical signal understanding approach provides any answers to these design issues.

The Helicopter Signal Tracker represents an application where even the lowest levels of processing require an interleaved application of classical numeric signal processing and classical heuristic search methods for finding perceptual features of interest. For example, the task of detecting harmonic components in a signal is divided into three interdependent tasks of peak detection, peak characterization, and peak grouping. The peak detection and peak grouping tasks utilize classical signal processing techniques. On the other hand, peak characterization utilizes heuristic search methods more commonly found in classical signal understanding systems. The focus in the design of HST was therefore on developing the extended spectrum representation for facilitating the application of heuristic and mathematically based techniques on a common signal representation. The need for such a representation does not arise in the classical system design approaches.

The above discussion illustrates the variety of reasons why classical signal processing and signal understanding approaches to system design are not suitable for particular applications. Instead, system architectures are desirable that permit specific techniques and data representations from signal processing and artificial intelligence to be integrated in a more sophisticated manner than allowed by classical signal understanding systems.

Finally, we emphasize that the KBSP systems described in this chapter are not the only ones to have been designed and/or implemented with the view of overcoming limitations of the classical approaches. There have been several other research efforts in the context of various applications. These include speech analysis and enhancement [6], seismic data interpretation [7], weather radar data interpretation [8], image interpretation [9], nonlinear image restoration [10] and automatic electroencephalogram (EEG) interpretation [11].

REFERENCES

[1] W. P. Dove, "Knowledge-Based Pitch Detection" RLE Technical Report 518, Cambridge Mass.: MIT (1986).

[2] L. R. Rabiner and R. W. Schafer, *Digital Processing of Speech Signals* (Englewood Cliffs, N.J.: Prentice Hall, 1978).

[3] H. Nawab and V. Lesser, "High-Level Adaptive Signal Processing," *NAIC Final Report*, 17 (October 1989).

[4] S. Albert Bregman, *Auditory Scene Analysis* (Cambridge: MIT Press, 1990).

[5] E. Milios, "Signal Processing and Interpretation Using Multilevel Signal Abstractions" RLE Technical Report 516, Cambridge Mass.: MIT (1986).

[6] C. Myers, A. Oppenheim, R. Davis, and P. W. Dove, "Knowledge Based Speech Analysis and Enhancement," *IEEE ICASSP-1984* (1984), 39A.4.1–39A.4.4.

[7] G. J. Cleary, L. L. Kramer, and P. M. Wingham, "Knowledge-Based Systems for the Interpretation of Seismic Data," *Coupling Symbolic and Numerical Computing in Expert Systems* (New York: Elsevier Science Publishers B.V., 1986).

[8] S. D. Campbell and H. S. Olson, "WX1—An Expert System for Weather Radar Interpretation," *Coupling Symbolic and Numerical Computing in Expert Systems* (New York: Elsevier Science Publishers B.V., 1986).

[9] C. A. Kohl, A. R. Hanson, and E. M. Reisman, "A Goal-Directed Intermediate Level Executive for Image Interpretation," *Proceedings of IJCAI-87* (1987), 811–14.

[10] S. D. Jeong and A. A. Beex, "Nonlinear Image Restoration Using a Segmentation-Oriented Expert System," *IEEE ICASSP-1988*, 2 (1988), 984–87.

[11] B. M. Dawant and B. H. Jansen, "Coupling Numerical and Symbolic Methods for Signal Interpretation," *IEEE Trans. on Systems, Man Cybernetics*, 21, no. 1 (January/February 1991), 115–24.

INDEX

A

Abstractions:
  array abstraction, 16
  data abstraction, 1, 237–39, 300–301
  signal abstractions, 2, 286–302
    in blackboard systems, 299–300
  system abstractions, 2
Abstract signal objects, 2–3, 20–26
  closure model for signals, 20–21
  See also Signal Representation Language (SRL)
Addition/multiplication cost, 161–62
ADE (Algorithm Design Environment), 31–32, 44–55, 73–74, 89, 134, 136, 178
  algorithm definition and manipulation examples, 45–55

contributions/limitations, 73–74
definition of, 45
overview of, 44–45
pruned FFT structure generated by, 50–54
sample interactive session in, 49
signal/system classes defined in, 48
Adjustable control parameters, SPA, 253–54
Adobe's *Photoshop,* 185
Adviced pitch rule, Pitch Detector's Assistant (PDA), 309
Algorithm Design Environment, *See* ADE (Algorithm Design Environment)
Algorithm design/rearrangement, 30–57
  integrated signal processing environment, capabilities/requirements for, 36–44

Algorithm design/rearrangement (*cont.*)
  ranking of equivalent algorithms:
    constrained derivation and, 62–73
    unconstrained derivation and, 56–62
  See also ADE (Algorithm Design Environment)
AOL language, 3, 17–19
Applications:
  knowledge-based signal processing (KBSP), 303–30
    Helicopter Signal Tracker (HST), 322–27
    Pitch Detector's Assistant (PDA), 45, 304–15
    Sound Understanding Testbed (SUT), 315–22
Array processing, 2, 16–19
  array abstraction, 16
  array operator model, 16–17
  deferred evaluation criterion and, 19
  languages, 17–19
  manifest typing criterion and, 19
Arrays, 2–3
  array abstraction, 16
  array aliasing, 18
  array operator model, 16–17

B

BB1, *See* Control blackboards
BDL language, 3, 11–15, 17
Binary signals, 144–46
Blackboard systems, 205–50
  aircraft monitoring example, 217–25
  components of, 212–13
  concept origin, 211
  control, 231–46
  control blackboards, 239–42
  data abstraction-based planning, 237–39
  future of, 246
  goal-directed blackboards, 235–37
  issues, 231–34
  RESUN models, 230–31, 242–46, 274–75
  Hearsay II and, 211, 215, 230, 232, 239, 299
  hypotheses, representation of, 214
  idealized, 214–16
  implementation of, 226–30
  and real-time problems, 247
  research in, 246–47
  signal abstractions in, 299–300
  structuring interpretation problems for, 225–31
  and understanding problems, 206
Block diagram languages, 9–15
  BDL, 11–15
BLODIB language, 9
BLODI language, 9
BLOSIM language, 9

C

CaminoReal, 202
Cascade, 167
Causal hierarchy, 208
CIRCUS language, 9
Classification problems, 210
CLOS, 88
Closed-loop operation, Helicopter Signal Tracker (HST), 326–27
Closure model for signals, 20–21
Common Lisp, 89, 143
Complex expressions, reducing into less complex forms, 148
Computational cost definitions, 161–65

Index

representing alternative architectures, 161
signal access cost, 163–65
union and maxima cost, 162–63
Computational equivalence, 40
Computation kernels, 2
Conflicts, types of, 260
Constrained derivation, and ranking of equivalent algorithms, 62–73
Constructive problem-solving techniques, 263–64
Continuous and discrete convolution, 128–29
Continuous convolution, 128–29
Control:
blackboard systems, 231–46
Integrated Processing and Understanding of Signals (IPUS), 274–77
Resolving Sources of Uncertainty (RESUN) models, 242–44
Control abstractions, 1
Control blackboards, 239–42
Control knowledge sources, 240–41
Cost definitions, 162
Cost weighting, 165

D

DARE language, 9
Data abstraction, 1, 300–301
data abstraction-based planning, 237–39
Data approximation, 278
Data types, abstract, 1
Deferred evaluation, 3, 8, 15
Design:
algorithms, 30–57
interactive signal processing documents, 196–202
Mathematica as design tool, 134–36

DFT, *See* Discrete Fourier transform (DFT)
Differential diagnosis, 209, 242, 265
Dilation operation, 144–45
Dimensional extensibility, 3, 7, 15
Discrepancies-to-distortions mapping, 273–74
Discrepancy detection, IPUS, 258–61
Discrete convolution, 128–29
Discrete Fourier transform (DFT), 103, 117–18, 134
Discrete-time Fourier transform (DTFT), 103, 115–17, 134
Discrete time signal (sequence), 4
Distributed Vehicle Monitoring Testbed (DVMT), 235, 237–39
DSP systems, 134–36, 176
tools, 176, 184–96
DTFT, *See* Discrete-time Fourier transform (DTFT)
Dual form generation, 160
DVMT, *See* Distributed Vehicle Monitoring Testbed (DVMT)

E

Electronic notebook, *Mathematica*, 174–203
Equivalent algorithms, ranking of:
constrained derivation and, 62–73
unconstrained derivation and, 56–62
Equivalent form generation, 157–58
Erosion operation, 144
E-SPLICE, 31–32, 44, 45, 134, 136
Evidence aggregation, 209, 265
Expression rewriting, 157–60
dual form generation, 160
equivalent form generation, 157–58
simplifications generation, 158–59
Extension KSs, 229

F

Fast Fourier transform (FFT), 23, 268, 270
Forward z-transforms, 107–12
Fourier transform, 102–3, 118–19, 178, 184
FSK-code detector, 44–46, 60, 62
Function closure, 20

G

GABRIEL language, 9
Global database, blackboard systems, 212–13
Goal-directed blackboards, 235–37
  limitations of, 237
Goal signal state, 280
Goertzel algorithm, 41–42
GOSPL language, 9
Greyscale signals, 146–47
GZP (Gain, Zeros, and Poles) structure, 187

H

Hanning window, 38, 40–41
Harmonic set abstraction level, 292–93
HASP/SIAP system, 299
Hearsay-II, 211, 215–16, 230, 232, 234, 239, 299
  policy KSs, 239n
  scheduling KSs, 239n
Helicopter Signal Tracker (HST), 322–27, 329–30
  objectives of, 323
  system details, 324–27
    closed-loop operation, 326–27
    open-loop operation, 325
Hypermedia, 179–81
  definition of, 179
  links, 180–81
  modifiability, 180
  multimedia, 180
  smart links, 181

I

Idealized blackboard systems, 214–16
IEEE Signal Processing Library, 184
ILS, 74
Immutability, 3, 7, 14–15
IMSL, 177, 184
Incremental hypothesis and test, 209, 242
In-place computation, 18–19
Input/output mapping, 4
Integrated Processing and Understanding of Signals (IPUS), 251
  adaptive control comparison, 266
  AI theories supporting, 274–83
    control issues, 274–77
    diagnostic reasoning, 280–81
    interleaved planning and execution, 281–82
    real-time issues, 277–80
  architecture:
    motivation for, 252–57
    overview of, 257–66
  diagnosis and reprocessing, 261–63
  discrepancy detection, 258–61
  evidence gathering in, 265–66
  interpretation process, 263–65
  signal-processing theories to support, 266–74
    data-model concept, 267–70
    discrepancies-to-distortions mapping, 273–74
    model variety analysis, 270–72

signal-processing distortions, 272–73
Integrated signal processing environment:
  capabilities/requirements for, 36–44
    algorithm definition support, 38
    automation of algorithm rearrangement, 40–42
    computational cost comparison, 422–24
    numerical signal processing support, 37–38
    signal property/transform analysis support, 38–39
Interactive signal processing documents, 173–204
  design issues, 196–202
  interactive document properties, 176–83
    hypermedia, 179–81
    Literate Programming, 183
    models, 181–83
    symbolic manipulation, 177–79
  problems with, 199–201
  research issues, 202–3
  writing, 196–99
Interleaved planning and execution, 281–82
Inverse z-transforms, 112–15
IPUS, *See* Integrated Processing and Understanding of Signals (IPUS)
ISPUD, 43

applications, 303–30
  Helicopter Signal Tracker (HST), 322–27
  Pitch Detector's Assistant (PDA), 304–15
  Sound Understanding Testbed (SUT), 315–22
Mathematica programming for, 93–96
Knowledge Manager (KM), Pitch Detector's Assistant (PDA), 314–15
Knowledge representation, *Mathematica*, 96–102
Knowledge source activation records (KSARs), 240–42
Knowledge source instantiations (KSIs), 216–17, 219, 225, 227–30, 237–38, 240–42
Knowledge sources (KSs):
  action component of, 229
  blackboard systems, 212–17, 227–30
  categories, 229
  grain size, 230
  parallel execution of, 247
  precondition of, 228–29

K

KBSP, *See* Knowledge-based signal processing (KBSP)
Knowledge approximation, 274
Knowledge-based signal processing (KBSP), 88–89

L

Laplace transform, 102, 118–19, 178, 184
Length computation, 18
Likely-burst rule, Pitch Detector's Assistant (PDA), 308
Linear transforms, 102
Links, hypermedia, 180–81
Lisp, 89, 184
Literate Programming, 177, 183
LOTUS language, 9

M

MACSYMA, 45, 74, 89, 91–92, 135, 184, 186, 300
Manifest typing, 3, 8–9, 15
Maple, 45, 89, 92, 134, 186
Mathematica, 88–89, 92–96, 129–36, 174–76, 184, 186–96
  as design tool, 134–36
    new applications, 138
      representing system properties, 136–37
      signals as objects, 137–38
  as educational tool, 129–31
  electronic notebook, 174–203
    animation capabilities, 194–95
    example of, 188–93
    features of, 186–87, 194–96
  in engineering curriculum, 132–34
  in freshman curriculum, 131–32
  help system, 195
  knowledge representation, 97–102
    common data structures, 97–98
    computational functions for signal processing, 98–100
    operators for signal processing, 100–102
  programming for KBSP, 93–96
  rules, 194
  user interface, 184
Mathematical equivalence, 40
Mathematical morphology:
  compared to traditional signal processing, 142–43
  definition of, 142
  signal properties used in, 150
MATLAB, 74, 177, 185, 187
Merge KSs, 229
MetaMorph environment, 143, 152, 153
Meta-systems, 167
Milo, 202

Minkowski subtraction/addition, 145–46, 150
M inputs, 4
MITSYN language, 9
Modulated filter banks, 33–36
Morphological algorithms:
  symbolic analysis of, 142–72
  symbolic manipulations, 148–57
Morphological equivalent form rules, 157
Morphological inclusion rules, 166
Morphological simplification rules, 159
Morphological system properties, 153
Morphological system theory:
  binary signals, 144–46
  greyscale signals, 146–47
Multimedia, 180
Multiple-outputs-synthesis, 262

N

Next-state simulation, 2, 14–15
Noninteger sampling rate conversion, 32–33
N!Power signal processing package, 20
N.s.s.-realizable systems, *See* Next-state simulation
NthPOWER (Signal Technology, Inc.), 43, 74
Number-theoretic transforms (NNTs), 135–36
Numerical signal inquiry operations, SRL, 22–23
Numeric spectrum level, 289

O

Objective C, 88

Index

Open-loop operation, Helicopter Signal Tracker (HST), 325
Operations, abstract, 1

P

Partially verified expectations, 260–61
PATSI language, 9
PDA, *See* Pitch Detector's Assistant (PDA)
Peak abstraction level, 290–92
Perspicuity, 9
Phonetic advice rule, Pitch Detector's Assistant (PDA), 309
Phonetic-verify rule, Pitch Detector's Assistant (PDA), 307
Photoshop (Adobe), 185
Pitch Detector's Assistant (PDA), 45, 304–15, 327–29
   adviced pitch rule, 309
   likely-burst rule, 308
   objectives of, 305–7
   phonetic advice rule, 309
   Phonetic F0 rules, 309
   phonetic-verify rule, 307
   power estimates rule, 308
   preliminary pitch assertion, 309–10
   similarity detection rule, 308–9
   system details, 307–15
      epoch system, 310–11
      Knowledge Manager (KM), 314–15
      rule system, 311–14
   voicing analysis rule, 307
Planning-based control framework, motivation for using, 244–46
P outputs, 4
Power estimates rule, Pitch Detector's Assistant (PDA), 308
Prediction KSs, 229–30
Pseudo-Wigner distribution, 282

Q

QM, 45, 89
QuickSig language, 4, 20

R

Realization, 4–5
Reduce, 186
Region of convergence (ROC), 89, 103, 121, 134
Resolving Sources of Uncertainty (RESUN) models, 230–31, 242–46, 266, 274–75, 316
   control in, 242–44
   refocusing mechanism, 244
Rules, *Mathematica*, 194

S

SCRATCHPAD, 136
Script-based planners, 244
Sensor interpretation, 207, 246
Sensor interpretation problems, 206
Sequential integration, 316
Short-time Fourier transforms (STFT), 33–36, 65, 253–56, 258–62, 268, 270–74, 278–80, 282–83, 318, 320–21
   data model for, 269
Signal abstractions, 2, 286–302
   auditory signal perception and, 298–99
   in blackboard architectures, 299–300
   data abstraction in signal processing, 300–301
   definition of, 287–89

Signal abstractions (cont.)
   extended spectrum, 289–93
      adaptive spectral estimation application, 293–98
      harmonic set abstraction level, 292–93
      numeric spectrum level, 289
      peak abstraction level, 290–92
      object recognition using decomposition and, 300
      spectral peaks and, 299
Signal access cost, 163–65
Signal generation process (SGP), 267
Signal processing, concepts/notation, 4–5
Signal processing algorithm (SPA), 253–57, 266–68, 280–81
Signal processing distortions, 272–73
Signal properties, used in mathematical morphology, 150
Signal property analysis, 148, 160–65
   addition/multiplication cost, 161–62
   computational cost definitions, 161–65
      representing alternative architectures, 161
      signal access cost, 163–65
      union and maxima cost, 162–63
   cost weighting, 165
Signal representation:
   deferred evaluation, 8
   dimensional extensibility, 7
   immutability, 7
   manifest typing, 8–9
   requirements, 3–9
   signal processing concepts/notation, 4–5
   uniform reference, 5–7
Signal Representation Language (SRL), 3, 20, 22–26, 34, 43, 184
   numerical signal inquiry operations, 22–23
   signal type definitions, 23–25
   symbolic signal inquiry operations, 26
Signal representations, 1–29
Signal reprocessing planner, 262
Signal types:
   definition of, 22
   implementation of, 23–25
Signal understanding, 253–54
   as interpretation, 207–11
   use of term, 205
Similarity detection rule, Pitch Detector's Assistant (PDA), 308–9
Simplifications generation, 158–59
Smart links, 181
SMP, 186
Sound Understanding Testbed (SUT), 315–22, 327, 329
   definition of, 315
   objectives of, 316–18
   system details, 318–22
Sources of uncertainty (SOUs), 243, 275–76
Spectral integration, 316
SPLICE language, 3, 4, 20, 23, 184
SRL, *See* Signal Representation Language (SRL)
Sspice, 136
State-space realization, 4–5
Streams, 2–3, 9, 10–11
SUT, *See* Sound Understanding Testbed (SUT)
Symbolic analysis, 88–141
   continuous and discrete convolution, 128–29
   differential/difference equations, solving, 119–21
   expression rewriting, 157–60
   knowledge representation, 96–103
   Mathematica, 88–89, 92–96, 129–36, 174–76, 184, 186–96
   of morphological algorithms, 142–72
   signal property analysis, 160–65

system property analysis, 166–67
transform-based analysis techniques, 121–27
transforms, 102–19
Symbolic discrete Fourier transforms, 115–19
Symbolic manipulation, 177–79
   complex expressions, reducing into less complex forms, 148
   manipulation requirements, 149
   morphological system information, 152–53
   new forms, generating from old expression, 148
   signal property analysis, 148
   signals/structuring elements, 149–52
   symbolic interferencing mechanism, 153–55
   system properties, finding of compound expressions, 148–49
   systems/signals, interaction between, 155–57
Symbolic manipulation programs, 174
Symbolic mathematics programs, 90–93
   MACSYMA, 45, 74, 89, 91–92, 135, 184, 186, 300
   Maple, 45, 89, 92, 134, 186
   Mathematica, 88–89, 92–96, 129–36, 174–76, 184, 186–96
      programming for KBSP, 93–96
Symbolic math programs, 178–79
Symbolics Common Lisp (Symbolics, Inc.), 45, 74
Symbolic signal inquiry operations, 26
Symbolic signal processing, 9
Symbolic z-transforms, 106–15
   forward z-transforms, 107–12
   inverse z-transforms, 112–15
Synthesis KSs, 229
System abstractions, 2

System properties, morphological, 153
System property analysis, 166–67

T

Time-resolution distortion, 272
Transform-based analysis techniques, 121–27
Transforms, 102–19
   discrete Fourier transform (DFT), 103, 117–18, 134
   discrete-time Fourier transform (DTFT), 103, 115–17, 134
   fast Fourier transform (FFT), 23, 268, 270
   Fourier transform, 102–3, 118–19, 178, 184
   linear transforms, 102
   number-theoretic transforms (NNTs), 135–36
   short-time Fourier transforms (STFT), 33–36, 65, 253–56, 258–62, 268, 270–74, 278–80, 282–83
   symbolic discrete Fourier transforms, 115–19
   discrete Fourier transform (DFT), 117–18
   discrete-time Fourier transform (DTFT), 115–17
   Fourier transform, 102–3, 118–19
   LaPlace transform, 102, 118–19
   symbolic z-transforms, 106–15
   forward z-transforms, 107–12
   inverse z-transforms, 112–15
   transform rule bases, structure of, 103–6

U

Unconstrained derivation, and ranking of equivalent algorithms, 56–62
Uniform reference, 3, 5–7, 14
Union and maxima cost, 162–63
Unverified expectations, 260–61
User interface, *Mathematica,* 184

V

Voicing analysis rule, Pitch Detector's Assistant (PDA), 307

W

Weighting, cost, 165
Writing, interactive signal processing documents, 196–99

Z

Z-domain representations, 39
Z-transforms, 178